Computational Mathematical Programming

NATO ASI Series

Advanced Science Institutes Series

A series presenting the results of activities sponsored by the NATO Science Committee, which aims at the dissemination of advanced scientific and technological knowledge, with a view to strengthening links between scientific communities.

The Series is published by an international board of publishers in conjunction with the NATO Scientific Affairs Division

A Life Sciences **B Physics**	Plenum Publishing Corporation London and New York
C Mathematical and Physical Sciences	D. Reidel Publishing Company Dordrecht, Boston and Lancaster
D Behavioural and Social Sciences **E Applied Sciences**	Martinus Nijhoff Publishers Boston, The Hague, Dordrecht and Lancaster
F Computer and Systems Sciences **G Ecological Sciences**	Springer-Verlag Berlin Heidelberg New York Tokyo

Series F: Computer and Systems Sciences Vol. 15

Computational Mathematical Programming

Edited by

Klaus Schittkowski

Institut für Informatik, Universität Stuttgart, Azenbergstraße 12
7000 Stuttgart 1, Federal Republic of Germany

Springer-Verlag Berlin Heidelberg New York Tokyo
Published in cooperation with NATO Scientific Affairs Division

Proceedings of the NATO Advanced Study Institute on Computational Mathematical Programming held at Bad Windsheim, Federal Republic of Germany, July 23–August 2, 1984

ISBN 3-540-15180-X Springer-Verlag Berlin Heidelberg New York Tokyo
ISBN 0-387-15180-X Springer-Verlag New York Heidelberg Berlin Tokyo

Library of Congress Cataloging in Publication Data.
NATO Advanced Study Institute on Computational Mathematical Programming (1984 : Bad Windsheim, Germany) Computational mathematical programming. (NATO ASI series. Series F, Computer and systems sciences ; vol. 15) "Proceedings of the NATO Advanced Study Institute on Computational Mathematical Programming held at Bad Windsheim. Federal Republic of Germany, July 23–August 2, 1984"–T. p. verso. "Published in cooperation with NATO Scientific Affairs Division." 1. Programming (Mathematics)–Data processing–Congresses. I. Schittkowski, Klaus, 1946-. II. North Atlantic Treaty Organization. Scientific Affairs Division. III. Series: NATO ASI series. Series F, Computer and system sciences ; no. 15. QA402.5.N365 1984 519.7 85-8055
ISBN 0-387-15180-X (U.S.)

Printing: Beltz Offsetdruck, Hemsbach; Bookbinding: J. Schäffer OHG, Grünstadt
2145/3140-543210

Preface

This book contains the written versions of main lectures presented at the Advanced Study Institute (ASI) on Computational Mathematical Programming, which was held in Bad Windsheim, Germany F.R., from July 23 to August 2, 1984, under the sponsorship of NATO. The ASI was organized by the Committee on Algorithms (COAL) of the Mathematical Programming Society. Co-directors were Karla Hoffmann (National Bureau of Standards, Washington, U.S.A.) and Jan Telgen (Rabobank Nederland, Zeist, The Netherlands). Ninety participants coming from about 20 different countries attended the ASI and contributed their efforts to achieve a highly interesting and stimulating meeting.

Since 1947 when the first linear programming technique was developed, the importance of optimization models and their mathematical solution methods has steadily increased, and now plays a leading role in applied research areas. The basic idea of optimization theory is to minimize (or maximize) a function of several variables subject to certain restrictions. This general mathematical concept covers a broad class of possible practical applications arising in mechanical, electrical, or chemical engineering, physics, economics, medicine, biology, etc. There are both industrial applications (e.g. design of mechanical structures, production plans) and applications in the natural, engineering, and social sciences (e.g. chemical equilibrium problems, christollography problems).

In most cases, a theoretical foundation which predicts the numerical performance of a mathematical programming algorithm, does not exist. Nor can one at the present time prove theoretically that one algorithm is best for a given application. As a consequence, the performance of optimization methods must be evaluated by experiments. By defining a sample, i.e. by searching for suitable test problems, and by designing a computational experiment, the efficiency, robustness, and reliability of an algorithm can be investigated. Such experimental optimization will become even more important in future years as modeling efforts attempt to solve problems related to evermore complex systems and consequently, the mathematical algorithms used to solve these problems become more and more sophisticated. The need to disseminate information about computational mathematical programming procedures was the main impulse for organizing the Advanced Study Institute.

Consequently, the general purpose of the Advanced Study Institute was to link new algorithmic developments for optimization models with practical applications. From the organizational point of view, the ASI consisted of tutorilas, research seminars, and a software fair. The tutorials covered all major types of optimization models currently used in practice. Leading experts gave an introduction to the subject, provided any mathematical background, presented suitable algorithms stressing information about their computer implementations, usage and numerical performance. The topics of the tutorials ranged from linear and integer programming, networks, and various aspects of nonlinear programming to optimal control, stochastic and nonsmooth optimization. They were accomplished by about 60 research seminars of other participants leading to a more detailed insight in special individual research projects. Moreover, information material about existing optimization software was displayed at the ASI e.g. in form of user's guides or code descriptions.

The purpose of this book is to compose the tutorials and to offer them to the mathematical programming community. It is expected that only minor knowledge of mathematical theory and terminology is required to understand the most important opinions of the individual authors. In addition, the book contains the main results of the software fair, i.e. a collection of about 50 code descriptions, with the intention to disseminate information about existing optimization programs.

The organizer of the ASI is indepted to the NATO Science Committee for the generous financial support. Other sponsors were the National Bureau of Standards of the U.S. Department of Commerce, and the Deutscher Akademischer Austauschdienst. The success of the meeting is based on the quality of the tutorials held at the ASI. Without any exception, the organizer acknowledges the extensive amount of time the invited lecturers had to spend for preparing the tutorials and, in particular, for preparing the manuscripts submitted for publication. Other individuals who deserve special thanks, are the co-directors Karla Hoffmann and Jan Telgen, furthermore Marlis and Jochem Zowe for their extensive assistance concerning all local organizational arrangements.

K. Schittkowski

Contents

INTEGER PROGRAMMING

E.M.L. Beale
Scicon Limited
Brick Close, Kiln Farm,
Milton Keynes, Great Britain.

SUMMARY

Following an introduction, which discusses the motivation for studying integer programming, the relevance of computational complexity and the relative merits of integer and dynamic programming, the branch and bound method is introduced in general terms. Various types of global entity to which it can be applied are introduced. These are integer variables, semicontinuous variables, special ordered sets and chains of linked ordered sets. A discussion of the algorithmic details follows. Finally, various approaches to automatic model reformulation are discussed: this seems to be the most important current area of integer programming research.

1. INTRODUCTION

The requirement that the variables must take integer values is a natural mathematical extension of linear programming, which is known as Pure Integer Programming. Alternatively, one may think of planning problems in which some questions require qualitative answers - Yes or No, while others require quantitative answers - How much. The Yes-or-No decisions can be coded as zero-one (or binary) variables. And many of these problems can be formulated as linear programming problems with the additional restriction that some variables must be either zero or one. This is the problem of Mixed Integer Programming if there are other variables that can take continuous values. And it is sometimes useful to give some integer variables upper bounds greater than one.

In any integer programming problem, a local optimum may not be a global optimum. This property is shared by nonlinear problems in which either the objective function to be maximized is not concave or the feasible region is not convex or both. Markowitz and Manne (1957) showed that some of these problems can be formulated as mixed integer programming problems. But it has been found better to introduce other types of global entity as described in Section 3 below, rather than to reduce these problems explicitly to mixed integer programming. So the topic of integer programming can be considered as including global optimization of other models using the methodology developed for integer programming.

NATO ASI Series, Vol. F15
Computational Mathematical Programming
Edited by K. Schittkowski

Integer programming is one approach to problems of combinatorial optimization. Elementary considerations can teach some important lessons, notably that it is unrealistic to expect to solve all problems by waiting for computing to become cheaper or quicker. Any method that requires the enumeration of all 2^n possible combinations of Yes and No decisions on n items may well work for up to 10 or even 20 items. But an increase of computer speed by a factor of 1000 only allows n to increase by 10 with this approach. The situation is even worse if one has to enumerate all n! permutations of n items. A value of n around 10 may then be acceptable, since 10! is only about 3.6×10^6, but each further increase in n by a single unit multiplies the amount of work by a factor of more than 10. Similar remarks apply to methods that improve on complete enumeration but save "only" 99.9% of the work.

These facts also illustrate two facts about model building. If the essence of a problem is to make a selection from n items, or to find an optimum ordering of them, then it may be important to simplify the representation of each individual item, even if this requires a fair amount of computing. And it may also be important to suppress minor details from a combinatorial model designed to suggest an overall strategy.

Most integer programming problems are such that the amount of work needed to produce a guaranteed optimum solution may increase exponentially with the problem size. The significance of this proposition is limited by two facts: we usually wish to solve a particular class of problems of a given size, and we can also often accept a "good" solution that is "probably nearly optimum" if a guaranteed optimum is unavailable. We should remember that many practical problems with thousands of integer variables can be completely solved in a few minutes. But the result about exponential growth warns us that some problems can become much harder when the problem size increases by an apparently modest amount. We have already noted that complete enumeration of all 2^n combinations requires 1000 times as much work when n increases from 20 to 30.

Before studying integer programming in detail, it is worth considering dynamic programming as an alternative approach to some problems. For most practical purposes, dynamic programming can be considered as the process of formulating and solving problems as shortest-route problems. Its scope is much more restricted than integer programming, but when it is applicable it is sometimes much more efficient, and the reason is illuminating.

If we are seeking the shortest route from A to B, then we may have to consider going through X or Y or Z. But if we go through X, then the best route from A to X will not depend on how we choose to go from X to B. Dynamic programming exploits this important fact, but current methods for integer programming do not. This may not be too serious if we are prepared to accept a good solution that is not a guaranteed optimum, but it can add significantly to the task of completing the integer programming search.

As an example of a problem where either approach is possible, but dynamic programming is preferable, consider planning the construction of capacity to exploit an expanding market for a new product. A version of this problem can be formulated as follows:

Let D_t be the demand in Year t. (tonnes).

Let P_t be the "profit", i.e. revenue less variable operating cost, in Year t (£/tonne).

Let $C_t(y_t)$ be the total cost of installing y_t tonnes of new capacity in Year t. (This includes fixed operating costs in future years.)

Let x_t be the quantity produced in Year t (tonnes).

Let y_t be the new capacity installed in Year t (tonnes).

Let z_t be the total capacity available in Year t (tonnes).

Then we wish to maximize

$$\sum_t P_t x_t - \sum_t C_t(y_t)$$

subject to the constraints

$$x_t \leq D_t \quad \forall t$$

$$x_t - z_t \leq 0 \quad \forall t$$

$$-z_t + z_{t-1} + y_t = 0 \quad \forall t$$

$$x_t \geq 0, \quad y_t \geq 0 \quad \forall t$$

$$z_0 = 0.$$

If the functions $C_t(y_t)$ are not convex, this problem is best solved by dynamic programming, treating z_t as the one state variable at time t. The formulation can then be extended in various ways without difficulty. But if a plant becomes obsolete after a fixed number of years, then z_t no longer defines all the information about past construction that is relevant to the future, and dynamic programming is no longer attractive.

2. BRANCH AND BOUND (TREE SEARCH) METHODS

From a computational point of view, the most fundamental fact about integer programming is that the conventional strategy for numerical optimization will not work. This conventional strategy is generally known as "hill climbing". It is to take a trial solution, see if any small change improves it and if so to modify the trial solution in the indicated direction and if not to stop. If suitably implemented, this strategy converges to a local optimum. The simplex method for linear programming is such a method. The linear programming problem is a special case of maximizing a concave function in a convex region, so we know from mathematical theory that a local optimum must then be a global optimum. But integer programming problems typically have many local optima that are much worse than the global optimum, and a hill-climbing method is not much use if we do not know which hill to climb. So what are we to do?

We could try to climb all the hills and see which is highest. But how can we be (reasonably) sure that we have located all the high hills? Rinnooy Kan and Timmer (1985) discuss the strengths and weaknesses of this approach. But, quite apart from the fact that they do not consider the specific problems of how to analyse integer variables, there is a fundamental difference between their stochastic approach and integer programming. Integer programming works with a globally valid explicit formulation of the objective function and constraints. It is clear that if, and only if, this can be done reasonably simply, the approach has more potential than one that makes no assumptions about the behaviour of the objective and constraint functions outside the immediate neighbourhoods of trial solutions.

Most methods for solving integer programming problems use the following strategy: we wish to maximize some function $f(\underset{\sim}{x})$ subject to the constraint that $\underset{\sim}{x} \varepsilon R$. Suppose that we can find a point $\underset{\sim}{x}^1$ that maximizes $f(\underset{\sim}{x})$ subject to the constraint that $\underset{\sim}{x} \varepsilon R_1$, where R_1 includes R. This is much easier than the original problem if $f(\underset{\sim}{x})$ is concave and R_1 is convex but R is not convex, since we can use a hill-climbing method such as the simplex method to find a local optimum, which is then necessarily a global optimum. The problem is obviously not much harder if R_1 consists of the union of a moderate number of convex sets. Now if $\underset{\sim}{x}^1 \varepsilon R$, it solves the original problem. Otherwise we modify R_1 so as to exclude $\underset{\sim}{x}^1$ but no point in R, and solve the modified problem. More generally, if we have found a point $\underset{\sim}{x}^0$ in R, called the <u>Incumbent Solution</u> by Geoffrion and Marsten (1972), then we need only insist that the modified R_1 does not exclude any points in R for which $f(\underset{\sim}{x}) > f(\underset{\sim}{x}^0)$. Whenever we find a point $\underset{\sim}{x}$ in R with $f(\underset{\sim}{x}) > f(\underset{\sim}{x}^0)$, this becomes the new incumbent. This process continues until we have a solution to the modified problem that is in R, or until R_1 is empty. Either situation means that the incumbent is a guaranteed optimum.

The original region R_1 is usually taken as the set of points satisfying all the equality and inequality constraints, but omitting the requirement that variables must take integer values. So in particular zero-one variables are treated as continuous variables with lower bounds of zero and upper bounds of one. For a normal integer programming problem the task of maximizing $f(\underset{\sim}{x})$ subject to $\underset{\sim}{x} \varepsilon R_1$ is then a linear programming problem. The solution to this problem is called the <u>Continuous Optimum</u>. The modifications to R_1 usually consist of adding linear inequality constraints. If a single inequality is added, it is known as a <u>Cutting Plane</u>. This approach was pioneered by Gomory (1958), and many alternative types of cutting plane have been developed since. But the apparently more pedestrian approach of adding two alternative linear inequalities has proved more reliable. This was called <u>Branch and Bound</u> by Little et al (1963). This means that the original linear programming problem is replaced by two linear programming subproblems. And at subsequent steps of the algorithm one subproblem is replaced by two new ones unless its optimum solution is either in R or has a value no better than $f(\underset{\sim}{x}_0)$.

In practice, the inequalities added are usually more restrictive lower or upper bounds on the integer variables. The size of the linear programming subproblems then stays constant, and indeed it is a considerable advantage from a data-processing point of view that the only differences between the subproblems are in the bounds on the integer variables. So R_1 is defined as the union of a set of subproblems. The following data are stored for each subproblem:

(a) the lower and upper bounds for each integer variable.

(b) the optimum basis for the LP subproblem from which the new subproblem was generated.

(c) a guaranteed upper bound U on the <u>value</u> of the subproblem. The value of a subproblem is defined as the maximum value of $f(\underset{\sim}{x})$ for any $\underset{\sim}{x} \varepsilon R$ that also satisfies the additional constraints of the subproblem.

and

(d) some estimate E of this value.

Let v_j denote the value of x_j in the optimum solution to any LP subproblem. So in particular v_0 denotes the value of x_0, the objective function being maximized. These quantities are available when deciding how to branch on a subproblem. They do not need to be stored.

The branch and bound process is often represented as a <u>Tree Search</u>. Each node of the tree represents a subproblem. The root represents the original linear programming relaxation of the problem; and when any node or subproblem is explored, branches (or arcs) are drawn from this node to the nodes representing any subproblems created from it. The list of the alternative subproblems to be solved is therefore called the <u>list of unexplored nodes</u>. This tree-structure representation may clarify the logic of the method. It does not need to be stored explicitly in the computer.

The implementation of the process is discussed further in Section 4 below, in a way that applies both to integer variables and the other types of Global Entity discussed in Section 3.

3. OTHER GLOBAL ENTITIES

The introduction of integer variables is a natural way to formulate a mathematical programming problem in which a local optimum is not necessarily a global optimum. But other concepts enable some of these problems to be tackled more efficiently. The variables, or sets of variables, introduced in this way are known as Global Entities. The name Global Entity perhaps needs some justification. The procedure for solving integer programming problems used to be called Integer, but the name was widely considered illogical when applied to problems containing no integer variables. So it was changed to Global. A Global Entity is therefore an entity recognized by the procedure Global.

A semi-continuous variable is a mathematically trivial but practically useful type of global entity described in a wide-ranging survey by Land and Powell (1979). It is a variable that may be either zero or else between one and some larger specified upper bound. It was introduced originally to simplify the solution of blending problems when materials must be excluded from the blend if they cannot be used in significant quantities. If the jth material must be used at a level between L_j and U_j, or else not at all, then the obvious formulation is to define x_j as a continuous variable defining how much is used, and δ_j as a zero-one variable to say if it is used. Then the conditions can be represented by introducing 2 constraints per material as follows:

$$x_j - L_j\delta_j \geq 0$$

$$x_j - U_j\delta_j \leq 0.$$

These ensure that if $\delta_j = 0$ then $x_j = 0$, while if $\delta_j = 1$, $L_j \leq x_j \leq U_j$.

But the need for these extra constraints, and the possibility of a trial solution with $x_j \geq L_j$ but $\delta_j < 1$, are avoided by replacing x_j and δ_j by a single semicontinuous variable y_j with an upper bound of U_j/L_j, where $L_j y_j$ now defines the amount of the $j^{\underline{th}}$ material used.

There is an important class of problems that are linear or integer programming problems except for the presence of some nonlinear nonconvex functions of continuous scalar arguments. This applies in particular to economies of scale being represented in the objective function, but it may also apply to constraints. Markowitz and Manne (1957) showed how such problems can be reduced explicitly to mixed integer programming, but this reduction has proved clumsy and inefficient. It is also unnecessary, because one can use a simple extension of the so-called λ-separable formulation due to Miller (1963). This assumes that any nonlinear function f(z) is adequately approximated by a piecewise linear function between $z = Z_0$ and $z = Z_K$, and that the argument z cannot lie outside this range. Then, if one introduces nonnegative variables λ_k and the constraints

$$\sum_{k=0}^{K} \lambda_k \qquad = \quad 1,$$

which is called the <u>convexity row</u>, and

$$\sum_{k=0}^{K} Z_k\lambda_k \;-\; z \qquad = \quad 0,$$

which is called the <u>reference row</u>, the nonlinear function f(z) can be replaced by the linear function

$$\sum_{k=0}^{K} f(Z_k)\lambda_k,$$

provided that no non-adjacent λ_k are nonzero. If the linear programming solution does not satisfy this condition, then the problem can be subdivided into two subproblems, with reduced ranges for z, by choosing a suitable value of k, say k_1, and saying that

<u>either</u> $\lambda_k = 0$ for all $k > k_1$

<u>or</u> $\lambda_k = 0$ for all $k < k_1$.

This is equivalent to choosing a Reference Row Entry Z_{k_1}, and setting $z \le Z_{k_1}$ in one subproblem and $z \ge Z_{k_1}$ in the other.

This formulation allows up to 2 members of this set to be nonzero, provided that they are adjacent. So the sets are known as Special Ordered Sets of Type Two, or S2 Sets. They were introduced by Beale and Tomlin (1970).

The same paper also introduced Special Ordered Sets of Type One, or S1 Sets, as sets in which at most one member may be nonzero. They can be used for multiple-choice problems, and in particular to restrict z to one of the specified values Z_k. The branching rules are almost the same: in one branch $\lambda_k = 0$ for all $k > k_1$, and in the other $\lambda_k = 0$ for all $k \le k_1$.

Each special ordered set is then a global entity.

The combination of special ordered sets with automatic interpolation of additional variables λ_k in relevant places proved to be reasonably straightforward in principle. It is described in Beale and Forrest (1976). This allows the optimization of problems containing genuinely nonlinear functions of scalar arguments.

A further generalization, known as Chains of Linked Ordered Sets, is described in Beale (1980). They were introduced to handle product terms. In particular, if the terms $x_i f_i(y)$ occur in different constraints with different variables x_i and possibly different functions $f_i(y)$ of the same scalar argument y, then all these terms are linear in the x_j once the value of y has been chosen. A chain of linked ordered sets allows one to start by considering a range of possible values of y and to narrow down this range by branch and bound operations.

The formal definition of a chain may be clarified by first considering its application to 2-dimensional separable programming. To represent some general function f(x,y) over the region

$$X_{MIN} \le x \le X_{MAX}, \qquad Y_{MIN} \le y \le Y_{MAX},$$

Miller (1963) suggests defining J+1 values of x, say X_j for $0 \le j \le J$ with $X_0 = X_{MIN}$, $X_{j-1} < X_j$ for $1 \le j \le J$

and $X_J = X_{MAX}$, and K + 1 values of y, say Y_k for $0 \le k \le K$ with $y_0 = Y_{MIN}$, $Y_{k-1} < Y_k$ for $1 \le k \le K$ and $Y_K = Y_{MAX}$. Then we can define $(J+1)\times(K+1)$ nonnegative variables λ_{jk} for $0 \le j \le J$, $0 \le k \le K$, representing weights on the points $x = X_j$ and $y = X_k$. This means that if we now introduce the constraints

$$\sum_j \sum_k \lambda_{jk} = 1 \tag{3.1}$$

$$\sum_j \sum_k X_j \lambda_{jk} - x = 0 \tag{3.2}$$

$$\sum_j \sum_k Y_k \lambda_{jk} - y = 0, \tag{3.3}$$

then the linear function $\sum_j \sum_k f(X_j, Y_k)\lambda_{jk}$ approximates the nonlinear function $f(x,y)$, provided that the function is adequately approximated by linear interpolation between all points (X_j, Y_k) for which the corresponding weights λ_{jk} are nonzero.

To control the combinations of variables that may take nonzero values, we can introduce further constraints. If we write

$$\sum_k \lambda_{jk} - \xi_j = 0 \quad \forall j \tag{3.4}$$

$$\sum_j \lambda_{jk} - \eta_k = 0 \quad \forall k, \tag{3.5}$$

and define the ξ_j and η_k as two S2 Sets, then in the final solution all the λ_{jk} will vanish except for those associated with the vertices of one of the elementary rectangles in the grid. In other words, all λ_{jk} will vanish except for $\lambda_{j_1 k_1}, \lambda_{j_1+1, k_1}, \lambda_{j_1, k_1+1}$ and λ_{j_1+1, k_1+1} for some j_1 and k_1.
The formulation will therefore correctly represent $f(x,y)$ to the extent that it is approximately linear within any rectangle defined by the constraints

$$X_{j_1} \le x \le X_{j_1+1} \qquad Y_{k_1} \le y \le Y_{k_1+1}$$

for some j_1 and k_1.

It will be noted that, unless $f(X_{j_1},Y_{k_1}) - f(X_{j_1+1},Y_{k_1}) - f(X_{j_1},Y_{k_1+1}) + f(X_{j_1+1},Y_{k_1+1}) = 0$, this formulation allows a range of fitted values for $f(x,y)$ within the rectangle. This may not be considered a serious objection; and if it is it can be overcome in various ways. One way is to restrict y to one of the specified values Y_k. This is appropriate if y represents say the diameter of a pipeline which must be chosen from a set of standard diameters supplied by the manufacturer, and it is in any case acceptable if the solution algorithm allows automatic interpolation of additional values Y_k where relevant.

The restriction of y to one of the values Y_k can be achieved by redefining the η_k as an S1 Set. But a more compact approach is to omit the η_k and the constraints (3.5) from the model and to define the λ_{jk} as a chain of J+1 sets of linked ordered sets. The definition of a chain is that in the final solution a single reference row entry must be selected for the chain, and all variables in all sets in the chain that do not have this reference row entry must vanish. So for this model, (3.3) would be chosen as the reference row for all sets in the chain, and the permitted variables λ_{jk} then all correspond to the same value of y. No such restriction is imposed at the continuous optimum, but during the branch and bound process the ranges of permitted values of the reference row entries are tightened between one subproblem and the next in the same way as for unlinked special ordered sets. The only difference is that the bounds for all sets in a chain are tightened in parallel.

Three further aspects of this formulation are worth noting:

If the same variable y occurs in several functions, say $f_i(x_i,y)$, each having a possibly different other argument x_i, then the sets of variables defining the possible values of y for any given value of the other argument can all be members of a single chain.

Although the approach can be used for general functions of two arguments by making the grid of values X_j fine enough, the approach is particularly attractive if $f(x,y)$ is linear in x for any given value of y, since the accuracy of the representation does not then depend on the fineness of the grid. So, as noted earlier, the approach is particularly appropriate to the representation of products of the form $x_i f_i(y)$.

Finally, this method of handling products of two arguments can in principle be extended to products of three or more arguments. For example $x_1 x_2 x_3$ can be written as zx_3 where $z = x_1 x_2$. But the best way to exploit these concepts in practice remains uncertain. Some formulations introduce an excessive discrepancy between the continuous optimum and the true optimum.

4. ALGORITHMIC DETAILS

The general description of the branch and bound operation in Section 2 left 6 important questions unanswered. These are:

(Q1) Which subproblem should be explored next?

(Q2) How should it be solved?

(Q3) On which global entity should we branch?

(Q4) How should we partition the possible values of this global entity between the new subproblems being created?

(Q5) How should the bound U be computed?

(Q6) How should the estimate E be computed?

These are addressed in this section.

Concerning Q1, there is a theoretical argument for always exploring the subproblem with the highest value of U, since this minimizes the number of subproblems created before completing the search. This is sometimes called a Breadth First Strategy, since it starts expanding the tree on a broad front. But the argument assumes that the branching procedure does not depend on the value $f(\underset{\sim}{x}^0)$ of the incumbent solution. The strategy makes no attempt to find any good integer solutions early in the search, and is subject to 5 objections.

(i) The branching procedure can sometimes be improved if we have a good value of $f(\underset{\sim}{x}^0)$.

(ii) A good value of $f(\underset{\sim}{x}^0)$ reduces the work on some subproblems, since a subproblem can be abandoned before it is solved as soon as we know that its value cannot exceed $f(\underset{\sim}{x}^0)$. The value of $f(\underset{\sim}{x}^0)$ can sometimes be used to refine the subproblem while it is being solved, as will be shown later.

(iii) We can often solve one of the subproblems that have just been created without recomputing the triangular factors of the basis.

(iv) The list of unexplored nodes may grow very long and create computer storage problems.

(v) No useful results may be obtained if the search has to be abandoned before a global optimum has been found.

To illustrate the first of these objections, note that if the reduced cost of any nonbasic integer variable exceeds $v_0 - f(\underset{\sim}{x}^0)$, then no solution better than the incumbent can be found if this variable changes by a single unit, so the variable can be fixed at its current trial value in this subproblem. Similar remarks apply to other types of global entity. The methods are very similar to those noted below for fixing nonbasic variables while solving a subproblem.

An extreme alternative to this strategy was used by Little et al (1963), and in early work on large-scale integer programming, see Beale (1968). This is "Last In First Out" or Depth First. In this strategy the last subproblem created is selected, provided only that the bound U exceeds $f(\underset{\sim}{x}^0)$: and otherwise the most recently created subproblem with a bound that is better than $f(\underset{\sim}{x}^0)$ is selected. (A subproblem with a smaller bound is deleted.) This strategy minimizes the number of subproblems in the list at any time, and if the other questions are answered suitably it usually finds a good solution early, which can be used to improve the rest of the search. It also allows the subproblems in the list to be stored sequentially, which was convenient in the 1960's when magnetic tape was the standard backing store. But the approach has been found inadequate and has been discarded.

There is not much wrong with Last In First Out when branching on one of the two subproblems just created, particularly if the more promising of the two is chosen. But when neither subproblem leads to any new subproblem, Last In First Out involves backtracking through all the most recently created subproblems, and this is often unrewarding.

Practical integer programming codes now contain many options, but the following answer to Q1 is often as good as any:

Divide the algorithm into stages. At each stage, consider a pair of subproblems that were created at the same time, and explore either or both of them. Choose the pair containing the subproblem with the highest value of E; but if any new subproblems were created in the previous stage restrict the choice to one of these. Having selected a pair of subproblems, explore the one with the higher value of E. Then explore the other (if it is still in the list) unless a new subproblem has just been created with at least as good a value of E.

Once an incumbent has been found, it is rather better to use the percentage error criterion due to Forrest et al (1974). This is to maximize $(U - f(\underset{\sim}{x}^0))/(U - E)$ rather than E when choosing subproblems. The motivation for this criterion is as follows. The value of the subproblem is less than or equal to U, and our estimate of the degradation from U in this value is $(U - E)$. If the true degradation is $\theta(U - E)$, this still improves on the incumbent if

$$\theta < (U - f(\underset{\sim}{x}^0))/(U - E).$$

So the subproblem for which this ratio is maximized is in a sense the one most likely to yield a better solution.

Question Q2 asks how to solve each LP subproblem. Since we start with a dual feasible basis, it is natural to use some dual method, particularly as this always provides an upper bound on the value of the subproblem and we can abandon the subproblem as soon as this upper bound ceases to be greater than $f(\underset{\sim}{x}^0)$. It is less clear whether one should use a standard dual method or parametric variation of the right hand side vector, which may include the lower and upper bounds on some variables.

Once we have an incumbent solution, there is scope for further refinement of the subproblem while it is being solved. If the variable x_q selected to enter the basis is not a continuous variable, we should check whether it can be fixed at its current trial value and rejected as an incoming basic variable. This check is based on a Lagrangian relaxation of the problem in which the constraints are multiplied by their shadow prices and added to the objective function, and all constraints other than convexity rows and bounds on the nonbasic variables are relaxed. If $\bar{a}_{00}$ denotes the current trial value of the objective function, and $\bar{a}_{0q}$ the reduced cost of x_q, then x_q can be fixed at its current trial value in the subproblem and rejected as the incoming basic variable under the following circumstances:

(i) x_q is an integer variable and $\bar{a}_{00} - |\bar{a}_{0q}| \leq f(\underset{\sim}{x}^0)$. (Note that $\bar{a}_{0q}$ will only be negative if x_q is nonbasic at its upper bound.)

(ii) x_q is a semi-continuous variable with current trial value zero and $\bar{a}_{00} - \bar{a}_{0q} \leq f(\underset{\sim}{x}^0)$.

(iii) x_q is a member of an unlinked S1 Set with a convexity row whose right hand side is B and whose slack has a reduced cost of d_s and

$$\bar{a}_{00} - B \min(\bar{a}_{0q}, d_s) \leq f(\underset{\sim}{x}^0).$$

(iv) x_q is a member of an unlinked S2 Set with neighbours x_{q-1} and x_{q+1} and a convexity row whose right hand side is B and whose slack has a reduced cost of d_s and

$$\bar{a}_{00} - B \min(\bar{a}_{0q-1}, \bar{a}_{0q}, \bar{a}_{0q+1}, d_s) \leq f(\underset{\sim}{x}^0).$$

(v) x_q is a member of a chain, and the Lagrangian relaxation suggests that all members of the chain with this reference row entry must be zero.

This brief discussion illustrates the fact that, while the procedures for handling different types of global entity vary in detail, the strategy can be carried out in a uniform way for all types.

Question Q3 is perhaps the most important and the most difficult. We must try to branch first on the most important global entities. To see this, note that a first branch on an irrelevant entity roughly doubles the size of the problem without making any progress towards solving it. Often the problem formulator can assess the relative importance of these entities. So he should have the option to assign them priorities, and the code will then always branch on some unsatisfied global entity with the highest priority.

If no priorities are assigned, or if there is more than one unsatisfied global entity with the highest priority, then some numerical criterion must be used to choose between them. This can be done by considering the possibility of branching on each in turn and estimating the resulting values of either U or E.

Early work on integer programming emphasized the bound U. Driebeek (1966) introduced the term penalty for a lower bound on the extent to which the value of an LP subproblem is reduced below v_0 by driving the value of an integer variable from its current trial value to the nearest integer in some particular direction. He showed how these penalties can be computed by implicitly carrying out one step of the dual simplex method. Tomlin (1971) showed how these penalties can sometimes be sharpened at a negligible computing cost by using the fact that no nonbasic integer variable can be increased to a value less than one. But practical problems often have several nonbasic variables with zero reduced costs, when these methods are largely useless.

It therefore seems best to concentrate on the estimates E. These can be derived in various ways, but a natural one is based on the Lagrangian relaxation discussed earlier.

A linear programming subproblem can be expressed as

Maximize x_0 subject to

$$\begin{aligned} x_0 + \sum_j a_{0j}x_j &= b_0 \\ \sum_j a_{ij}x_j &= b_i \quad \forall i>0 \end{aligned} \tag{4.1}$$

together with lower and upper bounds on individual variables.

In the course of solving this subproblem we will compute shadow prices π_i such that the optimum solution to the original subproblem also maximizes x_0 subject to

$$x_0 + \sum_j (\sum_i \pi_i a_{ij}) x_j = \sum_i \pi_i b_i \tag{4.2}$$

where $\pi_0 = 1$, together with the same bounds on individual variables.

Now if some global entity is unsatisfied at the solution to the LP subproblem, then we will probably have to change the contribution of the global entity to the left hand side of (4.1) in order to satisfy it. If we increase the left hand side of the $i^{\underline{th}}$ row of (4.1) by z_i for all i, then we can estimate the degradation D in the objective function value of the LP subproblem by

$$D = \sum_i \pi_i z_i, \tag{4.3}$$

since this is the resulting decrease in the value of x_0 defined by (4.2).

The computation of z_i depends on the global entity. For example if x_k is an integer variable with a trial value of $n_k + f_k$, with n_k integer, then it can be satisfied by increasing the trial value by $1 - f_k$, when $z_i = a_{ik}(1 - f_k)$.

But (4.3) may not give a realistic estimate of D. It defines a guaranteed lower bound, which will often be zero.

We can find an upper bound on the degradation in terms of the minimum and maximum shadow prices π_{MINi} and π_{MAXi}. Brearley et al (1975) show how to derive these bounds. In particular, note that $\pi_{MIN0} = \pi_{MAX0} = 1$, and $\pi_{MINi} \geq 0$ if the $i^{\underline{th}}$ constraint is a less-than-or-equal-to inequality.

If P denotes the set of rows for which $z_i > 0$, and N the set of rows for which $z_i < 0$, then

$$\sum_i \pi_i z_i \leq D \leq \sum_{i \varepsilon P} \pi_{MAXi} z_i + \sum_{i \varepsilon N} \pi_{MINi} z_i .$$

So we may write

$$D = \sum_i (\pi_i z_i + r_i),$$

where $0 \leq r_i \leq (\pi_{MAXi} - \pi_i) z_i$ for $i \varepsilon P$

and $0 \leq r_i \leq (\pi_{MINi} - \pi_i) z_i$ for $i \varepsilon N$.

But these upper bounds are not necessarily useful and may be infinite. It seems that only heuristic methods can provide realistic estimates for r_i. One approach is to write

$$r_i = (\min(|\pi_i| + T_{OL}, \pi_{MAXi} - \pi_i)) z_i \quad \text{for } i \varepsilon P$$

and

$$r_i = (\max(- (|\pi_i| + T_{OL}), \pi_{MINi} - \pi_i) z_i \quad \text{for } i \varepsilon N,$$

where T_{OL} is a small positive tolerance.

Given some such formulae for computing D from the z_i, it is then easy to estimate degradations from satisfying either integer variables or semi-continuous variables. Indeed we can easily estimate two degradations; one from driving the variable down to the integer part of its current trial value, and the other from driving it up to this number plus one.

For special ordered sets, and still more for chains of linked ordered sets, the computation of the estimated degradation is not so straightforward. But for any given reference row entry one can estimate the degradation, and hence one can find a reference row entry giving a (possibly local) minimum of the estimated degradation.

We then branch on the global entity giving the largest estimated degradation. For global entities giving two such estimates, there is an argument for choosing the larger. There is a possibly stronger argument for choosing the smaller. So this answers Q3.

Question Q4 is trivial for integer variables: on one branch the upper bound is made the integer part of the current trial value of the variable, while on the other branch the lower bound is made this number plus one. Similarly, semi-continuous variables are always satisfied unless the current trial value lies between zero and one, in which case the same rule applies.

For special ordered sets and for chains, we may define the current interval as the interval between the smallest and largest reference row entries for which any member of the set (or chain) takes a nonzero value. We must then branch somewhere in this interval. It is often appropriate to branch at or near the <u>Average Reference-Row Entry</u> $\bar{X}$.

For an unlinked set, this is defined by

$$\bar{X} = \sum_k X_k \lambda_k / \sum_k \lambda_k ,$$

where λ_k denotes the value of the $k^{\underline{th}}$ member of the set and X_k its reference row entry. For a chain,

$$\bar{X} = \sum_s \sum_k X_k \lambda_{sk} / \sum_s \sum_k \lambda_{sk} ,$$

where λ_{sk} denotes the value of the $k^{\underline{th}}$ member of the $s^{\underline{th}}$ set in the chain, and X_k its reference row entry.

There does not seem to be any simple rule for deciding when it is not appropriate to branch at the average reference row entry. On the other hand the number of extra nodes that have to be explored may be greatly increased by branching at an inapproriate point. So a plausible strategy is to try branching at the average reference row entry, to solve both of the resulting LP subproblems and to compute the estimated further degradations from satisfying the set (or chain) in both subproblems. If these estimated degradations are approximately equal, then the branch point is acceptable. But if one estimated degradation is much larger than the other, then the branch point should be moved so that it subdivides the subproblem giving the larger degradation.

Concerning Question Q5, the bound U is usually taken as the value of v_0 in the LP subproblem from which the new problem will be generated. But a more sophisticated approach based on some form of Lagrangian relaxation may be worthwhile.

Question Q6 is usually answered by summing the estimated degradations from v_0 required to satisfy each global entity.

5. AUTOMATIC MODEL REFORMULATION

Ever since integer programming was first applied to practical problems, some 20 years ago, it has been realized that some models with thousands of integer variables can be solved quite easily, while other models with less than 100 integer variables produce no useful answers. It soon became clear that some models can be made much more amenable to integer programming by reformulation. This can often be done without changing the physical significance of the model at all, by introducing constraints that become redundant when all the integer variables are restricted to integer values, but which are far from redundant without these restrictions. The importance of such reformulation has been stressed by various authors, notably Williams (1978) and Williams (1985). A good general strategy for solving such problems is to start with the "obvious" formulation and hope for the best, but if the solution is not found after a moderate number of branching operations to study the original continuous optimum solution to see if some simple additional constraints can be introduced to reduce the discrepancy between the continuous and integer optima. The next step is to automate this reformulation process. Major pioneering work was carried out in this direction by Crowder et al (1983) on pure integer programming models. This work is both surveyed and extended by Hoffman and Padberg (1985). These ideas have been extended to mixed integer programming, notably in unpublished work by L.A. Wolsey and T.J. Van Roy at CORE. Automatic model reformulation can reasonably be regarded as the most important current area of integer programming research.

A brief introduction to some relevant ideas follows:

So-called M coefficients can be tightened. This refers to constraints of the form $\sum_j A_j x_j - M\delta \leq 0$, where the A_j are positive coefficients, the x_j are nonnegative variables and

δ is a zero-one variable. If M is large enough, this constraint ensures that each $x_j = 0$ if $\delta = 0$, while imposing no restriction if $\delta = 1$. Mathematically, all sufficiently large M are equivalent. But computationally it is important to make M as small as possible. So, as part of the standard Reduce (or Presolve) procedure, one can compute upper bounds on the x_j say U_j, and M can be replaced by $\min(M, \sum_j A_j U_j)$.

New constraints can be generated from a constraint of the type considered above. Specifically $x_j - U_j \delta \leq 0$ for each j, and if this constraint is not satisfied at the continuous optimum it can usefully be added to the model.

Given any constraint of the form $\sum_j A_j x_j \leq B$, where the x_j are all restricted to integer values and the A_j are not all equal, one can solve a knapsack problem to find an upper bound on any function of the form $\sum_j N_j x_j$. Such a constraint may be worth adding to the model. The bound can sometimes be tightened if some of the x_j have a small upper bound, or if more than one belongs to the same S1 Set. In particular, Crowder et al (1983) emphasize the <u>cover constraint</u>

$$\sum_{j \varepsilon C} x_j \leq |C| - 1,$$

which must hold if the x_j are all zero-one variables, the A_j are all positive and $\sum_{j \varepsilon C} A_j > B$. Here C is any set containing $|C|$ elements.

When no further constraints can usefully be added in this way, the branch and bound process is started. During the process a Dynamic Presolve can be useful. Each constraint is turned into an equation with a right hand side of zero, by adding slack variables and transferring the constant term to the left hand side. Bounds (which may be infinite) are computed for each continuous variable, and hence upper and lower bounds are computed for the sum of the contributions of the continuous variables and the global entities to each row individually. If the lower bound is say L_{Bi}, then the problem would be infeasible if the contribution of any continuous variable or global entity exceeded its minimum possible value by more than L_{Bi}. This fact may enable us to tighten the bounds on some continuous variable or, more relevantly, on some global entity.

Similar remarks apply to the upper bounds, and each tightening operation can tighten the bounds on several rows.

Dynamic presolve is particularly helpful on multi-time-period models in which some nonlinear function occurs with an argument that is a nondecreasing function of time. For example, Beale (1983) describes a model for the long term development of an offshore gas field in which p_{SQt}, the reservoir pressure squared at the end of the tth time period, is a convex function of q_{CUMt}, the cumulative gas production. Now a high value of reservoir pressure squared facilitates the flow of gas, other things being equal, so the LP relaxation of any subproblem uses the largest available approximation to p_{SQt} for any assumed value of q_{CUMt}. If Q_{MINt} and Q_{MAXt} are the minimum and maximum permitted values of q_{CUMt}, and $p_{SQ}(Q_{MINt})$ and $p_{SQ}((Q_{MAXt})$ are the corresponding values of pressure squared, then the subproblem will use the equations

$$p_{SQt} = p_{SQ}(Q_{MINt})\lambda_{MINt} + p_{SQ}(Q_{MAXt})\lambda_{MAXt},$$

where

$$\lambda_{MINt} + \lambda_{MAXt} = 1$$

and

$$Q_{MIN}\ \lambda_{MINt} + Q_{MAX}\lambda_{MAXt} - q_{CUMt} = 0.$$

Now if we branch on the value $q_{CUMt} = Q_{VALt}$, so that in one new subproblem Q_{MINt} is replaced by Q_{VALt} and in the other new sub-problem Q_{MAXt} is replaced by Q_{VALt}, then the approximation to p_{SQt} will become more realistic. But if we could exploit the fact that q_{CUMt} is a monotonically nondecreasing function of t, we would automatically reduce Q_{MAXt_1} to Q_{VALt} for all $t_1 < t$ when Q_{MAXt} is reduced to Q_{VALt}, and we would automatically increase Q_{MINt_1} to Q_{VALt} for all $t_1 > t$ when Q_{MINt} is increased to Q_{VALt}. This obviously sharpens the branch and bound process. In fact we considered implementing this as a special facility. But then we realized that a dynamic presolve, generalized in the obvious way to special ordered sets, would take care of this situation automatically. So we implemented this instead. This generalization is reasonably straightforward in the usual situation when the convexity row has no slack, and a right hand side of B. Then if the kth member has an entry a_k in some row, the bounds on the contribution of the set to the row are $B \min_k a_k$ and $B \max_k a_k$.

But if the first member is nonzero, the contribution becomes Ba_0 if the set is an S1 Set, or between Ba_0 and Ba_1 if the set is an S2 Set. These values may demonstrate that the first member cannot be nonzero without making the whole problem infeasible.

6. ACKNOWLEDGEMENT

This paper is based in part on research sponsored by the BP Venture Research Unit, carried out in Scicon by O.S. Brooker and D. Kennedy under my supervision.

REFERENCES

E.M.L. Beale (1968) Mathematical Programming in Practice (Pitmans. London).

E.M.L. Beale (1980) "Branch and bound methods for numerical optimization of non-convex functions" COMPSTAT 80 Proceedings in Computational Statistics Edited by M.M. Barritt and D. Wishart pp 11-20 (Physica Verlag. Wien).

E.M.L. Beale (1983) "A mathematical programming model for the long-term development of an off-shore gas field" Discrete Applied Mathematics 5 pp 1-9.

E.M.L. Beale and J.J.H. Forrest (1976) "Global optimization using special ordered sets" Mathematical Programming 10 pp 52-69.

E.M.L. Beale and J.A. Tomlin (1970) "Special facilities in a general mathematical programming system for non-convex problems using ordered sets of variables" in Proceedings of the Fifth International Conference on Operational Research Ed. J. Lawrence pp 447-454. (Tavistock Publications. London).

A.L. Brearley, G. Mitra and H.P. Williams (1975) "Analysis of mathematical programming problems prior to applying the simplex method." Mathematical Programming 8 pp 54-83.

H. Crowder, E.L. Johnson and M.W. Padberg (1983) "Solving large-scale zero-one linear programming problems" Operations Research 31 pp 803-834.

N. Driebeek (1966) "An algorithm for the solution of mixed integer programming problems" Management Science 12 pp 576-587.

J.J.H. Forrest, J.P.H. Hirst and J.A. Tomlin (1974) "Practical solution of large mixed integer programming problems with UMPIRE" Management Science 20 pp 736-773.

A.M. Geoffrion and R.E. Marsten (1972) "Integer Programming Algorithms: A framework and state-of-the-art-survey" Management Science 18 pp 465-491.

R.E. Gomory (1958) "Outline of an algorithm for integer solutions to linear programs" Bulletin of the American Mathematical Society 64 pp 275-278.

A.H. Land and S. Powell (1979) "Computer codes for problems of integer programming" in Annals of Discrete Mathematics 5: Discrete Optimization Edited by P.L. Hammer, E.L. Johnson and B.H. Korte. pp 221-269 (North Holland Publishing Company. Amsterdam).

K.L. Hoffman and M.W. Padberg (1985) "LP-Based combinatorial problem solving". This volume.

J.D.C. Little, K.C. Murty, D.W. Sweeney and C. Karel (1963) "An algorithm for the traveling salesman problem" Operations Research 11 pp 972-989.

H.M. Markowitz and A.S. Manne (1957) "On the solution of discrete programming problems" Econometrica 25 pp 84-110.

C.E. Miller (1963) "The simplex method for local separable programming" in Recent Advances in Mathematical Programming Edited by R.L. Graves and P. Wolfe pp 89-100 (McGraw Hill. New York).

A.H.G. Rinnooy Kan and G.T. Timmer (1985) "Stochastic methods for global optimization" This volume.

J.A. Tomlin (1971) "An improved branch and bound method for integer programming" Operations Research 19 pp 1070-1075.

H.P. Williams (1978) Model Building in Mathematical Programming (Wiley. Chichester).

H.P. Williams (1985) "Model Building in Linear and Integer Programming" This volume.

MODEL BUILDING IN LINEAR AND INTEGER PROGRAMMING

H.P. Williams
Faculty of Mathematical Studies
University of Southampton, U.K.

Abstract

This paper surveys the topic of model building in mathematical programming discussing, (i) the systematisation of model building, including the use of Matrix Generating Languages, (ii) the use of Boolean Algebra for formulating 0 - 1 integer programming models and the efficient formulation of integer programming models considering both their facial structure and the desirability of creating meaningful dichotemies for the branch-and-bound tree search, (iii) the desirability and possibility of converting models to network flow models, (iv) the building of stable models.

Keywords, Model Building, Mathematical Programming.

NATO ASI Series, Vol. F15
Computational Mathematical Programming
Edited by K. Schittkowski

1. INTRODUCTION

Model Building in Linear and Integer Programming is a difficult subject to discuss in isolation from specific applications. Hence this paper will be illustrated at many points by particular examples. There is, nevertheless, a certain amount of theory and experience, of a general kind, that has been developed which it is worth organising into as coherent a form as possible.

Three objectives should apply when building a Mathematical Programming Model. These are (i) Accuracy and Relevance, (ii) Understandability, and (iii) Computational Tractability. Relatively little in general can be said about (i) since each application needs to be treated individually. It is, however, clearly important, through sensitivity analysis, to examine the potential effects of inaccuracy in data on computed solutions. Section 10 briefly examines this aspect. Also it is obviously necessary to exercise judgement in only modelling those aspects of a problem which are relevant to obtaining the solution required. While such observations may appear trite they are of major importance in the successful practice of the "art" of model building.

Most of this paper will be concerned with objectives (ii) and (iii) since this is where some degree of generality is possible. In Sections 2, 3 and 4 we pay attention to (ii) and the remaining sections apply to (iii).

Much of the discussion in this survey paper will refer to other peoples work. Appropriate references will be given as an alternative to a lengthy description.

2. SYSTEMATISATION IN BUILDING MODELS

As Mathematical Programming models have got larger the need for systematisation has got greater. The linking together of large amounts of information in a model demands systematic naming conventions for variables, constraints and objectives. Relatively little attention has been paid to these somewhat mundane considerations. The topic has, however, been given attention by Beale, Beare and Tatham (1974). Connected with such considerations is the use of Matrix Generators which is discussed in Section 2.

Most practical models contain a large Network Flow component. Building and understanding such a model is greatly eased if sign conventions are adhered to regarding flows into and out of nodes. A sensible convention (which accords with accounting practice) is to give flows out of a node a negative coefficient and flows into a node a positive coefficient. A further advantage of being systematic is that it then becomes possible to identify network structures in a model, often to computational advantage. This topic is considered in Section 9.

The Network Flow representation of a model can be generalised to model different types of process with a variety of input/output relations. For example, we can distinguish processes where inputs are in flexible proportions (e.g. blending) from processes where the chemistry demands inputs in fixed proportions. Similarly some processes (e.g. distillation and cracking) fix the proportions of the outputs while other processes allow flexible proportions of the outputs. The modelling of systems with a variety of interconnected processes of different types is greatly helped by different notational conventions which then become translated into algebraic relationships. A way of doing this has been developed by Müller-Merbach (1978).

Applicable to certain types of process is the idea of modal formulations. With these processes it is more satisfactory to let variables represent levels of extreme modes of operation rather than quantities of material flowing into and out of a process. Figure 1 illustrates a case of this (taken from Wagner (1969)).

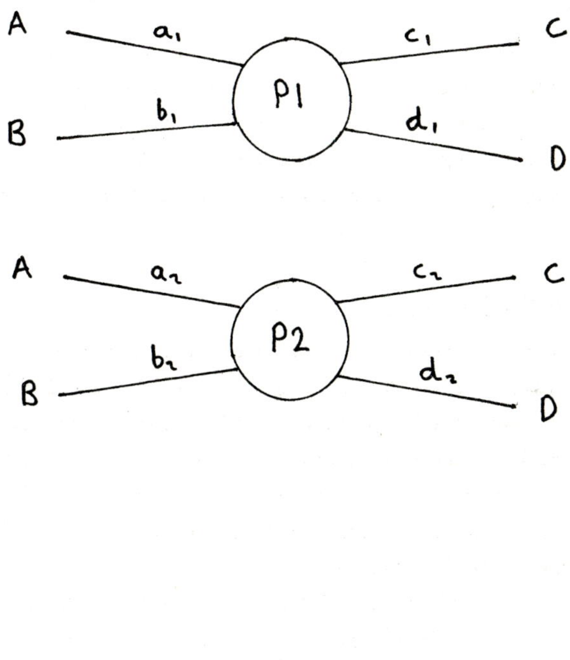

Figure 1.

P1 and P2 are two processes which demand inputs and outputs in fixed proportions $a_1 : b_1 : c_1 : d_1$ and $a_2 : b_2 : c_2 : d_2$ respectively. One possible model is to introduce 8 variables s, t, u, v, w, x, y, z representing the flows into and out of P1 and P2. The constraints are then

$$s + u \leq \text{Crude A availability}, \quad w + y = \text{Demand for C}$$
$$t + v \leq \text{Crude B availability}, \quad x + z = \text{Demand for D}$$

$$b_1 s - a_1 t = 0 , \quad c_1 s - a_1 w = 0 , \quad d_1 s - a_1 x = 0$$
$$b_2 u - a_2 v = 0 , \quad c_2 u - a_2 y = 0 , \quad d_2 u - a_2 z = 0$$

The modal formulation has 2 variables p_1 and p_2 representing levels of P1 and P2 and constraints

$$a_1 p_1 + a_2 p_2 \leq \text{Crude A availability}, \quad c_1 p_1 + c_2 p_2 = \text{Demand for C}$$
$$b_1 p_1 + b_2 p_2 \leq \text{Crude B availability}, \quad d_1 p_1 + d_2 p_2 = \text{Demand for D}$$

If the number of modes of operation of a process is few such formulations will be more compact as well as easier to interpret. Dantzig-Wolfe Decomposition can be regarded as the application of modal formulations to submodels.

3. THE USE OF MATRIX GENERATORS AND LANGUAGES

The structural similarities of Linear Programming (LP) and Integer Programming (IP) models across a wide variety of applications makes it efficient to use a Matrix Generator (MG) for the building of models. It is, in the author's opinion, better to think of such a system as a language (ML). For some other methods in Operational Research (e.g. Simulation) it has been common for a long time to use modelling languages. With Mathematical Programming the development has been more recent. The advantages of using an MG/L system can be summarised as

(a) The Input Format is usually more natural. Most Mathematical Programming Packages use the MPS fixed format which requires thinking in terms of columns instead of constraints. Although MG/L systems output this format they take a more natural input.

(b) The Data is usually separated from the MG/L Model allowing the modeller to think about the structure of the model independently of the data.

(c) Debugging of a model is made easier in an analogous way to the considerations which apply to computer programming.

(d) Modification is made easier. Models frequently undergo change and development.

(e) Repetition is automated. Models often involve similar products, multiple plants and time periods resulting in replication. MG/Ls automatically index row and column names to indicate this.

A variety of MG/L systems exist. There is an excellent survey by Fourer (1983) which also contains a comprehensive discussion of the subject.

The design of MG/L systems (as with programming languages) is to a large extent governed by subjective considerations. It is, however, the author's opinion that an MG/L should, as far as possible, mirror conventional mathematical notation. This is because such notation is both extremely well suited to modelling mathematical programmes and is also widely known by model builders (hence making the learning of such a language easier). In addition the following features are desirable.

(a) Automatic indexing of row and column names for repetition.
(b) Notation (Σ) for repeated sums.
(c) Relations between indices e.g. $I < J$.
(d) Separation of the data from the structure of the model.
(e) Arithmetic operations for generating coefficients within the language.
(f) Interactive syntax checking.
(g) Automatic formatting (e.g. MPS).
(h) Facilities for comment statements.
(i) User friendliness.

In order to illustrate the above considerations we give some examples taken from the language MAGIC (MAtrix Generator Input Convertor) described by Day (1984) and Day and Williams (1982). It is emphasised that this language is used here for illustrative rather than promotional purposes.

Example 3.1

The statement

```
DEM(T=1,MAXT) : SIGMA I = 1,MAX I : OUT(I,T)> = DEMAND(T) ;
```

generates constraints named DEM1, DEM2 , etc. involving the MP variables OUT11, OUT12 , etc.. DEMAND(T) is a one dimensional array of data defined previously in the MAGIC program or read from an external file.

Example 3.2

The statement

```
GUAR(T=1,MAXT) : SIGMA I = 1,MAXI :
    MAXLEVEL(I) * NUM(I,T)> = [1.15 * DEMAND(T)] ;
```

generates constraints GUAR1, GUAR2 , etc. involving the MP variables NUM11, NUM12 , etc. with coefficients defined in the array MAXLEVEL(I) .

Example 3.3

The statement

```
TCOST(MIN) : -SIGMA I = 1,NDEPTS :
    SIGMA J = 1,NCITIES - 1 : BENEFIT(I,J) * D(I,J)
  + SIGMA I = 1,NDEPTS : SIGMA J = 1,NCITIES :
    SIGMA K = 1,NDEPTS,K > I : SIGMA L = 1,NCITIES :
    [COMM(I,K) * DIST(J,L)] * G(I,J,K,L) ;
```

represents the following objective function (for a quadratic assignment problem).

$$\text{Minimise} \quad - \sum_{\substack{i=1,d \\ j=1,c-1}} B_{ij}\delta_{ij} + \sum_{\substack{i=1,d \\ j=1,c \\ k=1,d:k>i}} C_{ik}D_{j\ell}\gamma_{ijk\ell}$$

where the variables δ_{ij}, $\gamma_{ijk\ell}$ are represented by

D11, D12 , etc. G1121, G1122 , etc. .

In order to illustrate further features of MAGIC we present the full model from which the above three examples are taken. This model (and MAGIC) are discussed at greater length in Williams (1985). One small point should, however, be made here. The use of FOR ... END loops is an optional alternative to the incrementing inside brackets demonstrated in Examples 3.1 and 3.2. While convenient it could be objected that this was an undesirable option since it deviates from standard mathematical convention. This MAGIC model generates a file (TARIFILE) in MPS format.

```
NAME:TARIFF ;                              {Job name }

MAXI=3 ;                                   { Parameter assignments }
MAXT=5 ;

MINLEVEL(MAXI)=0.85,1.25,1.5 ;
MAXLEVEL(MAXI)=2,1.75,4 ;
COSTPH(MAXI)=1,2.6,3 ;
EXCOSTPH(MAXI)=2,1.3,3 ;
EXCTMINL(MAXI)=1.7,1.625,4.5 ;             { Coefficient assignments }
STARTCOST(MAXI)=2,1,0.5 ;
DEMAND(MAXT)=15,30,25,40,27 ;
NHOURS(MAXT)=6,3,6,3,6 ;
NUMBER(MAXI)=12,10,5 ;

OUT(MAXI:MAXT) ;
NUM(MAXI:MAXT) INTEGER ;                   { Variable definitions }
NST(MAXI:MAXT) INTEGER ;
                          { Objective function }

COST(MIN):SIGMA I=1,MAXI:SIGMA T=1,MAXT:
  ([EXCOSTPH(I)*NHOURS(T)]*OUT(I,T) + STARTCOST(I)*NST(I,T) +
  [COSTPH(I)*NHOURS(T)]*NUM(I,T)-[EXCTMINL(I)*NHOURS(T)]*NUM(I,T));

                          { Constraint statements }
FOR T=1,MAXT ;
  DEM(T):SIGMA I=1,MAXI:OUT(I,T) >= DEMAND(T) ;
  GUAR(T):SIGMA I=1,MAXI:MAXLEVEL(I)*NUM(I,T) >= [1.15*DEMAND(T)];
  FOR I=1,MAXI ;
    MIN(I:T):OUT(I,T) - MINLEVEL(I)*NUM(I,T) >= 0 ;
    MAX(I:T):OUT(I,T) - MAXLEVEL(I)*NUM(I,T) <= 0 ;
    ST(I:T,T>1):NST(I,T) - NUM(I,T) + NUM(I,T-1) >= 0 ;
    ST(I:T,T=1):NST(I,T) - NUM(I,T) + NUM(I,MAXT) >= 0 ;
    BOUNDS(I:T):NUM(I,T) <= NUMBER(I) ;
    BOUNDS(I:T):NST(I,T) <= NUMBER(I) ;
  END ;
END ;
MPS TARIFILE ;            { MPSX output format to file TARIFILE }
STOP ;
```

4. USING BOOLEAN ALGEBRA TO FORMULATE 0 - 1 INTEGER PROGRAMMING MODELS

For Integer Programming (IP) models there is much more scope for variation in possible representations than there is for Linear Programming (LP) models. In most IP models the integer variables can only take two values 0 or 1 . We therefore concentrate on the formulation of this type of model. The formulation of logical and non-linear conditions using 0 - 1 variables can be complex (and open to error). One possible way of helping systematise such modelling is to use Boolean Algebra.

Boolean Algebra (the Propositional Calculus) can be regarded as the algebra of statements which are either True or False. In order to make use of this algebra we must regard constraints as being true or false. In order to reason about such constraints we will associate with each of them a 0 - 1 variable δ .

$\delta = 1$ corresponds to the constraint being True.

$\delta = 0$ corresponds to the constraint being False.

In order to illustrate how this idea may be applied we use two examples. Further examples are given in Williams (1985).

Example 4.1

$$\delta = 1 \rightarrow \sum_j a_j x_j \leq b$$

where "$\rightarrow$" is the Boolean Algebra connective "implies" .
This condition may be re-expressed as

$$1 - \delta = 0 \rightarrow \sum_j a_j x_j - b \leq 0 .$$

If M is an upper bound for $\sum_j a_j x_j - b$ then the inequality

$$\sum_j a_j x_j - b \leq M(1-\delta)$$

clearly suffices to impose the condition.
This inequality is more conventionally written as

$$\sum_j a_j x_j + M\delta \leq M + b .$$

Clearly so long as M is large enough to be an upper bound for the expression $\sum_j a_j x_j - b$ then any values of M will suffice. In practice it is desirable to make M as realistic as possible for reasons which will be discussed in Section 7.

Example 4.2

$$\sum_j a_j x_j \leq b \rightarrow \delta = 1 \; .$$

Logically this statement is equivalent to its contrapositive

$$\delta = 0 \rightarrow \sum a_j x_j > b \; .$$

Converting this into the form of example 4.1 and introducing a small number $\epsilon > 0$ we can produce

$$\sum_j a_j x_j - (m-\epsilon)\delta \geq b + \epsilon$$

to model the condition where m is a lower bound for the expression $\sum_j a_j x_j - b$.

Examples 4.1 and 4.2 provide machinery which is sufficient to model any logical conditions among constraints. Once 0 - 1 "indicator" variables are introduced to represent the truth or falsity of constraints further constraints can be introduced, involving these variables, to model logical relations between them.

While such machinery may provide a systematic means of modelling it will not always provide the "best" model in terms of computational efficiency as discussed in Section 6. There is scope for further development in this area.

5. REDUCTION TECHNIQUES FOR LINEAR PROGRAMMES

A number of techniques exist for reducing the size of LP models by removing constraints and variables. These techniques can be automated to give preprocessors (known by names such as REDUCE, ANALYSE, PRESOLVE etc.) for MP systems. The main aim of such reductions is to reduce overall solution time although the recognition of redundancies should also clarify one's knowledge of the problem being modelled.

One method of reduction, taking account of PRIMAL and DUAL concepts, is described by Brearley, Mitra and Williams (1975). Developments of this method as well as other methods are given in Karwan, Lotfi, Telgen and Zionts (1983).

Recent research by the author has suggested that it is sensible to combine such procedures (which apply to a model) with CRASH procedures (which apply to the Simplex Algorithm). It is well known that the application of a CRASH procedure to find a good starting basis for the Simplex Algorithm can greatly reduce overall solution time. One of the most valuable functions of CRASH is to, as far as possible, replace artificial variables (in "=" constraints) by structural variables. If there are free (i.e. non sign-constrained) variables present in a model then these can be introduced into the basis. Clearly it would be desirable to replace artificial variables by free variables. In practice it is usually possible, by means of a REDUCE procedure, to make many variables free i.e. recognise that the conventional non-negativity conditions are automatically satisfied. For example suppose we have the sign pattern below in a model

x_1	x_2	x_3	x_4	
$-$	$+$			$\geqslant \oplus$
$\ominus$	$-$	$+$		$\geqslant \oplus$
$\ominus$	$\ominus$	$-$	$+$	$\geqslant \oplus$

$+$ indicates a positive coefficient.

$-$ indicates a negative coefficient.

$\oplus$ indicates a non-negative coefficient.

$\ominus$ indicates a non-positive coefficient.

It is easy to see that

$$x_1 \geqslant 0 \rightarrow x_2 \geqslant 0 \rightarrow x_3 \geqslant 0 \rightarrow x_4 \geqslant 0 .$$

Therefore x_2, x_3 and x_4 could be regarded as free variables since their non-negativity is implied by that of x_1 .

In order to detect such a structure it is desirable to "triangularise" the matrix as has been done above in order to show that it is necessary to look at the variables in the order x_1, x_2, x_3, x_4 . Triangularisation can also be used to avoid "circular arguments" which spuriously imply free variables. This complication is discussed by Tomlin and Welch (1982). Such a triangularisation can also be used by CRASH to bring quickly the variables x_2, x_3 and x_4 into the basis by avoiding updating their columns of coefficients.

The topic of modal formulations discussed in Section 2 could also have been considered under the heading of this section. Conversion of part of a model into a modal formulation can be automated by the dual of Fourier-Motzkin elimination as described by Williams (1984).

6. COMPUTATIONALLY EFFICIENT INTEGER PROGRAMMING FORMULATIONS

In view of their computational difficulty the efficient formulation of Pure and Mixed Integer Programmes is even more important than for LPs . Theoretically it might seem desirable to reformulate any IP model with constraints representing the facets of the convex hull of feasible integer solutions. This is illustrated by Figure 2.

The bold lines give the constraints corresponding to the LP relaxation. By replacing these constraints by those corresponding to the fainter lines we obtain a formulation which will always yield an integer solution to an optimal vertex solution of the LP relaxation.

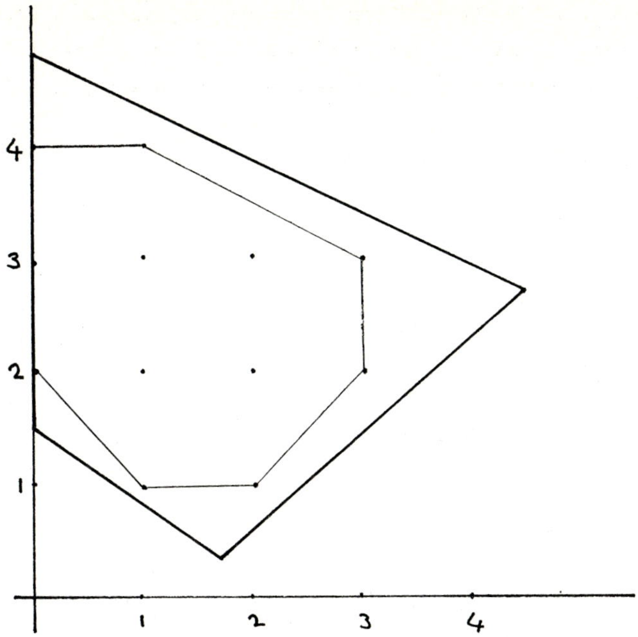

Figure 2.

In a general IP model, however, such a reformulation would be impractical since it would be (i) computationally very expensive and (ii) there would be an astronomic number of new constraints. There are, however, models where it is possible, either totally or partially, to carry out such a reformulation. We will therefore examine some circumstances where it is possible to "tighten" (or "sharpen") some constraints in order to reduce the size of the reasible region of the LP relaxation.

Example 6.1

Consider the constraint

$$\delta_1 + \delta_2 + \ldots + \delta_n - n\delta \leqslant 0$$

where δ_i, $\delta \in \{0,1\}$.

This can be reformulated by the constraints

$$\delta_1 - \delta \leqslant 0$$
$$\delta_2 - \delta \leqslant 0$$
$$\vdots$$
$$\delta_n - \delta \leqslant 0$$

The LP relaxation corresponding to the second formulation excludes fractional solutions admitted by the first.

If all constraints are of this form the matrix of coefficients can be shown to be totally unimodular and the constraints represent facets of the convex hull of feasible integer solutions.

Example 6.2 Multiperiod Planning

Formulation 1:

$\delta_{it} = 1$ if plant i is used in period t
$= 0$ otherwise

n_i = Number of periods for which the option to use plant i, in current, or future years is kept open.

At most N plants can be used in any year

$$t \in \{1,2,...,T\}$$

$$\sum_i \delta_{it} \leqslant N \quad \forall t$$

$$t\delta_{it} - n_i \leqslant 0 \quad \forall i,t \; .$$

Formulation 2:

Replace variables n_i by

$\gamma_{it} = 1$ if plant i "open" in year t
$= 0$ otherwise

$$\sum_i \delta_{it} \leqslant N \quad \forall t$$

$$\delta_{it} - \gamma_{it} \leqslant 0 \quad \forall i,t$$

$$-\gamma_{it} + \gamma_{it+1} \leqslant 0 \quad \forall i,t \neq T \; .$$

It is not immediately obvious why the second formulation is "tighter". This can be seen by adding the constraints of the second formulation together.

$$t[\delta_{it}-\gamma_{it}\leqslant 0]$$

$$+\sum_{\lambda=1}^{t-1} \lambda[-\gamma_{i} +\gamma_{i\gamma+1}\leqslant 0]$$

$$+\sum_{\mu=t+1}^{T} \mu(-\gamma_{i\mu}\leqslant 0)$$

to give

$$t\delta_{it} - \sum_{t=1}^{T} \gamma_{it} \leqslant 0 .$$

The original variable n_i in the first formulation represents the expression $\sum_{t=1}^{T} \gamma_{it}$ demonstrating that the desired sum of constraints above is equivalent to the last constraint of formulation 1. By adding constraints together we weaken their effect in the LP relaxation so demonstrating that the second formulation is tighter.

This model is discussed by Williams (1978).

Example 6.3 Piecewise Linear Approximations

Two standard methods exist for modelling a non-linear function of a single variable (a separable function). These are known as the "δ-formulation" and the "λ-formulation", and will be illustrated by the function y (of x) in Figure 3.

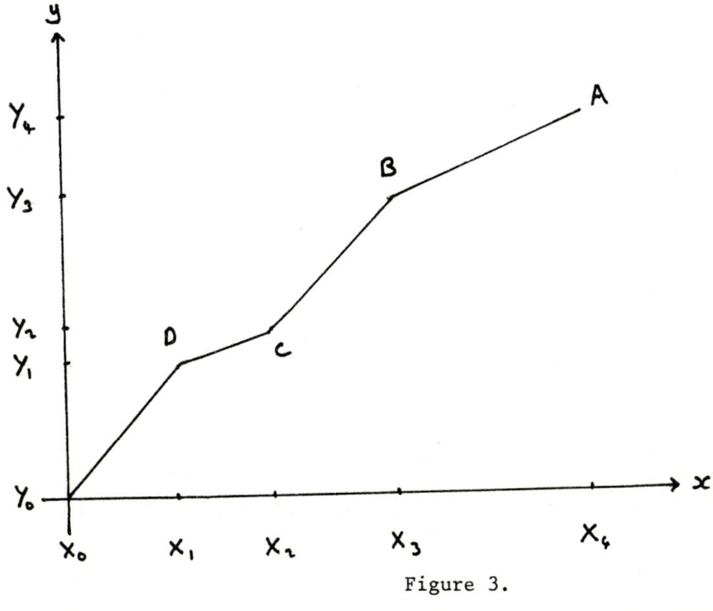

Figure 3.

The δ-formulation:

x and y are related indirectly through the new variables δ. The δ variables are not to be confused with integer variables. It is unfortunate that the symbol δ has become standard notation for this type of formulation.

$$x = (X_1-X_0)\delta_1 + (X_2-X_1)\delta_2 + (X_3-X_4)\delta_3 + (X_4-X_3)\delta_4$$
$$y = (Y_1-Y_0)\delta_1 + (Y_2-Y_1)\delta_2 + (Y_3-Y_2)\delta_3 + (Y_4-Y_3)\delta_4$$
$$0 \leqslant \delta_1,\delta_2,\ldots,\delta_4 \leqslant 1 .$$

If any $\delta > 0$ all preceding δ must be 1 and all succeeding δ must be 0.

This last condition is that which demands extra integer variables and logical relations between these variables. If the condition is ignored we obtain a relaxation in which (x,y) can represent any point in the rectangle $Y_0Y_4AX_4$ rather than points on this piecewise linear curve Y_0DCBA.

The λ-formulation:

x and y are now related indirectly through the new variables λ.

$$x = X_0\lambda_0 + X_1\lambda_1 + X_2\lambda_2 + X_3\lambda_3 + X_4\lambda_4$$
$$y = Y_0\lambda_0 + Y_1\lambda_1 + Y_2\lambda_2 + Y_3\lambda_3 + Y_4\lambda_4$$
$$\lambda_0 + \lambda_1 + \lambda_2 + \lambda_3 + \lambda_4 = 1 .$$

At most 2 adjacent λs can be non-zero.

This last condition is that which demands an integer programming formulation. The relaxation obtained by ignoring this condition allows (x,y) to represent points within the convex hull $ABDY_0C$.

Clearly the λ-formulation gives a more restricted relaxation than the δ-formulation making it computationally easier to solve.

7. REFORMULATION TECHNIQUES FOR INTEGER PROGRAMMES

Although it is impractical to generate facet constraints for general IP models there are procedures for partially doing this for special (simple) types of IP model. Attention has mainly been confined to Pure 0 - 1 IP models although Padberg, Van Roy and Wolsey (1982) have recently looked at a special class of Mixed 0 - 1 IP models.

More attention is given to this specialist topic in the paper by Hoffman and Padberg (this volume). Another relevant paper (besides those mentioned below) is Guignard and Spielberg (1981). Therefore in this paper we will restrict our attention to small numerical examples in order to explain some results.

Example 7.1 The 0 - 1 Knapsack Polytope

There is no loss of generality in considering a "$\leqslant$" constraint with all coefficients positive.

$$23\delta_1 + 21\delta_2 + 19\delta_3 + 17\delta_4 + 14\delta_5 + 13\delta_6 + 13\delta_7 + 9\delta_8 \leqslant 70$$

$$\delta_i \in \{0,1\} .$$

A Minimal Cover is a minimal set of indices (i.e. no proper subset has the property) for which the sum of the corresponding coefficients exceeds the right-hand-side,

e.g. {1,2,3,8},{1,3,4,5},...,{2,3,4,5},...,{1,5,6,7,8},...,{3,4,5,6,7}

are all examples of minimal covers. Hammer and Nguyen (1972) show how all such covers can be generated in order of increasing cardinality.

Clearly minimal covers could be used to produce corresponding constraints e.g.

$$\delta_1 + \delta_2 + \delta_3 + \delta_8 \leqslant 3$$

$$\delta_1 + \delta_3 + \delta_4 + \delta_5 \leqslant 3$$
$$\delta_2 + \delta_3 + \delta_4 + \delta_5 \leqslant 3$$
$$\delta_1 + \delta_5 + \delta_6 + \delta_7 + \delta_8 \leqslant 4$$

$$\delta_3 + \delta_4 + \delta_5 + \delta_6 + \delta_7 \leqslant 4 .$$

It has, however, been demonstrated by a number of authors Balas (1975), Hammer, Johnson and Peled (1975), Wolsey (1975) and Padberg (1979) that stronger (e.g. "tighter") constraints than this can be produced. In order to do this we extend minimal covers (where possible) to produce Extended Covers. The extension consists of adding indices to a minimum cover which correspond to coefficients greater-than-or-equal-to the largest coefficient in the minimum cover. This immediately implies stronger forms of some of the constraints above, e.g.

$$\{\delta_1\} + \delta_2 + \delta_3 + \delta_4 + \delta_5 \leqslant 3$$

$$\{\delta_1+\delta_2\} + \delta_3 + \delta_4 + \delta_5 + \delta_5 + \delta_7 \leqslant 4 ,$$

etc.

If a minimal cover gives rise to an extended cover which is not a proper subset of any other extended cover arising from a minimal cover of the same cardinality it is known as a strong cover.

The authors above demonstrate that all the constraints corresponding to extensions of all the strong covers encompass all the facets of the convex hull of solutions to the original constraints whose coefficients are 0 or 1 . Unfortunately there may be additional facets with coefficients larger than 1 . Nevertheless the constraints so generated are usually very strong (e.g. "tight"), cheap to generate, and therefore well worth adding to the original model in order to reduce the size of the feasible region of the LP relaxation.

Experience in automating these ideas is discussed by Crowder, Johnson and Padberg (1981).

In practice there will be a large number of facets even for the 0 - 1 knapsack polytope. A more modest aim is to reformulate the original constraint as another single constraint with smaller coefficients. Such a constraint will almost certainly be "tighter". In order to demonstrate this we use the same example as before e.g.

$$23\delta_1 + 21\delta_2 + 19\delta_3 + 17\delta_4 + 14\delta_5 + 13\delta_6 + 13\delta_7 + 9\delta_8 \leqslant 70$$

$$\delta_i \in \{0,1\} .$$

The coefficients have deliberately been placed in decreasing order.

Bradley, Hammer and Wolsey (1974) give the following method.

(i) Find "ceilings". These are lexicographically maximal sets of indices for which the sum of coefficients is less-than-or-equal to the right-hand-side, i.e. no set properly containing the set has this property nor does any set obtained by interchanging an index with one to the left in the ordering.
In this example the ceilings are {1,2,3},{1,2,4,8},{1,2,6,7},{1,3,5,6}, {2,3,4,6},{2,5,6,7,8} .

(ii) Find "roofs". These are lexicographically minimal sets of indices for which the sum of coefficients exceeds the right-hand-side and are minimal with respect to this property, i.e. no proper subset has this property nor does any set obtained by interchanging an index with one to the right in the ordering.
In this example the roofs are {1,2,3,8},{1,2,5,7},{1,3,4,7},{1,5,6,7,8}, {2,3,4,5},{3,4,6,7,8} .

The roofs and ceilings of the constraint give a complete characterisation of its logical import (i.e. solution set). We therefore seek another constraint

$$a_1\delta_1 + a_2\delta_2 + \dots + a_8\delta_8 \leqslant a_0$$

with the same roofs and ceilings. The above roofs and ceilings can be used to give constraints on the values of the a_i ,

i.e.
$$a_1 + a_2 + a_3 \leqslant a_0$$
etc.
$$a_1 + a_2 + a_3 + a_8 > a_0$$
etc.

and
$$a_1 \geqslant a_2 \geqslant \dots \geqslant a_8 .$$

These conditions can be regarded as the constraints of a linear programme with a suitable objective e.g. Minimise a_0 . The solution of this LP gives the coefficients for the new constraint.

It is worth pointing out that this method is easily extendable to reducing coefficient size for general integer constraints. This is explained in Williams (1985).

Some 0 - 1 IP models (e.g. the Quadratic Assignment Problem) are most naturally formulated with non-linear terms. There are a number of approaches to linearising such terms (and so making the model amenable to standard solution methods). These are illustrated in the following numerical example.

Example 7.1

Linearise $2\delta_1\delta_2 + 3\delta_1\delta_3 - 4\delta_2\delta_3$

$\delta_i \quad \{0,1\}$.

Approach 1:

Replace each term by a 0 - 1 variable and relate it to the original variables by suitable constraints.

$\delta_1\delta_2$ is replaced by δ_{12} with the constraints

$$\delta_1 + \delta_2 - \delta_{12} \leqslant 1$$

$$\delta_1 + \delta_2 - 2\delta_{12} \geqslant 0 .$$

(Following the discussion in Section 6 there is merit in replacing this last constraint by

$$\delta_1 - \delta_{12} \geqslant 0$$

$$\delta_2 - \delta_{12} \geqslant 0 .)$$

The terms $\delta_1\delta_3$ and $\delta_2\delta_3$ are treated similarly. Clearly such a reformulation (when the terms are products of 2 variables) will result in as many new variables as there are product terms and 2 (or 3) times this number of constraints.

Approach 2:

Define a continuous variable w_1 to represent $\delta_1(2\delta_2+3\delta_3)$ and relations

$$\delta_1 = 0 \rightarrow w_1 = 0$$

$$\delta_1 = 1 \rightarrow w_1 = 2\delta_2 + 3\delta_3 .$$

These relations can be modelled by

$$w_1 \leqslant 2\delta_2 + 3\delta_3$$

$$w_1 \leqslant 5\delta_1$$

$$5\delta_1 + 2\delta_2 + 3\delta_3 - w_1 \leqslant 5 .$$

The term $-4\delta_2\delta_3$ would be (trivially) treated similarly. This formulation would result in one new (continuous) variable for each original 0 - 1 variable and 3 new constraints for each 0 - 1 variables.

Which of the above approaches is most economical will clearly depend on the structure of the model although the second approach avoids the combinatorial explosion in number of constraints which can result if there are many product terms. A restricted form of the first approach is used by Beale and Tomlin (1972). The second approach is due to Glover (1975) and is used to good effect by Ayles et al. (1978) and Stecke (1983). It is also discussed by Goldman (1983).

A common device in IP models (demonstrated in Section 4) is to use a "big M" coefficient in constraints such as that below

$$\sum_j a_j x_j - M\delta \leqslant b$$

where $\delta \in \{0,1\}$.

So long as M is an upper bound for $\sum_j a_j x_j - b$ the required logical condition is implied. It is, however, desirable to make M as realistic as possible. The main reason for making M as small as possible is to reduce the size of the feasible region of the LP relaxation although it is, of course, always desirable to reduce the magnitude of coefficients to reduce rounding errors.

A more general constraint than that above is

$$\sum_{j \notin s} a_j x_j + \sum_{j \in s} a_j x_j \leqslant b$$

$$s = \{j / a_j < 0 \text{ and } x_j \text{ integer}\} .$$

If U is an upper bound for $\sum_{j \notin s} a_j x_j - b$ then any a_j such that $|a_j| > U$ can be replaced by $-U$ for $j \in s$. It could be worthwhile performing a special optimisation by maximising $\sum_{j \notin s} a_j x_j - b$ in the LP relaxation in order to find as small a value of U as possible. This idea is discussed by Oley and Sjoquist (1982) and Crowder, Johnson and Peled (1981).

The theoretical study of IP formulation has only recently been attempted by Meyer (1981) and Jeroslow and Lowe (1984). From a "Theory of MIP Representability" a number of alternative IP formulations to the conventional ones are produced. In particular an alternative formulation of Disjunctive Constraints is produced.

Example 7.2 Disjunctive Constraints.

$$D_1\underline{x} \leqslant \underline{b}_1 \text{ or } D_2\underline{x} \leqslant \underline{b}_2 \text{ or } \dots \text{ or } D_n\underline{x} \leqslant \underline{b}_n$$

D_i are positive matrices.

Formulation 1: (conventional)

$$D_1\underline{x} + (\underline{M}_1-\underline{b}_1)\delta_1 \leqslant \underline{M}_1$$
$$D_2\underline{x} + (\underline{M}_2-\underline{b}_2)\delta_2 \leqslant \underline{M}_2$$
$$\vdots$$
$$D_n\underline{x} + (\underline{M}_n-\underline{b}_n)\delta_n \leqslant \underline{M}_n$$
$$\delta_1 + \delta_2 + \ldots + \delta_n \geqslant 1 .$$

$\underline{M}_i$ is a vector of upper bounds for $D_i\underline{x}$.

Formulation 2:

$$\underline{x} = \underline{x}_1 + \underline{x}_2 + \ldots + \underline{x}_n$$
$$D_1\underline{x}_1 - \underline{b}_1\delta_1 \leqslant 0$$
$$D_2\underline{x}_2 - \underline{b}_2\delta_2 \leqslant 0$$
$$\vdots$$
$$D_n\underline{x}_n - \underline{b}_n\delta_n \leqslant 0$$
$$\delta_1 + \delta_2 + \ldots + \delta_n = 1 .$$

Favourable computational experience with such alternative formulations is reported in Jeroslow and Lowe (1985).

8. MODELLING DICHOTEMIES BY INTEGER PROGRAMMING

The initial implementations of the Branch and Bound Algorithm only allowed branching (the creation of dichotemies) on integer variables. It was demonstrated by Beale and Tomlin (1969) that it was possible and desirable also to allow dichotemies over sets of variables known as Special Ordered Sets. Two types of set were considered.

Type 1 (S1 sets): At most 1 of $\{x_1,x_2,\ldots,x_n\}$ can be non-zero.

Dichotemy: $x_1 = x_2 = \ldots = x_i = 0$ or $x_{i+1} = x_{i+2} = \ldots = x_n = 0$.

Type 2 (S2 sets): At most 2 adjacent members of $\{\lambda_1,\lambda_2,\ldots,\lambda_n\}$ can be non-zero.

Dichotemy: $\lambda_1 = \lambda_2 = \ldots = \lambda_{i-1} = 0$ or $\lambda_{i+1} = \lambda_{i+2} = \ldots = \lambda_n = 0$.

In many practical situations the above dichotemies are natural ones to make and lead to a more efficient search for the Optimal Solution. It is possible explicitely to introduce extra 0 - 1 integer variables in the formulation of the model for each potential dichotemy above. Then it is not necessary to extend the Branch and Bound Algorithm to deal with sets.

Situations do arise where there is no one clear ordering of a set of variables and we might wish to create dichotemies over different orderings. This idea has been discussed by Jeffreys (1976) who terms such sets "Special Unordered Sets".

Example 8.1 A Special Unordered Set.

$\{x_{nbt}, x_{nbh}, x_{nct}, x_{nch}, x_{sbt}, x_{sbh}, x_{sct}, x_{sch}\}$

x_{ijk} represents an activity which is either

in the North (i=n) or South (i=s)

uses a Production Process which is Batch (j=b) or Continuous (j=c)

with Power which is Thermal (k=t) or Hydro (k=h) .

At most one of x_{ijk} can be non-zero.

Possible dichotemies on which it would be desirable to branch are

(i) $x_{njk} = 0$ or $x_{sjk} = 0 \quad \forall_{j,k}$

(ii) $x_{ibk} = 0$ or $x_{ick} = 0 \quad \forall_{i,k}$

(iii) $x_{ijt} = 0$ or $x_{ijh} = 0 \quad \forall_{i,j}$.

The algorithmic implementation of such dichotemies over sets would be complex. A more satisfactory solution is to represent each dichotemy by a 0 - 1 variable in the formulation of the model.

9. NETWORK FLOW FORMULATIONS

A surprisingly large class of LP and (seemingly) IP problems can be directly formulated as, or converted into, Network Flow models. The correspondence between a directed network and an LP matrix is given by the node-arc incidence matrix illustrated in Figure 4. We adopt the convention, mentioned in Section 2, of giving the column corresponding to an arc a negative coefficient in the row corresponding to its predecessor node and a positive coefficient in the row corresponding to its successor node. Clearly there is a one-to-one correspondence between directed networks and matrices with exactly one -1 and one +1 in each column. In order to convert an LP model to a Network Flow model it is therefore necessary to convert the matrix of coefficients into this form. We give a number of examples where this is known to be possible.

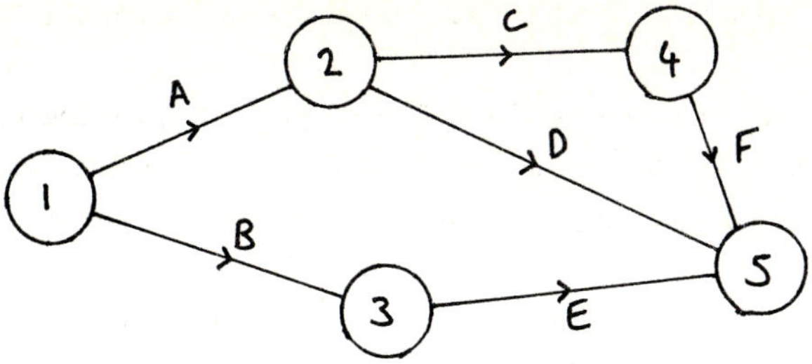

	A	B	C	D	E	F
1	-1	-1				
2	1		-1	-1		
3		1			-1	
4			1			-1
5				1	1	1

Figure 4.

<u>Example 9.1</u> Capacity Scheduling

x_{ij} = Number of items entering system at beginning of period i and leaving at beginning of period j subject to availabilities, capacities, demands, etc.

	x_{12}	x_{13}	x_{14}	x_{23}	x_{24}	x_{34}	u_1 (slacks)	u_2 (slacks)	u_3 (slacks)
A	1	1	1				1		
B		1	1	1	1			1	
C			1	1	1	1			1

By adding and subtracting constraints we can transform the model into the form below.

	x_{12}	x_{13}	x_{14}	x_{23}	x_{24}	x_{34}	u_1	u_2	u_3
A	1	1	1				1		
B - A	-1			1	1		-1	1	
C - B		-1				1		-1	1
- C			-1	-1	-1	-1			-1

demonstrating a network formulation.

Examples of such transformations are given by Dantzig (1969) and Veinott and Wagner (1973).

Example 9.2 Cyclic Rostering

This problem concerns manning over a repeating period of (say) 24 hours. If the period is divided into 6 intervals with each shift covering 2 intervals (say) we obtain the following model.

x_i = Number of people whose shift starts in interval i . We wish to maintain certain minimum manning levels in each interval.

							Surplus Variables					
	x_1	x_2	x_3	x_4	x_5	x_6	v_1	v_2	v_3	v_4	v_5	v_6
A	1	1					-1					
B		1	1					-1				
C			1	1					-1			
D				1	1					-1		
E					1	1					-1	
F	1					1						-1

This model can be transformed to a network form as indicated below.

	x_1	x_2	x_3	x_4	x_5	x_6	v_1	v_2	v_3	v_4	v_5	v_6
A - F		1				-1	-1					1
B - A	-1		1				1	-1				
C - B		-1		1				1	-1			
D - C			-1		1				1	-1		
E - D				-1						1	-1	
F - E	1				-1	1					1	-1

The transformation is possible if there are an even number of intervals in the period. This problem is given in Wagner (1969).

Example 9.3 Contractor Allocation

This problem is taken from Cheshire, McKinnon and Williams (1984) where the full details of the problem are discussed. We restrict our attention here to the essential details.

x_{ij} = Number of teams from Contractor i allocated to District j .

The constraints are

$\sum_{i \in I} x_{ij} = R_j$	Requirement of District j $\forall_j \in J$	(1)
$\sum_{j \in J} x_{ij} \leqslant A_i$	Available teams from Contractor i $\forall i \in I$	(2)
$x_{ij} - \delta_{ij} \geqslant 0$	$\delta_{ij} = 1$ implies Contractor i is represented in District j $\forall i \in I, j \in J$	(3)
$\sum_{i \in I} \delta_{ij} \geqslant N_j$	At least N_j contractors must be represented in District J $\forall j \in J$	(4)
$\sum_{i \in I' \subset I} \delta_{ij} \geqslant 1$	At least 1 "experienced" contractor must be present in District j $\forall j \in J$	(5)
$\delta_{ij} \leqslant 1$	and integer $\forall i \in I, j \in J$	

In order to convert this, apparently IP , model into a Network Flow model we can perform the following transformations:

(i) Introduce surplus variables u_{ij} into (3) and then use (3) to substitute x_{ij} out of (1) and (2).

(ii) Negate (2).

(iii) Introduce equations

$$\sum_{i \in I'} \delta_{ij} - y_j = 0 \qquad (6)$$

$$\sum_{i \in I} \delta_{ij} - z_j = 0 \qquad (7)$$

(iv) Use (6) to substitute δ_{ij} $\forall i \in I'$ out of (5).
Use (7) to substitute δ_{ij} $\forall i \in I$ out of (4) and (1),
Use (6) to substitute δ_{ij} $\forall i \in I'$ out of (7).

This results in the constraints

$$z_j + \sum_{i \in I} u_{ij} = R_j \quad \forall_j$$

$$- \sum_{j \in J} \delta_{ij} - \sum_{j \in J} u_{ij} \geq - A_i \quad \forall_i$$

$$\sum_{i \in I'} \delta_{ij} - y_j = 0 \quad \forall_j$$

$$\sum_{i \in I-I'} \delta_{ij} + y_j - z_j = 0 \quad \forall_j$$

$$y_j \geq 1 \quad \forall_j \, , \quad z_j \geq N_j \quad \forall_j \, , \quad \delta_{ij} \leq 1 \quad \forall_{i,j}$$

Apart from the simple bounding constraints on individual variables each variable appears in exactly two constraints with coefficients of -1 and +1 respectively. The transformed model therefore represents a Capacitated Network Flow model. A diagram of the network appears in the above reference. The key to this rather complex reformulation lies in the selection of subsets of the variables whose sums are to be represented by new variables through (iii) above. These new variables have simple bound constraints giving rise to capacitated arcs.

Another example of a transformation of an apparently IP model into a Network Flow model which requires the selection of subsets of variables to be represented by new bounded variables is given by Daniel (1973).

The great virtues of reducing models to Network Flow models are

(a) Specialist Network Flow algorithms allow such models to be solved very economically in space and time.

(b) The Network structure demonstrates that the model will always give integral vertex solutions (so long as the right-hand-side values are integer). For what might seemingly be IP models this has great virtue in demonstrating that computationally costly IP methods and software are not needed.

Three practical references to the subject are Glover and Klingman (1974), Glover, Hultz and Klingman (1978) and Bradley, Brown and Graves (1977).

An algorithm for either automatically converting an LP to a Network Flow model or showing that such a transformation is impossible is given by Bixby and Cunningham (1980).

Even if an LP (or IP) model is not transformable into a Network Flow model it is common to find models with large network components. The recognition or transformation into a model with network components can prove of great computational advantage. Daniel (1978) describes how the recognition of an embedded network structure was used to advantage in reformulating an otherwise intractible IP model. The optimisation of the first version of the model was abandoned after 4757 nodes of the tree search with no feasible solutions found. After reformulation the model was solved to (proved) optimality after 171 nodes.

Heuristic procedures exist for extracting large network components from LP (and IP) models. One such procedure with impressive computational results is described by Brown, McBride and Wood (1984).

It is also often worth examining the dual of a model for network structure as discussed by Williams (1982).

10. BUILDING STABLE MODELS

When the data of an LP or IP models change by a small amount it is often unralistic that the optimal solution should change radically. If this happens it can often be validly claimed that the model is a poor representation of reality. In practice, if for example, there is a small cutback in capacity it is more likely that extra capacity will be bought in than that the operating plan will be overturned. A good model should represent this. This can be done by the modelling of "soft" constraints with costed slack variables. A small example illustrates this.

Example 10.1 An Investment Problem

$$\text{Max} \quad 50x_1 + 40x_2 + 15x_3 + 10x_4 + 11x_5$$

$$\text{subject to:} \quad 10x_1 + 10x_2 + 5x_3 + 4x_4 + 3x_5 \leqslant B$$

$$x_i \in \{0,1\}$$

x_i represent 5 possible investments within a budget B.

If B is 20 the optimal solution is

$$x_1 = 1,\ x_2 = 1,\ x_3 = x_4 = x_5 = 0,\ \text{Objective} = 90\ .$$

If, however, B is reduced by a small amount the optimal solution becomes

$$x_1 = 1,\ x_2 = 0,\ x_3 = 1,\ x_4 = 0,\ x_5 = 1,\ \text{Objective} = 76\ .$$

In practice we would prefer to make

(i) The objective a continuous function of the right-hand-side.

(ii) The solution an upper semi-continuous function of the right-hand-side (i.e. continuous apart from changes at alternate solutions).

This may be done by introducing a costed slack variable into the model. The above example and further discussion of this important (and neglected) topic is given by A.C. Williams (1973).

11. REFERENCES

1. Ayles, P.S., E.M.L. Beale, R.C. Blues and S.J. Wild (1978). Mathematical Models for the Location of Government, Mathematical Programming Study 9, 59-74.

2. Balas, E. (1975). Facets of the knapsack polytope, Math. Prog., 8, 146-164.

3. Beale, E.M.L. and J.A. Tomlin (1969). Special facilities in a general mathematical programming system for non-convex problems using ordered sets of variables, in J. Lawrence (Ed.), Proc. 5th Int. Conf. of Operations Research, Tavistock, London.

4. Beale, E.M.L. and J.A. Tomlin (1972). An integer programming approach to a class of combinatorial problems, Maths. Prog. 1, 339-344.

5. Beale, E.M.L., G.C. Beare and P.B. Tatham (1974). The DOAE reinforcement and redeployment study: a case study in mathematical programming, in P.L. Hammer and G. Zoutendijk (Eds), Mathematical Programming in Theory and Practice, North Holland, Amsterdam.

6. Bixby, R.E. and W.H. Cunningham (1980). Converting Linear Programs to Network Problems, Mathematics of Operations Research, 5, 321-357.

7. Bradley, G.H., P.L. Hammer and L. Wolsey (1974). Coefficient reduction for inequalities in 0-1 variables, Math. Prog., 7, 263-282.

8. Brearley, A.L., G. Mitra and H.P. Williams (1975). Analysis of mathematical programming problems prior to applying the simplex algorithm, Math. Prog., 8, 54-83.

9. Brown, G.S., R.D. McBride and R.K. Wood (1984). Extracting Embedding Generalised Networks from Linear Programming Problems, Technical Report, Naval Postgraduate School, Monterey, California.

10. Cheshire, M., K.I.M. McKinnon and H.P. Williams (1984). The efficient allocation of private contractors to public works, Journal of the Operational Research Society, 35, 705-709.

11. Crowder, H., E.L. Johnson and M.W. Padberg (1981). Solving Large Scale Zero-One Linear Programming Problems, IBM Research Report RC8888, Yorktown Heights.

12. Daniel, R.C. (1978). Reducing computational effort in solving a hard integer programme, SIGMAP Association for Computing Machinery, Special Interest Group on Mathematical Programming, No. 25, December, 39-44.

13. Daniel, R.C. (1973). Phasing out capital equipment, Opl. Res. Q., 24, 113-116.

14. Dantzig, G.B. (1969), A Hospital Admission Problem. Technical Report N. 69-15, Stanford University, California.

15. Day, R.E. and H.P. Williams (1982). Magic: The Design and Use of an Interactive Modelling Language for Mathematical Programming, Dept. Business Studies, Working Paper 2/82, University of Edinburgh.

16. Day, R.E. (1984). MAGIC: User's Guide, Dept. Business Studies, University of Edinburgh.

17. Fourer, R. (1983). Modelling languages versus matrix generators for linear programming, ACM Transactions on Mathematical Software, 9, 143-183.

18. Glover, F. and D. Klingman (1974). Real world applications of network problems and breakthroughs in solving them efficiently, Research Report CC5159, Center for Cybernetic Studies, University of Texas at Austin.

19. Glover, F. (1975). Improved Linear Integer Programming Formulations of Non-linear Integer Problems, Management Science, 22, 455-459.

20. Glover, F., J. Hultz, D. Klingman and J. Stutz (1978). Generalised Networks: A Fundamental Computer-Based Planning Tool, Management Science, 24, 1209-1220.

21. Goldman, A.J. (1983). Linearisation on 0-1 variables. A clarification, Operations Research 31, 946-947.

22. Guignard, M. and K. Spielberg (1981). Logical reduction methods in zero-one programming. Minimal preferred inequalities. Operations Research 29, 49-74.

23. Hammer, P.L. and S. Nguyen (1972). "APOSS: A Partial Order in the Solution Space of Bivalent Programs, C.R.M.-163, Centre de Recherches Mathematiques, Universite de Montreal.

24. Hammer, P.L., E.L. Johnson and U.N. Peled (1975). Facets of regular 0-1 polytopes, Math. Prog., 8, 179-206.

25. Hoffman, K. and M. Padberg (1985), LP Based Combinatorial Problem Solving, This volume.

26. Jeffreys, M. (1976). The Next Generation of Branch and Bound Codes, Paper presented at IX International Symposium on Mathematical Programming, Budapest, Hungary, August.

27. Jeroslow, R.G. and J.K. Lowe (1983). Modelling with Integer Variables, Report No. 83270-OR, Institute für Okonometre und Operations Research, University of Bonn.

28. Jeroslow, R.G. and J.K. Lowe (1985). Experimental Results with the New Techniques for Integer Programming Formulations, Journal of the Operational Research Society, to appear.

29. Karwan, M.H., V. Lotfi, J. Telgen and S. Zionts (1983). Redundancy in Mathematical Programming: A State of the Art Survey, Springer-Verlag, New York.

30. Meyer, R.R. (1981). A Theoretical and Computational Comparison of 'Equivalent' Mixed Integer Formulations, Naval Research Logistics Quarterly, 28, 115-131.

31. Müller-Merbach, H. (1978). Entwurf von input-output-Modellen, in K. Brockoff et al. (Eds), Proz. Ops. Res., Physica-Verlag, Wurzberg, Wien, 7, 521-531.

32. Oley, L.A. and R.J. Sjoquist (1982). Automatic Reformulation of Mixed and Pure Integer Models to Reduce Solution Time in APEX IV, Paper presented at ORSA/TIMS Meeting, San Diego, California, October.

33. Padberg, M.W. (1979). Covering, Packing and Knapsack Problems, Annals of Discrete Mathematics, 4, 265-287.

34. Padberg, M.W., T.J. Van Roy and L.A. Wolsey (1982). Valid Linear Inequalities for Fixed Charge Problems, Core Discussion Paper No. 8232.

35. Stecke, K.E. (1983). Formulation and Solution of Nonlinear Integer Production Planning Problems for Flexible Manufacturing Systems, Management Science, 29, 273-287.

36. Tomlin, J.A. and J.S. Welch (1982). A Pathological Case in the Reduction of Linear Programs, Operations Research Letters, to appear.

37. Veinott, A.F. and H.M. Wagner (1962). Optimal capacity scheduling - I, Ops. Res., 10, 518-532.

38. Wagner, H.M. (1969). Principles of Operations Research, Prentice Hall, Englewood Cliffs.

39. Williams, A.C. (1973). Some Modelling Principles for MIP's, 8th International Mathematical Programming Symposium, Stanford, California.

40. Williams, H.P. (1978). The reformulation of two mixed integer programming problems, Math. Prog., 14, 325-331.

41. Williams, H.P. (1982). Models with Network Duals, J. Opl. Res. Soc., 33, 161-169.

42. Williams, H.P. (1984). Restricted Vertex Generation Applied as a Crashing Procedure for Linear Programming, Computers and Operations Research, Vol. 11, 401-407.

43. Williams, H.P. (1985). Model Building in Mathematical Programming, 2nd Edition, Wiley, New York.

44. Wolsey, L.A. (1975). Faces for a linear inequality in 0-1 variables, Math. Prog., 8, 165-178.

45. Bradley, G.H., Brown, C.G. and Graves (1977). Design and Implementation of Large Scale Primal Transhipment Algorithms, Management Science, 24, 1-34.

LP-BASED COMBINATORIAL PROBLEM SOLVING

by

Karla Hoffman†

and

Manfred Padberg††

June 1984

†Operations Research Division, National Bureau of Standards
Gaithersburg, Maryland

††Graduate School of Business Administration, New York University
New York, New York

The second author's work was supported in part by NSF Grant No. MCS 8304634. and by NBS contract #

NATO ASI Series, Vol. F15
Computational Mathematical Programming
Edited by K. Schittkowski

Abstract

A tutorial outline of the polyhedral theory that underlies linear-programming (LP)-based combinatorial problem solving is given. Design aspects of a combinatorial problem solver are discussed in general terms. Three computational studies in combinatorial problem solving using the polyhedral theory developed in the past fifteen years are surveyed: one addresses the symmetric traveling salesman problem, another the optimal triangulation of input/output matrices and the third the optimization of large-scale zero-one linear programming problems.

LP-BASED COMBINATORIAL PROBLEM SOLVING

K. L. Hoffman
Center for Applied Mathematics
National Bureau of Standards
Gaithersburg, Maryland 20899

Manfred Padberg
Graduate School of Business Administration
New York University
New York, New York 10006

1. Introduction

Several recent computational studies in combinatorial optimization have focused on applying the polyhedral theory developed in the past fifteen years to actual problem solving. The resulting computational advances in the state of the art have been considerable, especially as far as the exact optimization of large-scale combinatorial optimization problems is concerned. Indeed, a common aspect shared by all of the computational studies surveyed in this paper is that all of them address problems belonging to the class of NP-hard combinatorial optimization problems for which to date no technically good (or polynomially bounded) algorithms are known. Of course these empirical studies will not resolve the fundamental question of whether or not exponential worst-case behavior is the best that we can expect. However, it is entirely possible that - as it has been observed in several cases - a gulf exists between the empirical record and the theoretical (worst-case) analysis of numerical problem solving. For instance, the fact that the simplex method exhibits exponential performance on some artificially constructed problems has in no way discouraged the many users that rely daily on the simplex method for the optimization of linear programming problems having thousands of variables. Mounting empirical evidence indicates that both pure and mixed integer programming problems can be solved to optimality in economically feasible computation times, by methods derived by our improved understanding of the underlying polyhedral theory.

In Section 2 of this paper we give a tutorial outline of the polyhedral theory that underlies the computational studies referred to above. In the third section we discuss some design aspects of a combinatorial problem solver in general terms. Much of the implementational detail is left out. For a more detailed description see [Bernal, Hoffman & Padberg, 1984], [Padberg & Grötschel, 1983], and [Crowder, Johnson & Padberg, 1983]. In Sections 4, 5 and 6 we survey three computational studies concerning the symmetric traveling salesman problem, the optimal triangulation of input/output matrices and the optimization of large-scale sparse zero-one linear problems without special structure. Space and time limitations prohibit us from including the recent papers by [Martin & Schrage, 1983] and [Van Roy & Wolsey, 1984b] that address the optimization of large-scale mixed zero-one linear programming problems and are based on results on the facial structure of such problems developed in a series of recent papers by [Baranyi, Van Roy & Wolsey, 1983a and 1983b], [Padberg, Van Roy & Wolsey, 1982], [Van Roy & Wolsey, 1983 and 1984a].

2. Combinatorial Problem Solving and Linear Programming

Combinatorial optimization problems can be stated in the following way: Given a finite ground-set E, a family $\mathcal{F}$ of subsets of E and weights $c_e \in \mathbb{R}$ for all elements $e \in E$, we are interested in finding a member F^* of $\tilde{\mathcal{F}}$ such that

$$c(F^*) = \sum_{e \in F^*} c_e \qquad (2.1)$$

is as small (or alternatively, as large) as possible. To "model" combinatorial optimization problems one associates with very element $e \in E$ a variable x_e, i.e. a component of a vector $x \in \mathbb{R}^E$ indexed by e. (Note that for notational convenience we write $\mathbb{R}^E$ rather than $\mathbb{R}^{|E|}$.) With every subset $F \subseteq E$ one associates a vector $x^F \in \mathbb{R}^E$, called the <u>incidence</u> or <u>characteristic</u> vector of F, defined as follows:

$$x^F_e = \begin{cases} 1 \text{ for } e \in F \\ 0 \text{ for } e \in F. \end{cases} \qquad (2.2)$$

Thus to every subset $F \subseteq E$ there corresponds a unique zero-one vector in $\mathbb{R}^E$ and vice versa. For any family $\tilde{\mathcal{F}}$ we let $P_{\tilde{\mathcal{F}}}$ be the convex hull of the incidence vectors of all members of $\tilde{\mathcal{F}}$, i.e.

$$P_{\tilde{\mathcal{F}}} = \text{conv } \{ x^F \in \mathbb{R}^E \mid F \in \tilde{\mathcal{F}} \}. \qquad (2.3)$$

$P_{\tilde{\mathcal{F}}}$ is a polytope in $\mathbb{R}^E$ and every extreme point of $P_{\tilde{\mathcal{F}}}$ uniquely corresponds to a member of $\tilde{\mathcal{F}}$ and vice versa.

We can solve the combinatorial optimization problem (2.1) by solving the problem

$$\min \{c^T x \mid x \in P_{\tilde{\mathcal{F}}} \} \qquad (2.4)$$

since, by the linearity of the objective function of (2.4), every optimum solution to (2.1) corresponds to an optimum extreme point solution of (2.4) and vice versa.

By the theorem of [Weyl, 1935] there exists a finite (integer) matrix H and a vector h (of integers) such that

$$P_{\tilde{\mathcal{F}}} = \{ x \in \mathbb{R}^E \mid H x < h \} \qquad (2.5)$$

holds and consequently, our combinatorial optimization problem (2.1) becomes the linear program

$$\min \{c^T x \mid H x \leq h\}. \tag{2.6}$$

Evidently, if we know the description of $P_{\mathcal{F}}$ by way of linear inequalities (2.5) then we can use linear programming methods such as the simplex method to solve the combinatorial optimization problem (2.4).

The linear description (2.5) of the polytope $P_{\mathcal{F}}$ is, of course, in general not unique, as one can append any non-negative linear combination of the rows of $Hx \leq h$ to this system without changing the solution set. One is thus lead to consider minimal systems of linear inequalities that describe $P_{\mathcal{F}}$, i.e. linear inequalities that define facets of the polytope $P_{\mathcal{F}}$. We recall that the dimension of $P_{\mathcal{F}}$, denoted: $\dim P_{\mathcal{F}}$, is the affine rank of $P_{\mathcal{F}}$ minus one, where the affine rank of any set $S \subseteq \mathbb{R}^E$ is the cardinality of the largest affinely independent subset of S. An inequality $f^T x \leq f_o$ where $f \in \mathbb{R}^E$ and $f_o \in \mathbb{R}$ defines a facet of $P_{\mathcal{F}}$ if we have

(i) $P_{\mathcal{F}} \subseteq \{x \in \mathbb{R}^E \mid f^T x \leq f_o\}$ and

(ii) $\dim (P_{\mathcal{F}} \cap \{x \in \mathbb{R}^E \mid f^T x = f_o\}) = \dim P_{\mathcal{F}} - 1$.

Property (i) is referred to as the "validity" property of the inequality $f^T x \leq f_o$ with respect to the polytope $P_{\mathcal{F}}$ whereas property (ii) expresses the condition that the hyperplane $f^T x = f_o$ is generated by the (extreme) points of $P_{\mathcal{F}}$. In the linear description (2.5) of $P_{\mathcal{F}}$, we can drop all the rows of $H x \leq h$ that do not define facets of $P_{\mathcal{F}}$. Thus, a minimal system of linear inequalities defining $P_{\mathcal{F}}$ is obtained. If $P_{\mathcal{F}}$ is of full dimension, i.e. if $\dim P = |E|$, then such a minimal system is unique modulo multiplication by positive scalars. In the general case, i.e. if $\dim P_{\mathcal{F}} < |E|$, then such a minimal system is unique modulo multiplication by positive scalars and linear combinations of the equations defining the smallest affine subspace of $\mathbb{R}^E$ that contains $P_{\mathcal{F}}$.

There are several ways in which the family $\mathcal{F}$ of subsets of E arises in actual combinatorial problem solving. Typical for integer programming on graphs is the situation that one looks for a well-defined combinatorial configuration in a graph (e.g. a maximum-weight stable (or independent) node-set or edge-set, a minimum-weight hamiltonian cycle, or a minimum-weight clique-covering of all nodes). It is clear that in these cases the ground-set E corresponds to either all the nodes or all the edges (or in certain instances to both all nodes and edges) of the given graph. The combinatorial configurations that one looks for define the elements of the family $\mathcal{F}$.

In other situations, one starts out with a system of linear inequalities in n zero-one variables that correspond to the "formulation" of some underlying "real-world" problem arising in economics, the management of public or private enterprises, engineering, etc. In this case, one has $E = \{1,..., n\}$ and the family $\mathcal{F}$ of subsets F of E is given by the correspondence

$$F \in \mathcal{F} \iff x^F \in \{x \in \mathbb{R}^n \mid A x \leq b,\ x_j = 0 \text{ or } 1 \quad \text{for } j = 1,..,n\} \qquad (2.7)$$

where A is some given m x n matrix of rationals and b is a vector with m rational components. Of course, in the case of an integer program on a graph, one always can write down a "formulation" of the problem by way of linear inequalities in zero-one variables as well, thus obtaining a linear system $A x \leq b$ whose zero-one solutions correspond uniquely to the combinatorial configurations that one looks for in the graph.

We will assume now that - no matter what combinatorial problem is considered - we have a "formulation" of the optimization problem at hand and denote by

$$D = \{x \in \mathbb{R}^n \mid A x \leq b,\ x_j = 0 \text{ or } 1 \text{ for } j=1,.., n\} \qquad (2.8)$$

the set of feasible solutions. Indeed, any formulation will do and will be called the "user-supplied" formulation. The combinatorial optimization problem can then be stated as

$$\min \{c^T x \mid x \in D\}. \tag{2.9}$$

Letting

$$\tilde{\mathcal{F}} = \{F \subseteq \{1,..,n\} \mid x^F \in D\}$$

where x^F is the characterisitc vector of F, we have that

$$P_{\mathcal{F}} = \text{conv}\,(D) \tag{2.10}$$

where conv (D) means the convex hull of D in $\mathbb{R}^n$, i.e. the convexification of the discrete set D in $\mathbb{R}^n$.

The following concept of <u>relaxation</u> is typical for linear programming approaches to combinatorial problem solving. Denoting

$$P = \{x \in \mathbb{R}^n \mid A\,x < b,\ 0 < x_j < 1 \text{ for } j=1,..,n\} \tag{2.11}$$

we obtain a polytope that satisfies

$$P_{\mathcal{F}} \subseteq P \tag{2.12}$$

and thus we have the easy, but important consequence

$$\min \{c^T x \mid x \in P_{\mathcal{F}}\} > \min \{c^T x \mid x \in P\}. \tag{2.13}$$

Furthermore, one can readily prove that every extreme point of $P_{\mathcal{F}}$ is an extreme point of P, but of course <u>not</u> vice versa as P will, in general, have many more extreme points than $P_{\mathcal{F}}$. However, if the second problem in (2.13) produces a zero-one solution vector, then we have solved the combinatorial optimization (2.9).

There are only few instances when the preceding (naive) relaxation idea is proven to always work; this is, for instance, the case when the matrix A is totally unimodular or perfect, see [Padberg, 1974] for a related survey. But even if the relaxation procedure does not work for all possible choices of the objective function, it <u>may</u> still work for the particular objective function of the user-supplied formulation of the optimization problem at hand. Taking this chance seems worth while as we do know how to solve linear programming problems efficiently. Indeed - except for certain purely enumerative attempts at solving combinatorial optimization problems - all proven methods for combinatorial problem-solving use this relaxation idea as their point of departure. For instance, the only integer programming method implemented up to 1982 in the form of commercially available software - branch & bound - uses precisely this naive relaxation of the combinatorial optimization problem. The trouble (empirically observed in many instances of combinatorial optimization problems) is that the "integrality gap" in relation (2.13) may be very large. As a case in point, [Crowder, Johnson & Padberg, 1983] report on a zero-one problem with 548 variables where the left-hand side of (2.13) produced a value of 8691.0 while the right-hand side of (2.13) produced a value of 315.3 for the user-supplied formulation. Predictably, commercially available, state-of-the-art branch & bound codes failed to close this enormous integrality gap even when the computer (an IBM 370/168) was permitted to run for several hours of CPU time. The way out of this situation is to use improved formulations of the combinatorial optimization problem employing the polyhedral approach to combinatorial optimization outlined above.

Returning to the "ideal" situation where we know a complete linear description (2.5) of the underlying polytope, it is in general out of the question to generate and store all of the data of the linear program (2.6) in the computer. This is so, because - while H is guaranteed to be a finite matrix - it can still be expected to be extremely large. Indeed, in most cases of interest, the number of rows of H is known to grow exponentially and worse with the parameter n, i.e. with number of variables of the problems. An alternative to enumerating all known rows of H is to solve periodically the following problem:

(2.14) Facet-Identification Problem: Given a point $\bar{x} \in \mathbb{R}^E$ and a polytope $P_{\mathcal{F}}$ find a facet-defining linear inequality $f^T x \leq f_0$ for $P_{\mathcal{F}}$ which is violated by $\bar{x}$, i.e. such that $f^T \bar{x} > f_0$ holds, or prove that no such inequality exists, i.e. that $\bar{x} \in P_{\mathcal{F}}$ holds.

Using (2.14) as a subroutine in a linear programming code, we can generate violated facet-inequalities "on the fly" as we need them to cut off fractional linear programming optima. More precisely, we think of the following iterative procedure for solving combinatorial problems; see e.g. [Padberg & Grötschel, 1983].

(2.15) Relaxation Method for Combinatorial Problems

(2.15.1) (Initialization). Let (LP_o) be a valid linear programming relaxation of the combinatorial problem under consideration, (e.g. the user-supplied formulation without the integer restrictions). Set k=0 and go to (2.15.2).

(2.15.2) (LP-Solver). Solve (LP_k), let x^k be an optimal solution to (LP_k) and go to (2.15.3).

(2.15.3) (Facet-Identification). Solve the facet-identification problem (2.14) for x^k and the polytope P .

(2.15.3.1) If one or more violated facet-inequalities for P are found, define (LP_{k+1}) to be (LP_k) amended by the violated facet-inequalities, replace k+1 by k go to (2.15.2).

(2.15.3.2) If a violated facet-inequality does not exist, stop.

If a finitely convergent LP-solver is used in step (2.15.2) then the relaxation method (2.15) is finitely convergent. Moreover, if the simplex-method is used in step (2.15.2), then one shows readily that termination in (2.15.3.2) occurs at a zero-one feasible solution to the combinatorial optimization problem.

Unfortunately, the complete linear discription (2.5) of the associated polytope is known for only few classes of combinatorial optimization problems. For instance, for the problem of finding a maximum-weight matching in a graph, [Edmonds, 1965] has found a complete linear description of the polytope. An algorithm utilizing the facet-identification problem (2.14) as a subroutine was given by [Padberg & Rao, 1982]. This algorithm was implemented by [Grötschel & Holland, 1984] and found to have an equal or better empirical performance on large-scale problems than the theoretically proven polynomial algorithm of [Edmonds, 1965] as modified by [Burkard & Derigs, 1980]. There are other classes of problems where the above algorithmic idea works. In general, however, we know of many problems for which complete linear descriptions (2.5) of the associated polytope are unknown to date. Indeed, - in view of the NP-hard characteristic of the general zero-one problem - such complete characterizations may prove to be elusive forever for many practically important combinatorial optimization problems. Yet, just as branch & bound used only the most naive form of relaxation of the combinatorial optimization problem (and proved to be successful in solving sufficiently many practical problems), we can utilize the above polyhedral description to improve upon the naive relaxation of a combinatorial optimization problem by adding a partial list of facet-inequalities in the relaxation method (2.15). If it so happens that we do not find a violated facet-inequality in this method, either because it does not exist or because we just do not yet know the full description or because we do not yet know how to algorithmically generate a known facet-inequality then we can at termination of (2.15.3.2) still resort to branch & bound and have an improved chance of proving optimality by the latter method, due to a much "tighter" relaxation.

The challenge in combinatorial optimization thus is to find a sufficiently large sub-system $H^*x \leq h^*$ of the complete system $H x \leq h$ of the polytope $P_{\mathcal{Y}}$ that permits one to reduce the integrality gap in relation (2.13). For if we choose such a sub-system $H^* x \leq h^*$ carefully then we obtain

$$\min\{c^T x \mid Hx \leq h\} \geq \min \{c^T x \mid H^* x \leq h^*\} \geq \min \{c^T x \mid x \in P\}. \qquad (2.16)$$

The smaller the integrality gap between the "improved" linear formulation and the optimization problem in zero-one variables, the likelier the chances at optimizing the problem by branch & bound. As a case in point, in the example by [Crowder, Johnson & Padberg, 1983] referred to above, sufficiently many inequalities were generated automatically to yield an objective function value of 8643.5 for the "improved" formulation. As it so happened, with the remaining integrality gap being so small, the entire optimization (including a branch & bound phase) of this previously unsolvable problem took .91 minutes of CPU time on an IBM 370/168 starting from scratch with the user-supplied formulation.

We conclude this tutorial outline by giving a general example that illustrates the major points of this section.

(2.17) Example: Consider the problem of finding a maximum-weight stable (or independent) node-set in a finite undirected graph (without loops and multiple edges) $G = (V,E)$ with node-set V and edge-set E. Remember that $S \subseteq V$ is a stable set if $v, w \in S$ implies $(v,w) \notin E$, i.e. v and w are not connected by an edge of G. In terms of the general notation of this section, the ground set E is the node-set V of G, while the family $\mathcal{F}$ is the set of all stable subsets of V. $P_{\mathcal{F}}$ is the vertex-packing polytope. Introducing zero-one variables x_v for all $v \in V$ we can formulate the problem as follows:

$$\max \sum_{v \in V} c_v x_v$$

$$\text{(P1)} \qquad \text{subject to} \qquad \begin{aligned} & x_v + x_w \leq 1 \quad \text{for all } e = (v,w) \in E \\ & x_v = 0 \text{ or } 1 \quad \text{for all } v \in V \end{aligned}$$

where c_v for $v \in V$ are the node-weights for a given problem instance. The formulation (P1) correctly models the problem, but is rather "weak" from a linear programming point of view. For let C be the node-set of any clique of G (a node-set of a maximal complete subgraph of G) and $\mathcal{C}$ be any family of cliques of G that cover all edges of G. That is, $\mathcal{C}$ may (but need not) be the family of all cliques of G. Then we can formulate the same problem as follows:

$$
\text{(P2)} \qquad
\begin{array}{lll}
 & \max \sum_{v \in V} c_v x_v & \\
\text{subject to} & & \\
 & \sum_{v \in C} x_v < 1 & \text{for all } C \in \mathcal{C} \\
 & x_v = 0 \text{ or } 1 & \text{for all } v \in V
\end{array}
$$

Denote Z_G the optimal objective function value of (P1) and (P2), Z_{P1} the optimal objective function value of the (naive) linear programming relaxation of (P1) and Z_{P2} likewise for the formulation (P2). Noting that we are maximizing, we obtain as in (2.16) the following inequalities:

$$Z_G < Z_{P2} < Z_{P1} \qquad (2.18)$$

Moreover, [Padberg, 1973] has shown that the inequalities of (P2) define facets of the polytope $P_{\mathcal{F}}$, whereas the inequalities of (P1), in general, do not. Now consider the special case of finding a maximum-cardinality stable set in an odd anti-hole (the complement graph of an odd cycle without chords) having n nodes with $n > 7$, see [Padberg, 1977]. One shows readily that $Z_{P1} = n/2$. Taking $\mathcal{C}$ to be the family of all maximum-cardinality cliques of the odd anti-hole one finds $Z_{P2} = 2 + \frac{2}{n-1}$ while $Z_G = 2$. The maximum-cardinality cliques are, of course, a proper subset of all cliques of an odd anti-hole and their totality does not define the associated polytope $P_{\mathcal{F}}$, see [Padberg, 1973, 1977, 1979], [Trotter, 1976] and others. However, this example demonstrates the basic idea: A proper subset $H^*x < h^*$ of the totality of all facet-defining inequalities $Hx < h$ of the polytope $P_{\mathcal{F}}$ can have a dramatic effect upon the goodness of the resulting (naive) linear programming relaxation of the combinatorial optimization problem because the second formulation permits one to obtain Z_G by simply rounding down. From a user's point of view, of course, any formulation of the problem "will do". In terms of the relaxation method (2.15), formulation (P1) provides the first linear programming relaxation (LP_0). Then a zero-one problem solver based on (2.15) generates facet-defining inequalities to yield formulation (P2). Thus an improved formulation of the combinatorial problem is generated automatically. Branch & bound or any other linear-programming based enumeration method may be inevitable since at present, as for most integer programming problems of practical interest, the complete description of the vertex-packing polytope is unknown.

3. Design Aspects of a Combinatorial Problem Solver

As discussed in Section 2 we will assume that the combinatorial optimization problem is in the form

$$\min \{c^T x \mid Ax \leq b, x_j = 0 \text{ or } 1 \text{ for } j = 1,.., n\} \tag{3.1}$$

where c is a vector with n components, A is a mxn matrix and b is a vector with m components. An overall framework for a combinatorial problem solver is given in Figure 1. This flowchart is designed for a software package to solve large-scale zero-one problems having thousands of zero-one variables to optimality and can alternatively, be used as a heuristic procedure for obtaining "reasonable" feasible solutions and at the same time a true lower bound on the optimal objective function value of (3.1). The latter is, of course, a yardstick to measure the "reasonableness" of the solutions found.

In this design we start by reading the data of the user-supplied formulation into the computer. In a "pre-processing" phase the formulation is inspected automatically and permanent problem reductions are carried out. Such problem reductions concern the fixing of certain variables due to feasibility and/or optimality considerations to either zero or one, the clearing of rows of common divisors greater than one, the "cleaning up" of the user-supplied formulation such as reducing "Big M's", the "classification" of constraints and the detection of special structures. Some of the methods used in this preprocessing phase are discussed in [Crowder, Johnson & Padberg, 1983] and will be elaborated in a forthcoming paper [Bernal, Hoffman & Padberg, 1984]. After preprocessing, we thus obtain the definite version of the zero-one problem to be optimized, which we will call the "global problem" to be optimized.

In the next phase, heuristics are used to find an upper bound on the objective function of (3.1). While satisfactory heuristics exist for many specially structured zero-one problems (e.g. the set packing problem, the set-covering problem, the traveling salesman problems), quick and efficient heuristics for general zero-one problems are scarce. One such heuristic is the pivot-and-complement heuristic due to [Balas & Martin, 1980]. It is important to note that a good upper bound permits the "fixing" to zero or one of certain variables permanently once a good lower bound is also known.

Since this framework is to be used to solve zero-one problems having possibly thousands of variables, one proceeds next with the selection of a subproblem for actual optimization. How to select a subproblem, is clear in the context of integer programming on graphs: In the case of a traveling salesman problem with several hundred cities, for instance, one has to optimize over tens of thousands of variables (corresponding to the edges of the graph). Clearly the bulk of these variables will be zero in an optimal solution; thus, it will be sufficient to consider the problem on a sparse subgraph of the entire problem induced by the "short" edges of the graph. To prove the optimality of a solution of the sparse problem to the global problem, one needs to "price out" the variables temporarily set equal to zero. If their reduced cost exceed the gap between the linear programming optimum and the integer optimum, then they were correctly set to zero. Otherwise, the subproblem is redefined. By a judicious choice of the subproblem selected, one can assure that the process is convergent. In general, the subproblem selection can be done on the basis of a normalized cost criterion, i.e. we can order the variables by increasing ratios

$$c_j/(1 + \sum_{i=1}^{m} a_{ij}^2)^{1/2} \tag{3.2}$$

and use a cut-off value for determining the variables of the actual subproblem to be considered. At later stages of the calculation, such selection may be based on normalized reduced cost, rather than the original cost. Of course, except for

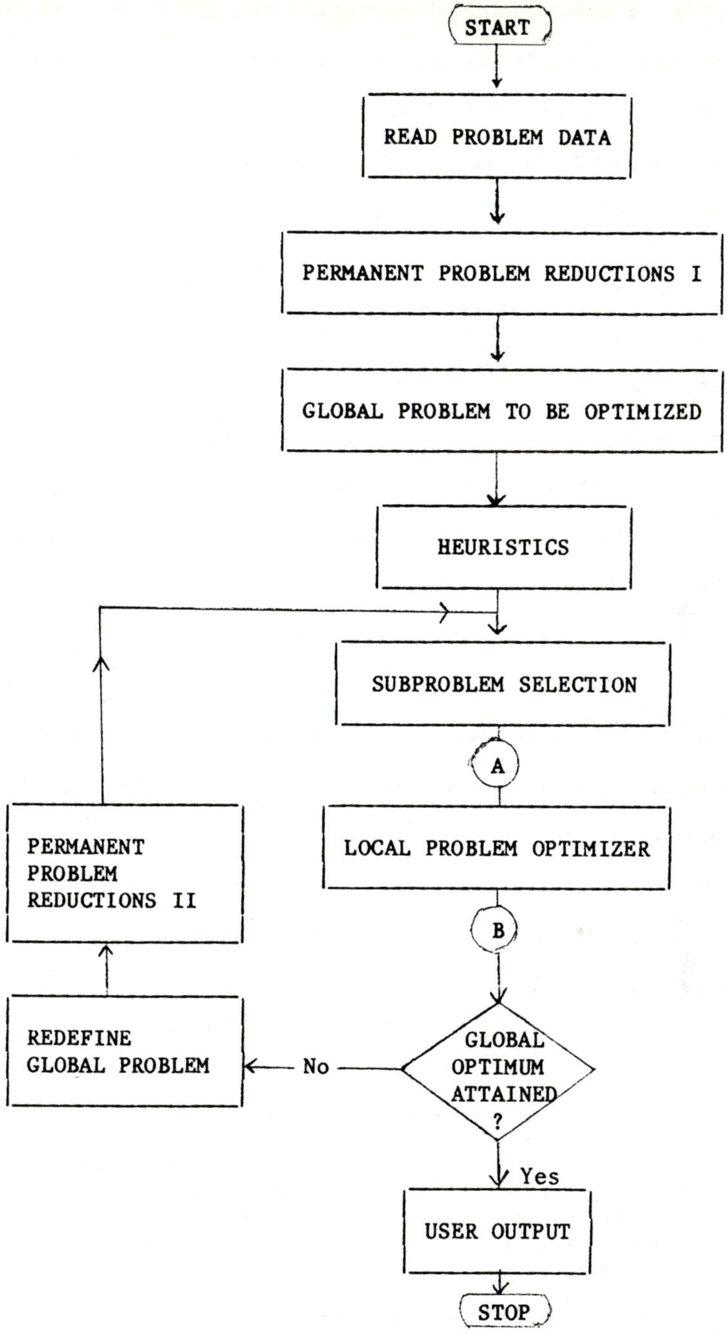

Figure 1. The Overall Flowchart

integer programs on graphs, it is not clear at present how to select a suitable subproblem so as to maintain zero-one feasibility (provided one knows that problem (3.1) is feasible). Of course, if a (good) feasible zero-one solution has been found heuristically, one can always include the variables that are in a feasible solution in each of the sub-problems, thus ensuring the feasibility of all selected subproblems. We note that in the case of a traveling salesman problem, criterion (3.2) is identical to the one based on the original cost c_j and was successfully used by [Crowder & Padberg, 1980] and [Padberg & Hong, 1980] to optimize a 318-city problem involving 50,403 zero-one variables.

The most important phase of the zero-one problem solver is the phase called the "local problem optimizer". Included in this subsystem is a problem set-up routine, a linear program optimizer, a package of problem-dependent subroutines for the constraint generation as well as a branch & bound routine. In other words, this is the relaxation method (2.15) of Section 1 amended by a branch & bound procedure. Examples of how the constraint generation (2.14) of Section 2 is carried out for different problem structures are discussed in later sections of this paper. Figure 2 shows the flow chart of the "local problem optimizer".

Having obtained an optimal solution for the subproblem considered in the local problem optimizer one proceeds to extend ("lift") the newly generated constraints to include all zero-one variables of the global problem. Such an extension is theoretically and computationally possible using the various "lifting" procedures described in the literature, see [Padberg, 1975] and [Wolsey, 1975], for the case of general zero-one problems. More specific results apply to specially structured zero-one problems, (e.g. the traveling salesman problem, set packing, set covering). One then computes a lower bound on the optimal objective function value of (3.1) e.g. by optimizing the linear programming relaxation of the global problem.

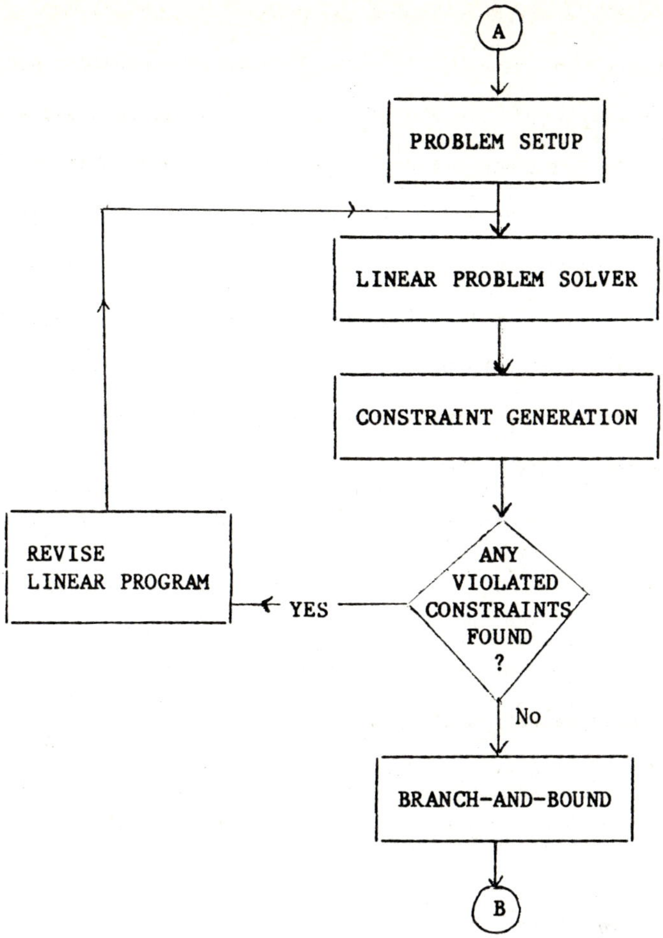

Figure 2. The Local Problem Optimizer

Assuming that a feasible zero-one solution was found (either by a heuristic or by the local problem optimizer) we have then an upper and a lower bound on the optimal objective function value of (3.1). At this point one applies permanent problem reductions based on the reduced costs of the linear program, see [Crowder, Johnson & Padberg, 1983] and selects a new subproblem among those variables that were not fixed at either zero or one. Of course, the latter has to be done in such a manner so as to guarantee convergence (the strategies for the selection will depend on the structure of the actual problem on hand). For general zero-one problems at present we use always the global problem, i.e. we bypass the subproblem selection.

In the next three sections we describe facets of the polytope $P_{\mathcal{F}}$ and the status of the respective facet-identification problem (2.14) for three classes of combinatorial problems: the traveling salesman problem, the triangulation of large input/output matrices and large sparse zero-one problems having no special structure. These three types of problems are chosen because computational studies have proven that cutting planes related to the facets of the underlying polytope are an indispensable tool for the exact solution to these problems. A summary of the computational results of these studies will be also provided in these sections.

4. The Symmetric Traveling Salesman Problem

In the symmetric traveling salesman problem (hereafter abbreviated to TSP) one is interested in finding a shortest roundtrip or tour through n cities such that every city is visited exactly once. So let $G = (V,E)$ denote the complete, undirected graph having n nodes (representing the cities) and let the vector c with components c_e for all edges $e \in E$ denote the symmetric distance table of the graph (i.e., if edge e connects nodes i and j of the graph then c_e is the (symmetric) distance between city i and j). The assumption that G is complete is for notational convenience only since, in a minimization context, missing edges of the original graph can be adjoined by assigning arbitrarily large distances to them.

A roundtrip or tour through the n cities correspondes to a hamiltonian cycle in the graph G and thus the family $\mathcal{F}$ of Section 2 is the family of all subsets of edges of G that form a hamiltonian cycle. The ground-set of Section 2 is the set E of all edges of G and the zero-one vectors of interest to us are those that correspond to subsets of edges forming a hamiltonian cycle in G. In an undirected complete graph with n nodes there are exactly 1/2 (n-1)! hamiltonian cycles and thus in $\mathbb{R}^m$ where $m = 1/2n\,(n-1)$ we have exactly that many zero-one points of interest to us. Taking the convex hull of these 1/2n (n-1)! points we obtain the traveling salesman polytope over which we wish to minimize the linear form given by the vector c.

To formulate the TSP as an integer program denote by A the n-by-m node-edge incidence matrix of G where as before $m=1/2n\,(n-1)$. Furthermore, for any vector $x \in \mathbb{R}^E$ with components x_e for all $e \in E$, we denote for any subset $W \subseteq V$

$$x(W) = \sum_{e \in E(W)} x_e \tag{4.1}$$

Where $E(W)$ is the set of edges in G with both ends in the node-set W. The TSP is the following minimization problem:

$$\text{Minimize} \quad cx \tag{4.2}$$

subject to

$$Ax = \underline{2} \tag{4.3a}$$

$$x(W) \leq |W|-1 \text{ for all nonempty proper subsets } W \subseteq V \tag{4.3b}$$

$$\left.\begin{array}{ll} 0 \leq x_e \leq 1 & (4.3c) \\ x_e \text{ integer} & (4.3d) \end{array}\right\} \text{ for all } e \in E$$

where $\underline{2}$ is the vector with n components equal to 2 and $|W|$ is the cardinality of W. The constraints (4.3a) express the condition that every node must be met by exactly two edges in every tour. Conditions (4.3b) express the conditions that cycles having less than n edges are of no interest to us as they do not correspond to hamiltonian cycles in the graph and are called <u>subtour-elimination constraints</u>, see [Dantzig, Fulkerson & Johnson, 1954]. (Note that the upper bound constraints are duplicated for notational convenience in (4.3c)). It is important to note that the requirement (4.3d) is essential; for <u>if</u> it could be dropped, then the traveling salesman problem would be solvable in polynominal time! This means, of course, that additional linear inequalities are needed to describe the traveling salesman polytope.

A first (naive) linear programming relaxation of the traveling salesman problem as used in the relaxation procedure (2.15) is given by (4.2), (4.3a), (4.3b), (4.3c). If the optimal solution of this linear program is integer, we have solved the traveling salesman problem. Otherwise we need more (facet-defining) inequalities to eliminate (chop off) the fractional linear programming optimum. If we run out of known facet defining inequalities to do this job before obtaining a tour, we then resort to branch & bound. In the next section we discuss classes of inequalities that define facets for the traveling salesman polytope.

The astute reader will have noticed that this naive relaxation involves exponentially many constraints and thus -*prima facie* -it is not implementable. Indeed, historically, it has been for this reason that the above formulation has been ignored in the literature. There are several formulations reported in the literature that replace the above inequality system by one that is polynomial in the number n of nodes of the graph (typically, however, at the expense of introducing additional variables). Whatever the merits of these other formulations, we know today that half of the constraints of (4.3b) define facets of the traveling salesman polytope. Thus, they are essential in defining the polytope. The way out of the dilemma is to include the constraints (4.3b) into those that are generated "on the fly". The real question therefore is not how many constraints we have, but whether or not we can find violated (facet-defining) inequalities efficiently. The answer to this question is positive for subtour elimination constraints as well as other inequalities.

4.1 Facets of the Symmetric Traveling Salesman Polytope

The set of linear inequalities (4.3a), (4.3b) and (4.3c) defines a polytope that contains (properly) the traveling salesman polytope. We now wish to append to this set of constraints, facet-inducing inequalities which will "tighten" the above formulation when the requirement "x integer" is ignored. There are many types of facet-inducing inequalities known for the problem, see [Grötschel & Padberg, 1983], of which will discuss three types because they can all be described by the inequality

$$\sum_{i=0}^{k} x(W_i) < |W_o| + \sum_{i=1}^{k} (|W_i|-1)-\langle k/2\rangle \qquad (4.4)$$

where <·> denotes rounding up. The sets W_i are proper subsets of V satisfying the following conditions for $i = 0, 1, \ldots, k$:

$$|W_0 \cap W_i| \geq 1, \qquad i = 1, \ldots, k, \tag{4.5}$$

$$|W_i \setminus W_0| \geq 1, \qquad i = 1, \ldots, k, \tag{4.6}$$

$$W_i \cap W_j = 0, \qquad 1 \leq i < j \leq k, \tag{4.7}$$

$$k \text{ odd.} \tag{4.8}$$

The edge-set $C = \bigcup_{i=0}^{k} E(W_i)$ is called a comb in G (see [Grötschel & Padberg, 1983] and [Padberg & Grötschel, 1983]). The inequalities (4.4) are called comb constraints. For $k = 1$ and $|W_0| = 1$, we retrieve the subtour-elimination constraints (4.3b) introduced above. A comb is a 2-matching constraint [Edmonds, 1965] if the inequalities in both (4.5) and (4.6) hold as equalities. A comb is a Chvatal-comb [Chvatal, 1973] if the requirement (4.7) is dropped and the inequality (4.5) is required to hold as an equality. A thorough discussion of comb inequalities can be found in [Grötschel & Padberg, 1983] where it is proven that comb inequalities belong to the class of linear inequalities that occur in any minimal linear constraint system for the TSP (modulo a multiple of the rows of A). The linear programming formulation of the symmetric traveling salesman problem provided by (4.2)-(4.4) is still only an approximation, however, since the associated linear constraint set is not complete (i.e., it admits basic solutions that are noninteger). Furthermore, as indicated above, one would not wish to list all known facets since even for very small TSP problems, computers are not available nor will ever be available which would be capable of handling systems of linear equalities of the size needed (see TABLE 1 below). But, to stress this important point again, it is not the number of constraints needed to linearly describe the polytope that matters in computation, rather it is the question of whether or not we are able to find violated constraints efficiently, i.e. to solve the facet-identification problem (2.14) in polynomial time.

Nr. of cities	Nr. of Subtour Elimination Constraints	Nr. of Comb Constraints
6	25	60
7	56	2,100
8	119	42,840
9	246	667,800
10	501	8,843,940
15	16,368	1,993,711,339,620
20	$0.5 \cdot 10^6$	$1.5 \cdot 10^{18}$
30	$0.5 \cdot 10^9$	$1.5 \cdot 10^{31}$
40	$0.5 \cdot 10^{12}$	$1.5 \cdot 10^{45}$
50	$0.5 \cdot 10^{15}$	10^{60}
120	$0.6 \cdot 10^{36}$	$2 \cdot 10^{179}$

Table 1*

*Extracted from [Padberg & Grotschel 1983].

Using the above classification of all comb constraints, we are faced with the following facet-identification problems (2.14):

4.9 Identification of subtour elimination constraints (SEC):

Given a point $\bar{x} \in \mathbb{R}^E$ satisfying $0 \leq \bar{x}_e \leq 1$ for all $e \in E$ find a set $W \subseteq V$, $2 \leq |W| \leq n-1$, such that $\bar{x}(W) > |W| - 1$ holds or prove that no such $W \subseteq V$ exists.

4.10 Identification of 2-matching constraints (2C):

Given a point $\bar{x} \in \mathbb{R}^E$ satisfying (1.6) find a set $W \subseteq V$ and a set T of edges $e_1, \ldots, e_{|T|}$ in E with the properties

(a) e_i has one endpoint in W $\qquad i = 1, \ldots, |T|$

(b) $e_i \cap e_j = 0$ $\qquad 1 \leq i < j \leq |T|$

(c) $|T|$ odd

such that $\bar{x}(W) + \bar{x}(T) > |W| + |T|/2$ holds or prove that no such $W \subseteq V$ and $T \subseteq E$ exist.

4.11 Identification of Comb Constraints (Co):

Given a point $\bar{x} \in \mathbb{R}^E$ satisfying (1.6) find sets $W, W_1, \ldots, W_k \subseteq V$ with the properties

(a) $|W \cap W_i| \geq 1$ $\quad i = 1,\ldots,k$

(b) $|W_i \setminus W| \geq 1$ $\quad i = 1,\ldots,k$

(c) $W_i \cap W_j = \emptyset$ $\quad 1 \leq i < j \leq k$

(d) k odd

such that $\bar{x}(W) + \sum_{i=1}^{k} \bar{x}(W_i) > |W| + \sum_{i=1}^{k} (|W_i|-1) - \langle\frac{k}{2}\rangle$ holds or prove that no such sets $W, W_1,\ldots,W_k \subseteq V$ exist.

In [Padberg & Hong, 1980] a heuristic is provided for the identification of violated subtour elimination constraints. The algorithm requires a total computational effort which is polynomially bounded. For a complete answer to the question of whether or not there exists a subtour elimination constraint which chops off some feasible point x, the algorithm of [Gomory & Hu, 1961] must be applied to the graph obtained by the refined shrinking procedure described by [Padberg & Grotschel, 1983]. If all calculations are performed carefully in both algorithms then the identification of subtour elimination constraints can be carried out by a polynomially bounded algorithm, i.e. problem (4.9) is solvable in polynomial time.

[Padberg & Rao, 1982] have shown that a modification of the minimum-cut algorithm of [Gomory and Hu, 1961] can be used to identify violated 2-matching constraints, and that this algorithm solves problem (4.10) in polynomial time. The Padberg & Rao algorithm as well as a reduction heuristic for 2-matching constraints are described in [Padberg & Grötschel, 1983].

For the general comb inequalities no polynomial exact algorithm is known to date. [Padberg & Grötschel, 1983] give a heuristic procedure whose effectiveness remains to be tested in numerical problems.

A complete description of the methodology for the facet-identification problem (2.14) for symmetric TSPs (heuristics and exact algorithms to the extent known to date) can be found in [Padberg & Grötschel, 1983]. The known results indicate that for a large class of facets of the traveling salesman polytope the facet-identification problem (2.14) is solvable in polynomial time.

4.2 Computational Results for Large-Scale Symmetric Traveling Salesman Problems

In this section we give a brief summary of the papers by [Crowder & Padberg, 1980], [Grötschel, 1980] and [Padberg & Hong, 1980]. There are, however, several other studies (unpublished) that utilized polyhedral information in the actual solution of symmetric traveling salesman problems such as the solution of a 125-city problem done by students of Professor W. R. Pulleyblank at the University of Grenoble (France) and the solution of a 68-city problem done by students of Professor L. Wolsey at the Catholic University of Louvain-la-Neuve (Belgium).

Related studies concerning the use of cutting-planes in the solution of symmetric traveling salesman problems have been carried out by [Land, 1979] and [Miliotis, 1976 and 1978], [Fleischmann, 1981 and 1982], and earlier by [Hong, 1972]. While these studies reported good results for medium-sized traveling salesman problems their methodology did not focus on testing the polyhedral theory described here which is the intent of our brief survey.

Figure 3 depicts the overall flow of calculations to be performed in implementing the relaxation method (2.15) as applied to traveling salesman problems. The three studies that we will survey used flowcharts differing in minor details. While [Grötschel, 1980] carried out the constraint-identification visually, both [Crowder & Padberg, 1980] and [Padberg & Hong, 1980] used fully automated procedures (no manual intervention) in their respective software.

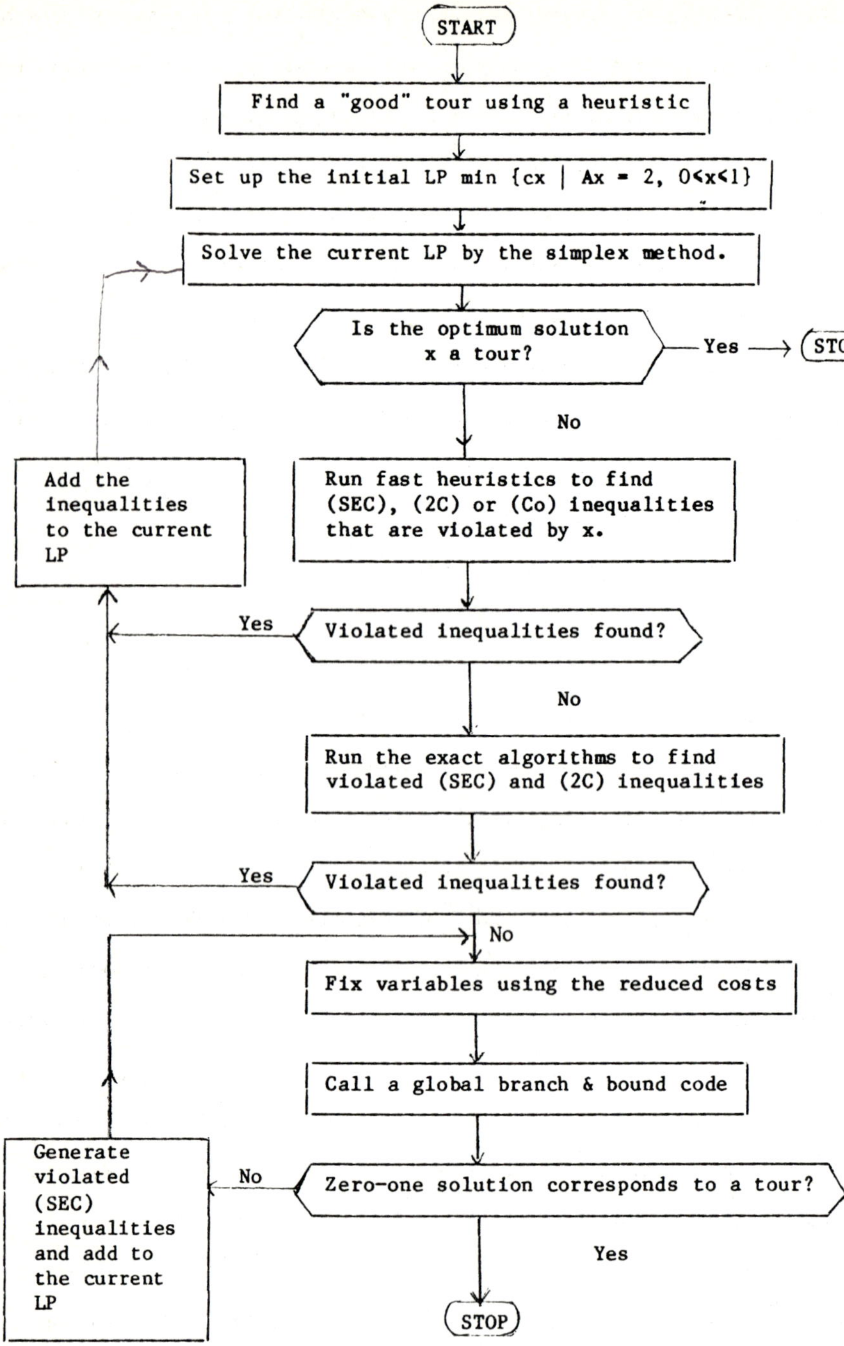

Figure 3. Flow Chart

The first instance of a symmetric traveling salesman problem involving more than 100 cities that was solved using only subtour elimination, 2-matching and comb constraints is a 120-city problem given by the data in the Deutscher General Atlas (Mairs Geographischer Verlag, Stuttgart, 1967/68). In [Grötschel, 1980] no recourse to branch & bound or other enumerative methods was needed in order to obtain an optimal tour, i.e. the relaxation method (2.5) for combinatorial optimization problems was used and nothing else.

In order to find and to prove optimality of this problem 13 iterations of the relaxation method (2.5) were necessary. In this process only 96 subtour elimination and comb constraints of the total universe of 10^{179} such constraints were generated to establish optimality of the tour. More precisely,

36 subtour elimination constraints

25 2-matching constraints and

35 (other) comb constraints

were needed to find and to prove optimality of the tour.

A comprehensive study using facet-defining inequalities for the TSP problem was performed by [Padberg & Hong, 1980]. They sampled 74 symmetric traveling salesman problems in order to broadly assess the computational value of facet-defining linear inequalities for the solution of traveling salesman problems. We note that out of the total 74 problems only 54 were solved (to optimality), but in all cases excellent lower bounds were obtained when compared to the solutions found by using the heuristic due to [Lin & Kerninghan, 1973].

In order to evaluate the value of facet-defining inequalities towards the goal of proving optimality Padberg & Hong took the following approach: Given a heuristically obtained solution, they solve the L.P. relaxation problem (4.2, 4.3a, 4.3c), i.e. the 2-matching relaxation of the symmetric TSP, with the heuristically obtained tour as a starting point. Using the same heuristically obtained solution they then ran the problem a second time generating facet-defining inequalities:

This run either terminates with an optimal tour or in case of an inability to identify a suitable new constraint, defaults with a lower bound for the problem obtained by solving the, by now, enlarged linear programming problem. This approach yields two values: VALUE1 is the objective function value without cuts and VALUE2 is the objective function value with cuts. If TOUR denotes the minimum length tour of the problem, then the following ratio is a good proxy for measuring the added value of the additional work:

$$\text{RATIO} = (\text{VALUE2} - \text{VALUE1})/(\text{TOUR}-\text{VALUE1}).$$

Note that RATIO is zero if no improvement is obtained (e.g. if no constraint was generated--this was never the case), while RATIO is one if the constraint-generation procedure terminates with the optimal tour. RATIO is, of course, always between zero and one and due to taking both differences and a ratio, the measure is invariant under scaling and translating the data. As the exact value of TOUR is known only _a posteriori_, it is worth pointing out that RATIO is a _conservative_ valid measure of the improvement (if TOUR is a possibly _suboptimal_ value obtained heuristically). It should be noted further that for the computation of both VALUE1 and VALUE2 the same starting solution for the respective linear programs is used, which is known to greatly impact the performance of simplex methods.

There are several parts to the computational study by Padberg & Hong of which we will discuss two. In one part they ran 55 randomly generated Euclidean TSPs using the pseudo-random number generator of [Lin & Kernighan, 1973] which generates coordinates x_i and y_i with values between 1 and 1000 for $i = 1,\ldots,n$. Due to the fact that coordinates are generated as pairs, the same random seed for $n + m$ cities produces a graph that is properly contained in the graph on the first n cities, thus permitting one to study how increasing n affects the added value of the facet-inducing inequalities. 10 different problems were run for n = 15, 30, 45, 60 and 75, respectively, using 10 different seeds for the random number generator. Furthermore, 5 Euclidean problems with n = 100 due to [Krolak, Felts & Marble, 1971] were included in this statistical part of the study, though they are different from the other ones.

n		15	30	45	60	75	100
Ratio	μ	1.0	0.99	0.93	0.92	0.88	0.92
	σ	0.0	0.03	0.11	0.10	0.09	0.02
TOUR	μ	3555	4738	5566	6297	6878	21507
	σ	383	314	224	181	224	525
GAP1	μ	224	352	379	452	387	1507
	σ	121	100	149	133	93	313
GAP2	μ	0.0	5	24	38	50	120
	σ	0.0	15	57	44	44	43
PIVOT2	μ	11	34	47	76	87	167
	σ	2	7	13	30	23	40
ΔPIVOT	μ	2	13	17	36	37	97
	σ	2	5	12	29	22	38
TIME2*	μ	0.33	1.37	4.46	14.47	30.52	108.74
	σ	0.03	0.26	1.27	6.82	10.81	39.97
ΔTIME*	μ	0.07	0.46	1.39	6.25	11.64	50.4
	σ	0.03	0.23	1.23	6.63	10.63	31.7
CUTS	μ	3	12	15	26	28	72
	σ	2	5	8	13	12	18
OPTIM		10	9	5	4	3	0

*Seconds, IBM 370/168 MVS

Table 2. 55 Euclidean TPS's

Table 2 contains all the relevant statistics for this part of the study. The entries in Table 2 were obtained by averaging the respective individual figures and their mean μ is given with σ being the standard deviation. The top row of Table 2 contains the value RATIO. As it is to be expected, RATIO declines with increasing n. TOUR is the tour length obtained by the heuristic.

The bottom line of Table 2 (OPTIM) specifies the number of times the linear program terminated by proving optimality of the heuristically obtained tour. GAP1 measures the average difference between TOUR and the objective function value of the (initial) linear program (4.2), (4.3a) and (4.3c), i.e. the 2-matching relaxation of the symmetric TSP. GAP2 is the crucial measure in evaluating the constraint-generation procedure. It is the difference between TOUR and the objective function value of the amended linear program. PIVOT2 is the average of the total number of pivot operations carried out by the constraint-generating program. ΔPIVOT is the average increment of the pivot count over what it takes to solve the initial linear program. Likewise, TIME2 specifies the total CPU-time of the constraint-generating program, ΔTIME the average increment over the respective times for the initial linear program. Thus to terminate with an average lower bound of 21,387 for $n = 100$ it took on average 108.74 seconds of CPU-time, while it took on average 50.4 seconds of CPU-time less than 108.74 to obtain an average lower bound of 20,000. (The number in the respective rows labelled σ are the respective standard deviations of the sample problems.) Finally, CUTS is the average number of constraints that were generated and amended to the original linear program. Thus for 100-city problems the initial linear program has 100 rows and 4950 variables, while at termination of the constraint-generation procedure the linear program increased on average to 172 rows and 5022 variables, a truly modest increase given the complexity of the problem and the goodness of the bound obtained. For the individual figures for all of the sample problems the reader is referred to the original paper [Padberg & Hong, 1980].

Another part of this study was done in order to permit a limited comparison of the performance of the constraint-generation procedure vis-a-vis other approaches. Therefore, a number of test problems that have been used by other researchers were solved. The results are summarized in Table 4. The heading "without cuts" refers to the solution of the (initial) linear program: TIME1 is the CPU-time in seconds, PIVOT1 the pivot count, VALUE1 the objective function.

TOUR refers to the minimum length tour or the value of the best tour found by the heuristic or during earlier runs with the TSP code and is the initial solution for the TSP code in the run reported in Table 4. The heading "with cuts" refers to the constraint-generation procedure : VALUE2 is the objective function value of the linear program with cuts, the first column under PIVOT2 refers to the total number of pivots, the second column under PIVOT2 refers to the number of pivots carried out after the default in the constraint-generation procedure (i.e. the second column is counted already in the first). CUTS specifies the total number of cuts generated in the run, its second column the number of cuts that were dropped after defaulting. TIME 2 is total execution time to termination in CPU-seconds. RATIO is the value discussed in the introduction to this section.

	Without cuts				With cuts						
Problem	TIME1*	PIVOT1	VALUE1	TOUR	VALUE2	PIVOT2		CUTS		TIME2	RATIO
DAN42	2.57	30	641	699	699	37	0	9	0	3.10	1.0
GRO48	4.09	33	4769	5046	5031[illegible]	83	9	32	8	9.16	0.95
HEL48	3.69	33	11197	11461	11461	38	0	10	0	4.30	1.0
TOM57	7.79	44	12633½	12955	12940	61	4	22	1	10.40	0.95
KROL70	16.33	53	623½	675	673[illegible]	120	8	44	10	31.91	0.98
				6942	6938[illegible]	243	4	75	17	221.50	0.99
GRO120	111.20	69	6662½								
				6951	6928[illegible]	166	17	49	4	171.44	0.92
KNU121	4.54	45	328	349	343½	74	13	10	1	7.25	0.76
LIN318	670.8	251	38765½	41349	41236[illegible]	578	70	171	64	1751.46	0.96

* Seconds, IBM 370/168 MVS

Table 4. Eight Problems from the Literature

DAN42 is the 42-city version of the 49-city problem due to [Dantzig, Fulkerson & Johnson, 1954]. The solution was proven to be optimal in 3.10 seconds of CPU-time after adding nine constraints. (These constraints include one type of constraints not discussed here, see [Padberg & Hong, 1980].) GRO48 is a 48-city problem due to Grötschel (48 cities with distances given in Shell's Roadatlas of Germany). After 9.16 seconds of CPU-time the program terminated with a lower bound of 5032 for the optimum tour; the best tour found by the heuristic has a length of 5046. HEL48 is the 48-city problem due to [Held & Karp, 1970]. The solution was proven to be optimal in 4.30 seconds of CPU-time after adding ten constraints. TOM57 is the 57-city problem due to [Thompson & Karp, 1964]. After 10.40 seconds of CPU-time a lower bound of 12940 for the optimum length tour of 12,955 was obtained. (Optimality was proven by [Held & Karp, 1970].) KROL70 is a 70-city problem due to [Krolak, Felts & Marble, 1971]. After 31.91 seconds of CPU-time a lower bound of 674 on the heuristically obtained best tour of length 675 was obtained. GRO120 is the 120-city problem due to Grötschel described above. KNU121 is a supersparse 121-city problem due to [Knuth, 1976]. 7.25 seconds of CPU-time were used to obtain a lower bound of 344 on the optimum tour length of 349 published in the New York Times.

LIN318 is a 318-city problem the data of which are published by [Lin & Kernighan, 1973]. The data come from an actual problem involving the routing of a numerically controlled drilling-machine through three identical sets of 105 points each plus three outliers. As the drilling is done by a pulsed laser, drilling time is negligible and the problem becomes a standard traveling salesman problem. The only exception from the standard form is that a particular start and end point are to be used; the resulting Hamiltonian path problem can, however, easily be accommodated within the linear programming framework by assigning a negative distance to the particular arc. The distance-table of the complete graph on 318 points (with the exception of one edge) was computed from the coordinates published by [Lin & Kernighan, 1973).

The most surprising outcome of the computational study by Padberg & Hong is that only very few additional facet-defining inequalities are needed in order to obtain an excellent lower bound on the minimum tour length, and in some cases, to also prove optimality. The results obtained in the statistical part of the computational study are consistently at most 1/2% off the optimal tour length and the standard deviations are consistently small as well. The results for the test-problems from the literature including the - by 1980 standards - truly large-scale traveling salesman problems with 120 and 318 cities, respectively, generally out-performed the results that one might expect based on the statistical part of the study. In particular, the bound for the 120-city problem obtained this way indicates that the solution is within 0.04% of the optimum tour and the bound for the 318-city problem indicates that the solution is within 0.26% of the minimum length Hamiltonian path through the 318 points. With the resulting (remaining) gap between the best tour found and the bound obtained by the use of facet-inducing inequalities being so relatively small, it was entirely realistic to expect that any good branch & bound procedure would enable one to solve large-scale traveling salesman to optimality and this is what was done by [Crowder & Padberg, 1980].

Crowder & Padberg selected the same ten problems from the literature as those chosen in the previous study and are shown in Table 5, with KRO124 through KRO128 being the five 100-city problems due to [Krolak, Felts & Marble, 1971]. Several changes in the code used by Padberg & Hong were implemented to improve the lower bounds. These changes concerned the identification of additional subtour-elimination constraints and the determination of the "hard core" of the respective TSP's that remained to be optimized. To this end a subroutine was written to fix non-basic variables of the last LP optimum at either zero or one using the available upper and lower bound on the optimal objective function value. The basic idea for this "variable fixing" can be found in [Dantzig, Fulkerson & Johnson, 1954]. It is very effective and on average a reduction of the n(n-1)/2 variables of a TSP with n cities to 5% of that number was obtained The reduction in the case of the 318-city problem was even more impressive and was between 2% - 3% of the original variables.

In Table 5, n is the number of cities and m = n(n-1)/2 is the number of variables (the number of edges of the graph) except in the case of KNU121 which is a supersparse problem having 222 edges only and 121 nodes. n_R is the number of rows after running the problem with the changed TSP code using the best available tour as a starting solution for the linear program; m_R is the number of variables that were either basic at the final linear programming optimum or that could not be fixed at their respective non-basic value. TIME is the total execution time to obtain LPVALUE at termination. RATIO is the ratio described in this section. Three runs were made for problem LIN318, labelled A, B and C. LIN318A is a run of LIN318 started with the suboptimal tour of length 41349 found by [Padberg & Hong, 1980] using the revised software system just described. The run LIN318B used the optimal tour of length 41345 found by Crowder & Padberg as a starting point--(a check to ensure the validity of the software output). The third run, 318C used the same information as 318B but with a factor of 10 in the data to increase the precision of the distance calculation.

PROBLEM	n	m	n_R	m_R	TIME*	RATIO	LPVALUE
GRO48	48	1,128	86	104	9	0.95	5,031.06
TOM57	57	1,596	89	91	7	0.98	12,948.5
KRO124	100	4,950	187	248	66	0.97	21,225.31
KRO125	100	4,950	170	446	122	0.91	21,978.00
KRO126	100	4,950	183	185	52	0.98	20,730.08
KRO127	100	4,950	192	220	114	0.97	21,257.48
KRO128	100	4,950	203	334	213	0.93	21,970.83
GRO120	120	7,140	199	239	149	0.97	6,934.89
KNU121	121	222	139	182	9	0.90	346.5
LIN318A	318	50,403	496	1,372	2,231	0.97	41,269.83
LIN318B	318	50,403	495	1,144	2,283	0.97	41,269.00
LIN318C	318	50,403	495	1,208	2,295	0.97	412,820.0

*Seconds, IBM 370/168 MVS

Table 5. 10 Large-Scale TSP's

Table 6 gives a breakdown of the facet-inducing linear inequalities that were generated in the respective runs with the (changed) TSP code to get the linear programs of Table 5.. As noted earlier, the constraint identification for (general) comb inequalities was rather ineffective.

Problem	Subtour	2-Matching	Comb	Total
GRO48	21	16	1	38
TOM57	16	16	0	32
KRO124	54	32	1	87
KRO125	40	28	2	70
KRO126	43	37	3	83
KRO127	55	36	1	92
KRO128	45	57	1	103
GRO120	51	28	0	79
KNU121	18	0	0	18
KNU318	157	20	0	177
TOTAL	500	270	9	779
PERCENT	64.3	34.6	1.1	100

Table 6. Breakdown of Constraints by Type

Finally, Table 7 presents the results of using the IBM-package MIP/370 as the branch & bound package to prove optimality for each of these problems. The first two columns display the problem size : n_R is the number of rows and m_R is the number of columns of the (pure) zero-one problem to be optimized. The column labelled MIP gives the number of times the MIP/370 program was called, i.e. the number of passes through the bottom loop of the flow chart in Figure 3 in which branch & bound calculations where required to prove optimality of a tour. The first of the two columns labelled NODES gives the number of nodes of the branch & bound tree generated to find the optimum zero-one solution and the second column gives the total number of nodes generated in MIP/370 to prove optimality. Likewise, the first of the two columns labelled TIME gives the time spent in the branch & bound procedure to find the optimum zero-one solution, whereas the second one gives the total time spent in the MIP/370 program to prove optimality. (Thus in problem GR048 it took 17 nodes and 2 seconds of CPU-time to find the optimum and one additional node and an additional second to prove optimality.) The respective times have been rounded to the nearest second; thus in some cases the MIP times add up to the total execution times indicating that MPSX/370 solved the linear program in less than one second. The total CPU-time for the MPSX/370 computation of solving the linear program (plus constraint-generation and re-optimization where applicable) as well as the time spent in the MIP/370 program are reported in the last but one column labelled TOTAL TIME.

NAME	n_R	m_R	MIP	NODES		TIME		Δn	NODES		TIME		Δn	NODES		TIME		TOTAL TIME*	OPT TOUR LENGTH
GRO48	86	104	1	17	18	2	3											5	5,046
TOM57	89	91	1	2	3	0	1											6	12,955
KRO124	187	248	1	8	15	2	7											13	21,282
KRO125A	170	445	2	133	147	59	70	1	115	146	51	79						159	22,141
KRO125B	170	445	2	187	201	98	110	1	160	191	80	108						220	22,141
KRO126	183	185	1	5	5	1	2											9	20,749
KRO127	192	220	2	4	12	2	7	1	2	12	0	7						21	21,294
KRO128	203	334	1	6	22	5	11											24	22,068
GRO120	199	239	2	17	19	5	8	1	18	20	5	8						24	6,942
KNU121	139	182	3	14	25	1	2	1	15	23	1	2	3	15	23	1	2	16	349
LIN318A	496	1,372	3	22	22	50	52	5	35	36	81	86	1	21	71	75	163	319	41,345
LIN318B	495	1,144	2	41	41	123	125	4	54	58	129	146						321	41,345
LIN318C	495	1,208	2	188	188	349	359	4	150	186	368	445						818	413,589
LIN318D	495	1,208	2	216	216	365	375	4	200	230	408	483						862	413,589

*Seconds, IBM 370/168 MVS

Table 7

A first glance at Table 7 shows that the computation times as well as the total branch & bound effort spent on optimizing the zero-one linear programming problems are extremely low. To optimize the (previously unsolved) problem GR048, a total of 18 nodes had to be generated and the entire execution time of the MPSX-MIP/370 routine took 5 seconds of CPU-time. The "gap" between the linear program value of the corresponding TSP output and the optimal tour is 14.94, i.e. roughly 0.3% of the optimal tour length, and this fact together with the high cutting power of the facet-defining subtour-elimination and comb constraints is responsible for the low computation time. The same can be said about virtually every other problem.

Computational and theoretical research on traveling salesman problems continues along the lines described here and we quote from [Padberg & Grötschel, 1983]: "318 cities should not be the end of the story. The codes we have reported on here have used only some of the heuristics known to date. Currently new LP-based cutting plane procedures are being developed which contain exact subroutines for the identification of subtour elimination and 2-matching constraints as well as several new heuristics and new implementation details to speed up various communication procedures between subroutines and to overcome certain problems with respect to space. These methods may lead to another jump in the range of solvable problem sizes."

5. <u>The Triangulation of Input/Ouptut Matrices</u>

The results described in this section are summarized from papers by [Grotschel, Junger and Reinelt, 1982a, 1982b, 1983]. The triangulation of input/output matrices is an interesting application of a NP-hard combinatorial optimization problem known as the <u>linear ordering problem</u> and is described as follows: Following [Leontief, 1936], the economy of a region or country is divided into n sectors, and a n-by-n input/output matrix A with the n columns and n rows is constructed where the entry a_{ij} denotes the amount of flow of goods from sector i to sector j in a certain year. The problem known as the "triangulation" problem consists of permuting the rows and columns of A simultaneously such that the sum of the entries above the main diagonal of the permuted matrix is as large as possible. As to the problem's economic interest and significance, Leontief writes in a 1963 article in <u>Scientific American</u>, "The triangulation of a real input-output table - that is, the discovery of its peculiar structural properties and its interindustry dependence - is a challenging task. It is complicated by the fact that one must take into account not only the distinction between zero and non-zero entries but also the more important difference between their actual numerical magnitudes," see also [Leontief, 1951 and 1966].

Let α_o denote the <u>largest</u> sum of the entries above the main diagonal of the permuted matrix that can be achieved by simultaneous row and column permutations and let α be the sum of all off-diagonal elements of the input/output matrix A. Then the number

$$\lambda (A) = 100 \frac{\alpha_o}{\alpha}$$

is called the <u>degree of linearity</u> of the economy represented by the input/output matrix A and is used to measure the mutual dependence or circularity of the sectors of the economy under consideration. To know α_o, it is necessary to solve the <u>triangulation</u> or <u>linear ordering problem</u>.

The linear ordering problem is described as follows: A linear ordering of a finite set V with $|V|=n$ is a one-to-one mapping σ from $\{1,\ldots,n\}$ to V. For $i, j \in V$ we say that i precedes j if $\sigma^{-1}(i) < \sigma^{-1}(j)$. In economic applications a "profit" (or a "cost") can be associated with a linear ordering in the following way: For every pair of elements $i, j \in V$ two values c_{ij} and c_{ji} are given that are interpreted as the profit that accrues if i precedes j or if j precedes i, respectively, in a linear ordering σ. Assuming the additivity of profits, the total profit of a linear ordering σ is

$$\sum_{(i,j)\in T(\sigma)} c_{ij} \tag{5.0}$$

where $T(\sigma) = \{(i,j) \mid \sigma^{-1}(i) < \sigma^{-1}(j)\}$, and (i,j) denotes the ordered pair i and j. In the linear ordering problem one is interested in finding a linear ordering σ that maximizes the objective function (5.0). Of course, in the triangulation problem the profits are given by the flow values a_{ij} of the input/output matrix A.

Graphically, a linear ordering σ induces an acyclic orientation of the edges of the complete graph K_n having n nodes corresponding to the elements of V: An edge connecting nodes i and j is oriented (or directed) from i to j if $(i,j) \in T(\sigma)$. The induced orientation of the edges of K_n is acyclic in the sense that the resulting directed graph does not contain any directed cycle since in a linear ordering no element can precede itself. On the other hand, every acyclic orientation of the edge-set of K_n induces a linear ordering on the nodes of K_n. We can thus consider the complete directed graph $D_n = (V,A_n)$ having n nodes and arc-set A_n. A linear ordering σ corresponds uniquely to a partial graph (V,T) of D_n such that (V,T) induces an acyclic orientation of the undirected graph K_n and vice versa. In the terminology of graph theory, an acyclic orientation of K_n is called an acyclic tournament in the directed graph D_n. Then the family $\mathcal{F}$ of Section 2 is the following set:

$$\mathcal{T}_n = \{T \subseteq A_n \mid T \text{ is an acyclic tournament in } D_n\}$$

where for notational simplicity we identify "tournament" with "arc set of a tournament". Since $|A_n| = n(n-1)$ the <u>linear ordering polytope</u> P^n_{LO}, i.e. the convex hull of the incidence vectors of all acyclic tournaments in D_n, is the polytope

$$P^n_{LO} = \text{conv}\,\{x^T \in \mathbb{R}^{n(n-1)} \mid T \in \mathcal{T}_n\},$$

over which we wish to maximize the linear form given by the profits c_{ij} for all $i \neq j$.

Having described the combinatorial objects we are looking for as certain subsets of the arc set A of the complete directed graph D_n we can now proceed to formulate the linear ordering problem as a linear program in zero-one variables. Let $x_{ij} = 1$ if i precedes j, $x_{ij} = 0$ otherwise. The linear ordering problem is the following maximization problem:

$$\max \sum_{i,j} c_{ij}\, x_{ij} \tag{5.1}$$

subject to

$$x_{ij} + x_{ji} = 1 \qquad \text{for all } i \neq j \tag{5.1a}$$

$$\left.\begin{array}{l} x_{ij}+x_{jk}+x_{ki} \leq 2 \\ x_{ik}+x_{kj}+x_{ji} \leq 2 \end{array}\right\} \qquad \text{for all } 1 \leq i < j < k \leq n \tag{5.1b}$$

$$x_{ij} \geq 0 \tag{5.1c}$$

$$x_{ij} \text{ integer.} \tag{5.1d}$$

Note that variables x_{ii} are not defined since we don't need them. The inequalities (5.1a) express the condition that either i precedes j or j precedes i for all $i \neq j$. Inequalities (5.1b) exclude directed cycles (dicycles) of length 3 and are all that are needed to exclude <u>all</u> directed cycles in conjunction with (5.1a). The condition (5.1d) is essential since otherwise the linear ordering problem could be solved in polynomial time! This, of course, means that additional linear inequalities are needed to describe the linear ordering polytope P^n_{LO} by way of linear inequalities.

5.1 Facets of the Linear Ordering Polytope

The following theorem from [Grötschel, Jünger & Reinelt, 1982] summarizes the partial description of the linear ordering polytope P^n_{LO} known to date. Related interesting structural properties of this polytope can be found in [Young, 1978] who studied this problem because of its interest in voting theory.

Theorem: Let $D_n = (V,A_n)$ be the complete directed graph having $n \geq 6$ nodes.

(i) The dimension of P^n_{LO} is $n(n-1)/2$.

(ii) The inequalities $x_{ij} \geq 0$ for all $(i,j) \in A_n$ define the (trivial) facets of P^n_{LO}.

(iii) The 3-dicycle inequalities (5.1b) define facets of P^n_{LO}.

(iv) Let $U=\{u_1,\ldots,u_k\}$, $W=\{w_1,\ldots w_k\} \subseteq V$ be disjoint sets of nodes of D_n of cardinality $3 \leq k \leq n/2$ and call the arc set $A=\{(u_i,w_i) \mid i=1,\ldots,k\} \cup \{(w_i,u_j) \mid i \neq j \in \{1,\ldots,k\}\}$ a k-fence. Then every k-fence inequality

$$x(A) \leq k^2-k + 1 \tag{5.2}$$

defines a facets of P^n_{LO} that is distinct from the facets of part (ii) and (iii). (Note that k-fences are particular orientations of the complete bipartite graph $K_{k,k}$).

(v) Let $M \subseteq A_n$ be an arc set (called Möbius-ladder) which is the union of $k \geq 3$, k odd, dicycles $C_1,\ldots,C_k$ satisfying the following properties:

(1) The length of C_i is three or four, $i = 1,\ldots,k$.

(2) The cycles C_i,C_{i+1}, $i = 1,\ldots,k-1$, resp. C_1,C_k have exactly one arc, say e_i resp. e_k, in common. No other pair of cycles has a common arc.

(3) Each node u in the vertex-set of M is contained in at least three arcs of M.

(4) If two dicycles C_i, C_j, $2 \leq i+1 < j \leq k$, have a node, say v, in common then v belongs to all dicycles $C_i, C_{i+1}, \ldots, C_j$ or to all dicycles $C_j, \ldots, C_k, C_1, \ldots, C_i$.

(5) Given any dicycle C_j, $j \in \{1, \ldots, k\}$,
the set $M \setminus \{e_i \mid i \in J\}$ contains exactly one dicycle, namely C_j, where $J = \{1, \ldots, k\} \cap (\{j-2, j-4, \ldots\} \cup \{j+1, j+3, \ldots\})$,

Then the Möbius-ladder inequality

$$x(M) \leq |M| - \frac{k+1}{2} \qquad (5.3)$$

defines a facet of P^n_{LO}. The Möbius-ladder inequality for $M = C_1 \cup C_2 \cup C_3$ where the C_i are dicycles of length four is a 3-fence inequality. In all other cases no inequality (5.3) is equivalent to any of the inequalities of part (ii), (iii) and (iv).

To use this theorem algorithmically, one notes first that because of the simplicity of the constraints (5.1a) one can eliminate half the number of the variables of the problem. One then rewrites the inequalities of the theorem accordingly. To begin the calculation one starts by considering the problem

$$\max \sum_{i<j} \bar{c}_{ij} x_{ij}.$$

$$\text{subject to} \quad 0 \leq x_{ij} \leq 1 \text{ for all } 1 \leq i < j \leq n$$

where $\bar{c}_{ij} = c_{ij} - c_{ji}$. This problem can be solved trivially.

One then adds the (transformed) 3-dicycle inequalities constraints (5.1b). The number of constraints of this type while polynomial in n is too large to list and consequently they are generated as cutting planes "on the fly". That is, all 3-dicycle inequalities are enumerated, but only those that are violated by the solution to the current linear programming relaxation are added.

On the other hand, there is no polynomial method to solve the facet-identification problem (2.14) for the k-fence inequalities described in (5.2) known to date. Heuristics for finding violated constraints can be found in [Grötschel, Jünger & Reinelt, 1982b].

The Möbius ladders (5.3) are even more difficult to handle, and a general approach for obtaining such constraints is unknown at present. Heuristics have been devised for three special Möbius-ladder inequalities, again see [Grötschel, Jünger & Reinelt, 1982b].

5.2 Computational Results for the Triangulation of Input/Output Matrices

Given the current state of information regarding the polyhedral structure of the linear ordering problem, we now summarize the computational results Grötschel, Jünger & Reinelt had with an algorithm using the facet-cuts described above. All experiments were run on the IBM 370/168 of the Rechenzentrum der Universität Bonn. The authors state that the range of matrices considered in this study is representative of almost all input-output matrices that have been compiled in Europe so far. The study comprises three sets of input/output matrices: one set is provided by the European Community, the second one by the West German Institut fur Wirtschaftsforschung and the third one by the West German Statistisches Bundesamt.

The European Community compiles (44x44)-input-output matrices for all members of the EC. This input-output program is of considerable importance since here, for all countries, all sectors are defined in a (more or less) identical way, so that structural comparisons between these countries can be made. These I/O tables can be obtained from the Office Statistique des Communautés Européennes (EUROSTAT, Boite postale 1907, Kirchberg, Luxembourg) free of charge on tape. In this study the authors chose the (most important) matrices which are the (44x44)-matrices of intermediate consumption at current prices for the years 1959, 1965, 1970, 1975. EUROSTAT has 30 of these matrices. A list of the matrices together with their degree of linearity (the computation of which requires the optimization of the linear ordering problem for the given input/output matrix), and the respective running times can be found in Table 8. The running times in Table 6 include the total time used including input and output of data, paging etc. In every cutting plane generation step all cutting planes that were found were added to the LP. In 28 of the 30 matrices an integral solution was obtained after the initial phase, and only for the matrices BELGIUM 59 and GERMANY 70 was it necessary to enter the branch & bound stage. Both problems were solved with only one branching operation. Thus, the branch & bound tree consisted of three nodes (including the root).

Input Output Table		Degree of linearity	CPU time in (min : sec) (IBM 370/168)
T59I11XX.B	(ITALY 59)	88.443	1 : 02.10
T59D11XX.B	(GERMANY 59)	88.148	0 : 45.84
T59F11XX.B	(FRANCE 59)	85.516	0 : 44.02
T59B11XX.B	(BELGIUM 59)	83.818	1 : 12.84
T59N11XX.B	(NETHERLANDS 59)	82.991	0 : 51.83
T65L11XX.B	(LUXEMBURG 65)	90.805	0 : 28.95
T65I11XX.B	(ITALY 65)	85.812	1 : 34.07
T65B11XX.B	(BELGIUM 65)	85.030	1 : 15.41
T65F11XX.B	(FRANCE 65)	84.395	1 : 10.87
T65D11XX.B	(GERMANY 65)	83.007	1 : 29.05
T65W11XX.B	(EUR-6 65)	82.921	0 : 56.78
T65N11XX.B	(NETHERLANDS 65)	82.585	1 : 15.86
T70L11XX.B	(LUXEMBURG 70)	89.808	0 : 58.87
T69R11XX.B	(IRELAND 69)*	87.862	0 : 56.85
T70K11XX.B	(DENMARK 70)	85.893	0 : 54.56
T70I11XX.B	(ITALY 70)	85.613	1 : 13.68
T70F11XX.B	(FRANCE 70)	85.115	1 : 16.51
T70B11XX.B	(BELGIUM 70)	84.132	0 : 54.96
T70W11XX.B	(EUR-6 70)	83.401	0 : 54.72
T70D11XX.Ba	(GERMANY 70a)*	83.181	1 : 12.06
T70N11XX.B	(NETHERLANDS 70)	82.810	1 : 08.82
T70X11XX.B	(EUR-9 70)	82.668	0 : 53.76
T70D11XX.Bb	(GERMANY 70b)*	82.199	1 : 36.69
T70U11XX.B	(UNITED KINGDOM 70)	80.636	0 : 36.86
T75E11XX.B	(SPAIN 75)	87.715	1 : 04.01
T75K11XX.B	(DENMARK 75)	86.054	1 : 04.47
T75I11XX.B	(ITALY 75)	85.136	1 : 10.56
T75N11XX.B	(NETHERLANDS 75)	82.943	0 : 52.04
T75D11XX.B	(GERMANY 75)	82.773	1 : 13.87
T75U11XX.B	(UNITED KINGDOM 75)	82.678	1 : 12.05

minimum CPU time: 0 : 28.95
maximum CPU time: 1 : 36.69
average CPU time: 1 : 04.09

Table 8

The second class of input/output matrices have the size 56x56 and are compiled by the Deutsches Institut für Wirtschaftforschung (DIW), Berlin, for the Federal Republic of Germany during the years 1954 -1972. The degrees of linearity and the running times are listed in Table 9. The version of the code used was the same as for the EUROSTAT-matrices. All nine triangulation problems could be optimized without entering the branch & bound stage.

Input Output Table	Degree of linearity	CPU time in (min : sec) (IBM 370/168)
DIW56N54	81.299	4 : 56.71
DIW56N58	81.605	4 : 24.14
DIW56N62	81.435	4 : 56.12
DIW56N66	81.791	4 : 47.15
DIW56N67	81.063	5 : 01.19
DIW56N72	79.812	5 : 10.75
DIW56R54	80.785	4 : 47.61
DIW56R58	81.061	4 : 50.81
DIW56R66	81.941	4 : 33.77
DIW56R67	81.801	5 : 08.58
DIW56R72	79.966	4 : 57.03

minimum CPU time: 4 : 24.14
maximum CPU time: 5 : 10.75
average CPU time: 4 : 52.16

Table 9

Thirdly, the authors ran the same version of the code on three (60x60) input/output matrices compiled for the years 1970, 1974 and 1975 by the Statistisches Bundesamt (Postfach 5528, D-6200 Wiesbaden, W. Germany) for the Federal Republic of Germany. The three tables "Input-Output Tabelle zu Ab-Werk-Preisen -Inländische Produktion 70, 74, 75" are available from this government agency. The results can be found in Table 10. The 1974 and 1975 problems could be solved using nothing but 3-dicycle inequalities. For the 1970 matrix only one branching operation was necessary to obtain the optimum solution.

Input Output Table	Degree of linearity	CPU time in (min : sec) (IBM 370/168)
Input-Output-Tabelle 1970 zu Ab-Werk-Preisen-inländische Produktion	83.186	13 : 25.01
Input-Output-Tabelle 1974 zu Ab-Werk-Preisen-inländische Produktion	83.965	9 : 32.95
Input-Output-Tabelle 1975 zu Ab-Werk-Preisen-inländische Produktion	83.799	9 : 57.76

Table 10

Up to this point, the authors used a version of the code where only 3-dicycle inequalities are generated. In a second version of the code, other heuristics were used only if no violated 3-dicycle inequalities were found. The only problems where the new code produces results different from the one described above are those input/output tables for which the branch & bound phase had to be called, i.e. for which 3-dicycle inequalities did not suffice to solve the problem.

For each of these problems, one violated Möbius-ladder constraint together with many 3-dicycle cuts was sufficient to get the optimal solution. Perhaps luckily, in every real-world application of this study, the authors were able to solve the linear ordering problem purely by LP cutting plane-techniques without resorting to branch & bound.

Thus, the linear ordering problem appears to be another class of NP-hard problems which can be optimized in computationally reasonable times using facet-defining linear inequalities in a linear-programming-based approach to combinatorial problem solving.

6. Large-Scale Zero One Linear Programming Problems

In this section, we describe two computational studies [Crowder, Johnson & Padberg, 1983] and [Bernal, Hoffman & Padberg, 1984], which apply the theory of polyhedral combinatorics to the solution of general large-scale zero-one programming problems. The zero-one programming problems that are considered here have the following form:

$$\text{(ZOP)} \qquad \text{minimize}\{cx \mid Ax \leq b,\ x_j = 0 \text{ or } 1 \text{ for } j = 1,\ldots,n\}$$

where A is an m-by-n matrix with arbitrary rational entries, and b and c are vectors of length m and n, respectively, with rational entries. In all test problems, the costs c_j are nonnegative numbers. The problems have a sparse contraint matrix A, i.e., the total number of nonzero elements a_{ij} of A divided by the product of m and n is typically less than 0.05. Furthermore, while A is permitted to have rows with entries equal to +1, -1 and 0 only (e.g. special ordered set constraints) the problems considered here primarily have a substantial number of rows in which the nonzero elements of A have a significant variance within each row. Indeed, if A has only +1, -1 and 0 entries, or if all but a few rows of A have this property, then neither problem preprocessing nor constraint generation as done in these studies can be expected to be very effective. This, indeed, was the case for several of the test problems having this property.

The approach taken to solve these problem consists of a combination of problem preprocessing, cutting planes and clever branch & bound strategies. The preprocessing and overall flow of the procedure was discussed in Section 3. We now provide the details of constraint generation and branch & bound for this class of problems. Computational results will then be presented.

6.1 A Polyhedral Approximation for Sparse Zero-One Linear Programs

The zero-one problem (ZOP) with a single linear constraint, i.e, the case where m = 1, is called the knapsack problem and the facial structure of the associated polytope has been studied thoroughly, even though a complete list of the linear inequalities that define the knapsack polytope remains unknown. Let (a_i, b_i)

denote the ith row of (A,b) and assume for simplicity that all constraints of (ZOP) are general constraints having arbitrary rational entries. Let

$$P_I^i = \text{conv}\{x \in \mathbb{R}^n \mid a^i x \leq b_i,\ x_j = 0 \text{ or } 1 \text{ for } j = 1, \ldots, n\} \tag{6.1}$$

denote the convex hull of the zero-one solutions to the single inequality $a^i x \leq b_i$ where $i \in \{1,\cdots,m\}$. P_I^i is the <u>knapsack polytope</u> associated with constraint i of problem (ZOP). Likewise, we let

$$P_I = \text{conv}\{x \in \mathbb{R}^n \mid Ax \leq b,\ x_j = 0 \text{ or } 1 \text{ for } j = 1, \ldots, n\} \tag{6.2}$$

denote the convex hull of zero-one solutions to the entire constraint set of problem (ZOP). P_I is the zero-one polytope associated with problem (ZOP) and clearly we have

$$P_I \subseteq \bigcap_{i=1}^{m} P_I^i \tag{6.3}$$

i.e., P_I is contained in the intersection of all the knapsack polytopes P_I^i, i = 1, ..., m. Equality in (6.3) does not, in general, hold, but does hold if, for example, problem (ZOP) decomposes totally into m knapsack problems. This is the case if every variable x_j appears in exactly one of the m constraints $a^i x \leq b_i$ for i = 1, ...,m. If we have a large-scale zero-one programming problem with a <u>sparse</u> matrix A and with no apparent special structure, it is reasonable to expect that the intersection of the m knapsack polytopes P_I^i provides a fairly good approximation to the zero-one polytope P_I over which we wish to minimize a linear objective function. This is a working hypothesis and the computational results confirm that it is a reasonable assumption. Practically speaking, this assumption permits one to concentrate on the individual rows of the constraint set of problem (ZOP) when one derives valid inequalities for the polytope P_I. While the preceeding approach is clearly valid for any zero-one problem, it cannot be expected to lead to substantial computational gains if, for instance, the matrix A is rather dense or if almost all entries in A equal +1, -1 and 0. In these two cases the constraint generation must be done differently, see, for example, the survey by [Padberg, 1979].

6.2 Facets of the Knapsack Polytope

Now consider a single inequality $a^i x \leq b_i$ of the constraint system of problem (ZOP). Using the variable substitution $x_j' = 1 - x_j$ where necessary, one brings the inequality into a form where all nonzero coefficients are positive. Dropping the index i for notational convenience, one therefore considers a linear inequality

$$\sum_{j \in K} a_j x_j \leq a_o \tag{6.4}$$

where the coefficients a_j are positive rational numbers and the variables x_j assume the values zero or one. Let $S \subseteq K$ be such that

$$\sum_{j \in S} a_j > a_o \quad \text{and} \quad \sum_{j \in S} a_j - a_k \leq a_o, \quad \text{for all } k \in S \tag{6.5}$$

hold. Then S is called a <u>minimal cover</u> with respect to (6.4) and obviously every zero-one solution to (6.4) satisfies the inequality

$$\sum_{j \in S} x_j \leq |S| - 1 \tag{6.6}$$

where $|S|$ denotes the cardinality of the set S. Suppose, next, that $S^* \subseteq K$ and $t \in (K \setminus S^*)$ satisfies

$$\sum_{j \in S^*} a_j \leq a_o \text{ and} \tag{6.7}$$

$Q \cup \{t\}$ is a miminal cover for every $Q \subseteq S^*$ with $|Q| = k$

where k is an integer number satisfying $2 \leq k \leq |S^*|$. Due to the one-element role of the index t and because k is some integer number, the set $S^* \cup \{t\}$ is called a (1,k)-configuration and the following inequalities are valid for the 0-1 solutions to (6.4):

$$(r - k + 1)x_t + \sum_{j \in T(r)} x_j \leq r \tag{6.8}$$

where $T(r) \subseteq S^*$ varies over all subsets of cardinality r of S^* and r varies over all integers from k to $|S^*|$, inclusively. If $k = |S^*|$ holds in (6.7), then a (1,k) configuration is a minimal cover, thus, in general, the class of inequalities associated with (1,k)-configurations properly contains the class of inequalities associated with minimal covers. Both inequalities (6.5) and (6.7) are best possible

ones if $K = S$ holds in the first case, or if $K = S^* \cup \{t\}$ holds in the second case, i.e., in these cases the respective inequalities define facets of the associated knapsack polytope; see Padberg [1979].
In general, however, these inequalities must be "lifted" to obtain facets of the knapsack polytope associated with (6.4), i.e., they must be extended appropriately to the variables x_j with index j in $K \setminus S$, or with index j in $K \setminus S^* \setminus \{t\}$, respectively.

The extension is done over the respective variables by the following recursive lifting procedure: Initially one sets $f_j = 1$ for all $j \in S$, $f_o = |S|-1$ or $f_j = 1$ for all $j \in S^*$, $f_t = (r - k + 1)$, $f_o = r$ and $S = S^* \cup \{t\}$, respectively. For the iterative step, let $k \in K \setminus S$ and determine

$$z_k = \max \{\textstyle\sum_{j \in S} f_j x_j \mid \sum_{j \in S} a_j x_j \leq a_o - a_k, \qquad (6.9)$$
$$x_j = 0 \text{ or } 1 \text{ for all } j \in S\}.$$

Define $f_k = f_o - z_k$, redefine S to be $S \cup \{k\}$, and repeat until $K \setminus S$ is empty. The resulting inequality $f_x \leq f_o$ defines a facet for the knapsack polytope associated with (6.4). Obviously, the lifting procedure requires the solution of several zero-one problems. But by relaxing (6.9) to a linear program one can approximate the lifting procedure and thus produce "almost" facet-defining inequalities for the knapsack polytope efficiently.

We note that the support of an inequality obtained by lifting (6.6) or (6.8) is contained in the support of the inequality (6.4), i.e., $a_j = 0$ implies that $f_j = 0$ holds. Therefore, the inequalities that are generated preserve the sparsity of the constraint matrix. This is an additional difference between these methods and the traditional cutting planes described in the textbooks on integer programming. The traditional cutting planes are typically rather dense and thereby lead to explosive storage requirements.

Since the number of possible minimal cover and (1,k)-configurations for (6.4) is exponential in the number of variables of the constraint (6.4) one cannot list a priori all possible minimal covers and (1,k) configurations for each row of the problem. One must therefore generate such constraints "on the fly," i.e. generate them in the course of computation as they are needed. To this end, one starts by solving the linear program

$$\min\{cx \mid Ax \leq b,\ 0 \leq x_j \leq 1 \text{ for } j = 1, \ldots, n\} \tag{6.10}$$

and obtains an optimal solution $\bar{x}$. If $\bar{x}$ is a zero-one solution, one can stop: $\bar{x}$ solves the problem (ZOP). Otherwise, taking any knapsack row of A (i.e. a non SOS row) one solves the following problem in approximation to the facet-identification problem (2.14):

CONSTRAINT IDENTIFICATION PROBLEM: Given $\bar{x}$ find a minimal cover inequality (6.6) or a (1,k)-configuration inequality (6.8) that chops off $\bar{x}$, if such an inequality exists.

The constraint identification problem is solved, in turn, for each row of the original constraint matrix A that qualifies, the resulting inequalities that are identified are "lifted" and appended to the linear programming problem, the augmented problem is reoptimized and the procedure is repeated until a zero-one solution is found, until no more constraints are found, or until the gain in the objective function value after a number of such iterations becomes too small, e.g., is less than one unit in terms of the units of the objective function. And, of course, when one of the latter two possibilities occur, one resorts to branch & bound.

The question is then to devise a procedure that solves the constraint identification problem. Consider the zero-one knapsack problem

$$\min\{\textstyle\sum_{j\in K} (1 - \bar{x}_j)s_j \mid \sum_{j\in K} a_j s_j > a_o, s_j = 0 \text{ or } 1 \text{ for all } j \in K\} \tag{6.11}$$

where the inequality is strict in the knapsack constraint. If follows that there is a minimal cover inequality (6.6) that chops off $\bar{x}$ if and only if the optimal objective function value of (6.11) is less than one. As can easily be verified, problem (6.11) is constructed in such a manner that its solution finds a most violated minimum cover inequality.

The formulation of the constraint identification problem for minimal cover inequalities involves the solution of a zero-one knapsack problem for which, to date, no technically good algorithm is known, but which can be approximated efficiently by its linear relaxation. For (1,k)-configurations, at present, a formulation of the constraint identification problem in a tractable form such as (6.11) is not known. Rather,both computational studies reported here use ad-hoc procedures for finding (1,k)-configurations. For details about these procedures see [Crowder, Johnson & Padberg, 1983] and [Bernal, Hoffman & Padberg, 1984].

We note that facets for the knapsack problem with special-ordered sets were described by [Johnson & Padberg, 1982]. The study by [Crowder, Johnson & Padberg, 1983] does not use these results, but rather a surrogate constraint is formed from the original knapsack constraint and a non-overlapping subset of SOS constraints. [Bernal, Hoffman & Padberg, 1984] compare results of using the surrogate strategy to that of using the theoretical results directly.

Both studies resort to branch & bound procedures when it is no longer possible to generate constraints which cut off the current LP solution. Crowder, Johnson & Padberg use IBM's MIP/370 package with an ad-hoc upper bound procedure. Bernal, Hoffman and Padberg use Marsten's IP83 code (an extension of XMP to integer programming) see [Marsten, 1983]. This code is an all-FORTRAN portable code which attempts to incorporate both breadth-first and depth-first strategies (via parametric analysis) into one code. For a complete description of the approach see [Marsten & Morin, 1978] and the dissertation of [Singhal, 1982]. Included in IP83 is the heuristic of [Balas & Martin, 1980] for determining a feasible integer solution to the problem.

6.3 Computational Results for Sparces Zero-One Programs

[Crowder, Johnson & Padberg, 1983] used a set of real-world large-scale zero-one linear programming problems for their tests. Table 11 summarizes the main characteristics of the test problem set. The columns headed VARS, ROWS and NONSOS contain the number of variables, the total number of constraints, and the number of those constraints that are not of the special ordered set type. The column headed VARIANCE indicates the average standard deviation of the absolute values of the nonzero elements of the non-SOS rows of the constraint matrix after preprocessing; the number in parentheses is the standard deviation of this aggregate statistic.

TEST PROBLEM SUMMARY

NAME	VARS	ROWS	NON-SOS	VARIANCE*	DENS	Z_{lp}	Z_{ip}
P0033	33	16	11	49.6 (33.9)	17.4	2520.6	3089.0
P0040	40	24	13	39.6 (92.8)	11.3	61796.5	62027.0
P0201	201	134	107	0.7 (1.0)	5.0	6875.0	7615.0
P0282	282	242	44	34.3 (16.2)	2.0	176867.5	258411.0
P0291	291	253	14	3.5 (2.3)	1.9	1705.1	5223.75
P0548	548	177	94	123.9 (88.3)	1.8	315.3	8691.0
P1550	1550	94	2	8.4 (8.4)	7.4	1706.5	1708.0
P1939	1939	109	2	3.5 (3.5)	4.8	2051.1	2066.0
P2655	2655	147	2	2.3 (2.3)	3.4	6532.1	6548.0
P2756	2756	756	386	149.2 (89.4)	0.4	2688.7	3124.0

*After problem preprocessing

Table 11

Thus, for example, P2756 has 756 rows out of which 386 are non SOS rows. Computing for each non SOS row of P2756 the standard deviation of the absolute values of its nonzero elements and averaging over all 386 rows yields an average intra-row standard deviation of 149.2, i.e., a substantial variation of the nonzero coefficients; the number 89.4 indicates that the variation is substantial among the non-SOS rows of the constraint matrix, but is somewhat less on average than the variation within each non-SOS row. In problems P1550, P1939 and P2655 these figures indicate that only one of the two non-SOS rows contributes to the standard deviation, the other row having identical nonzero elements. Indeed, these problems are set partitioning problems with two additional constraints: one constrains the sum of all variables to be a certain number, the other one is a knapsack constraint with positive coefficients which, however, have little variation. The column headed DENS specifies the density of the original constraint matrix plus the cost row; i.e. the total number of nonzero elements of A and c divided by the product of m and n and multiplied by 100. Finally, Z_{lp} denotes the optimal objective function value of the linear programming relaxation of problem (ZOP) in user-supplied form, while Z_{ip} denotes the optimal zero-one objective function value.

Crowder, Johnson & Padberg performed all the computational experiments on the IBM 370/168 computer, running the MVS operating system, at the Thomas J. Watson Research Center in Yorktown Heights, New York. FORTRAN programs were compiled using the FORTRAN H Extended Compiler. PL1 programs were compiled using the PLI Optimizing Compiler.

PROBLEM PREPROCESSOR SUMMARY

NAME	ORIGINAL PROBLEM		ELIMINATED		TIME*
	ROWS	VARS	ROWS	VARS	
P0033	16	33	0	0	0.01
P0040	24	40	0	0	0.02
P0201	134	201	0	6	0.01
P0282	242	282	20	0	0.02
P0291	253	291	47	1	0.02
P0548	177	548	20	21	0.03
P1550	94	1550	0	0	0.19
P1939	109	1939	0	0	0.18
P2655	147	2655	0	0	0.30
P2756	756	2756	17	22	0.50

*CPU minutes, IBM 370/168 - MVS

Table 12

The zero-one preprocessor is the initial computational phase. As indicated, the effect of the preprocessor is to tighten the user-supplied linear programming formulation of the zero-one problem using the standard "tricks" of integer programming to eliminate inactive constraints, and to fix variables at either zero or one. Table 12 summarizes the results of the initial preprocessor step on the test problem set. The column headed ORIGINAL PROBLEM specifies the number of rows and variables in the problem before preprocessing. The column headed ELIMINATED gives the number of rows and variables that the preprocessor successfully eliminated from the problem. The time for the initial preprocessor phase, in CPU minutes is given under the TIME column.

The constraint generation procedure is the second computational phase: It operates on the preprocessed problem to produce and solve a linear programming problem with a better (greater in the minimization case) optimal continuous objective function value. Table 13 summarizes the results of the first constraint generation step on the test problem set. The columns headed PREPROCESSED PROBLEM specify the number of constraints, the number of variables, the density of the problem, and optimal continuous objective function value Z_{lp}^{*} of the problem produced by the preprocessor, before constraint generation. The AUGMENTED PROBLEM section of the table indicates the number of constraint generation "passes," i.e., the number of intermediate linear programming problems required, the number of constraints in the final linear programming problem produced, the density of its constraint matrix including the cost row, and its optimal continuous objective function value Z_{lp}^{**}. The time for the initial constraint generation phase, in CPU minutes, is given under the heading TIME. This time includes all linear programming calculations in the constraint generation phase, including the solution of the original linear program.

CONSTRAINT GENERATION SUMMARY

NAME	PREPROCESSED PROBLEM				AUGMENTED PROB				TIME*
	VARS	ROWS	DENS	Z_{lp}^{*}	PASS	ROWS	DENS	Z_{lp}^{**}	
P0033	33	16	17.4	2819.4	6	36	12.1	3065.3	0.12
P0040	40	24	11.3	61829.1	4	29	13.8	61862.8	0.12
P0201	195	134	5.0	7125.0	1	139	5.0	7125.0	0.16
P0282	282	222	1.5	176867.5	14	462	1.3	255033.1	1.00
P0291	290	206	1.3	1749.9	6	278	2.5	5022.7	0.33
P0548	527	157	1.9	3125.9	9	296	1.3	8643.5	0.57
P1550	1550	94	7.4	1706.5	1	94	7.4	1706.5	0.81
P1939	1939	109	4.8	2051.1	1	110	4.8	2051.1	0.74
P2655	2655	147	3.4	6532.1	2	149	3.4	6535.0	2.04
P2756	2734	739	0.4	2701.1	8	1065	0.4	3115.3	3.06

*CPU minutes, IBM 370/168 - MVS

Table 13

The real measure of the effectiveness of the constraint generation procedure is determined by how much it closes the "gap" between the optimal value of the objective function of the linear programming relaxation and the optimal zero-one objective function value. Table 14 summarizes these data for the first application of the constraint generation procedure. The column headed ΔROWS specifies the number of constraints generated by the procedure. The GAP is the difference between the optimal zero-one objective function value and the optimal continuous objective function value before constraint generation, but after problem preprocessing, i.e., the GAP is $Z_{ip} - Z^{*}_{lp}$. The figure RATIO is computed as RATIO = $(Z^{**}_{lp} - Z^{*}_{lp})/(Z_{ip} - Z^{*}_{lp})$. Table 15 summarizes the time, in

EFFECT OF CONSTRAINT GENERATION

NAME	ΔROWS	GAP	RATIO
P0033	20	269.6	0.92
P0040	5	197.9	0.17
P0201	5	490.0	0.00
P0282	240	81543.5	0.96
P0291	72	3473.8	0.94
P0548	139	5565.1	0.99
P1550	0	1.5	0.00
P1939	1	14.9	0.00
P2655	2	15.9	0.19
P2756	326	422.9	0.98

Table 14

PIPX EXECUTION SUMMARY

NAME	PIPX Passes	TIME*			
		Pre-processor	LP/Constraint Generation	Branch-and-Bound	Total
P0033	1	0.01	0.12	0.56	0.69
P0040	2	0.03	0.13	0.02	0.18
P0201	1	0.01	0.16	9.90	10.07
P0282	1	0.02	1.00	11.70	12.72
P0291	2	0.04	0.51	0.30	0.85
P0548	2	0.12	0.70	0.09	0.91
P1550	3	0.50	0.90	0.10	1.50
P1939	3	0.49	0.86	13.64	14.99
P2655	3	0.82	3.23	2.70	6.75
P2756	2	1.07	4.03	49.32	54.42

*CPU minutes, IBM 370/168 - MVS

Table 15

CPU minutes, for the entire system execution on the test problem set. PIPX Passes is the number of major iterations performed by the system as a whole. The columns headed TIME specify the aggregate times spent in each of the three main computational phases, and the total time required to solve the test problems.

All test problems considered during the course of this study were solved to optimality. The methodology described above for the solution of large-scale zero-one linear programming problems produced--by 1983 standards--impressive computational results, particularly on sparse problems having no apparent special structure. The authors obtained the test problems from various sources within and outside of the IBM Corporation and, while they claim to have no comparative data on previous solution attempts, most were originally considered not amenable to exact solution in economically feasible computation times. The computational results reported contradict this sentiment and strongly confirm the hypothesis that a combination of problem preprocessing, facet-defining cutting planes, and clever branch & bound technique permits, in reasonable computation times, the optimization of sparse large-scale zero-one linear programming problems, even when no apparent special structure is present.

Computational results from the Bernal, Hoffman & Padberg study will not be available until the end of 1984. The authors report preliminary results that good preprocessing and the [Balas & Martin, 1980] heuristic do further improve the results obtained by Crowder, Johnson & Padberg. This study is significant in that it is the only study known to these authors that attempts to use constraint generation within an all-FORTRAN machine-portable non-proprietary code. The test effort took place on a UNIVAC 1100 Series Machine which allows only 262K words of storage. Within this significant machine memory restriction, the authors were still successful in solving the first six problems reported by Crowder, Johnson and Padberg, as well as a variety of other difficult zero-one problems not included in the Crowder, Johnson, & Padberg study.

7. Conclusions

The computational results reported in the last three sections of the report point to the suitability of facet-defining cutting-planes for the purpose of proving optimality in difficult, large combinatorial optimization problems. These constraints--generated only when needed--are able to reduce the gap between the linear-programming relaxation and the integer program substantially, thereby making branch & bound calculations tractable.

Codes which are LP-based and use cutting planes methods are not restricted to these three classes of problems. Facets for many other specially structured problems are known, e.g. plant location problems, see [Cho, (Johnson), Padberg & Rao, 1983a,b], set covering, packing and partitioning problems, see [Padberg 1979], and mixed 0/1 Linear Programming, see [Martin & Schrage, 1983] and the papers of the C.O.R.E. group mentioned in the introduction. Code development for such problems will require procedures that generate the facet-defining inequalities via polynomially-bounded algorithms or at least approximate them by efficient procedures. The codes are likely to consist of a preprocesser, a variety of heuristics for generating good feasible solutions, algorithms for generating facet-inducing cutting planes and a branch & bound algorithm, i.e. to follow the general design outlined in Section 3 of this paper. These codes will certainly be rather complex, but are likely to lead to methods for solving to optimality--with reasonable computational effort--many of the classes of difficult combinatorial problems for which only heuristic "guesses" at the optimal solution are currently possible. The excellent computational results of the most recent study by [Van Roy & Wolsey, 1984] shows that the validity of this statement is not restricted to the (pure) zero-one combinatorial optimization problems considered in this survey, but extends as well to the case of mixed zero-one linear programming problems.

References

Balas, E. & R. Martin [1980]: "Pivot and Complement--A Heuristic for 0-1 Programming" Management Science 26 (1980) 86-96.

Baranyi, I., T. Van Roy, & L. Wolsey, [1983a]: "Strong Formulations for Multi-Item Capacitated Lot Sizing," CORE Report No. 8313, Universite Catholique, Louvain-la-Neuve, Belgium.

Baranyi, I., T. Van Roy, & L. Wolsey, [1983b]: "Uncapacitated, Lot-Sizing: The Convex Hull of Solutions," CORE Report No. 8314, Universite Catholique, Louvain-la-Neuve, Belgium.

Burkard, R. & U. Derigs, [1980]: Assignment and Matching Problems: Solution Methods with FORTRAN-Programs, Springer Lecture Notes in Economics and Mathematical Systems, No. 184, Berlin.

Cho, D. Chinhyung, E.L. Johnson, M. Padberg & M.R. Rao [1982]: "On the Uncapacitated Plant Location Problem I: Valid Inequalities and Facets," Mathematics of Operations Research 8 (1983) 579-589.

Cho, D. Chinhyung, M. Padberg & M.R. Rao, [1983]: "On the Uncapacitated Plant Location Problem II: Facets and Lifing Theorems," Mathematics of Operations Research 8 (1983) 590-612.

Chvatal, V. [1973]: "Edmonds Polytopes and Weakly Hamiltonian Graphs," Mathematical Programming, 5(1973) 29-40.

Crowder, H. & M. Padberg [1980]: "Solving Large-Scale Symmetric Traveling Salesman Problems to Optimality," Management Science, 26 (1980) 495-509.

Crowder, H., E. Johnson, & M. Padberg [1983]: "Solving Large-Scale Linear Zero-One Programming Problems," Operations Research, 31 (1983) 803-834.

Dantzig, G., D.R. Fulkerson & S. Johnson [1954]: "Solution of a Large-Scale Traveling Salesman Problem" Operations Research 2(1954) 293-410.

Edmonds, J. [1965]: "Maximum Matching and a Polyhedron with 0,1 Vertices," Journal of Research, National Bureau of Standards, 69B (1965) 125-130.

Fleischmann, B. [1981]: "The Traveling Salesman Problem on a Road Network," Working paper, Fachbereich Wirtschaftswissenschaften, Universitat Hamburg, revised July 1982.

Fleischmann, B., [1982]: Linear-Programming Approaches to Traveling Salesman and Vehicle Scheduling Problems," paper presented at the XI. International Symposium on Mathematical Programming, Bonn, FRG.

Gomory, R. & T.C. Hu [1961]: "Multi-terminal Network Flows," J. SIAM 9 (1961) 551-570.

Grötschel, M. & O. Holland, 1984]: "Solving Matching Problems with Linear Programming," Preprint No. 37, Mathematisches Institut, Universitat Augsburg, FRG.

Grötschel, M. & M. Padberg [1983]: "Polyhedral Aspects of the Traveling Salesman Problem I: Theory," to appear in Lawler, E. et al. (eds.): The Traveling Salesman Problem, North-Holland, 1985.

Grötschel, M., M. Junger & G. Reinelt [1982a]: "On the Acyclic Subgraph Polytope," WP 82215-OR, Institut fur Operations Research, Universitat Bonn. (1982).

Grötschel, M., M. Jünger & G. Reinelt [1982b]: "Facets of the Linear Ordering Polytope. WP 82217-OR, Institut fur Operations Research, Universitat Bonn (1982).

Grötschel, M., M. Jünger, & G. Reinelt [1983]: "Optimal Triangulation of Large Real World Input-Output Matrices," Preprint No. 9, Mathematisches Institut, Universitat Augsburg (1983).

Held & Karp [1970]: "The Traveling Salesman & Minimum Spanning Trees," Part 1 Operations Research 18 (1970) 1138-1162, Part II, Mathematical Programming 1 (1971) 6-26.

Hong, S., [1972]: "A Linear Programming Approach for the Traveling Salesman Problem," Ph.D. Thesis, Johns Hopkins University, Baltimore.

Hammer, P.L., M. Padberg & U.N. Peled [1975]: "Constraint Pairing in Integer Programming INFOR - Canadian Journal of Operations Research, 13 (1975) 68-81.

Johnson, E., & M. Padberg [1981]: "A Note on the Knapsack Problem with Special Ordered Sets" Operations Research Letters, 1 (1981) 38-43.

Johnson, E. & M. Padberg [1982]: "Degree-Two Inequalities, Clique Facets and Biperfect Graphs," Annals of Discrete Mathematics, 16 (1982) 169-187.

Knuth, D. [1976]: "The Traveling Salesman Problem," illustrative example in H. Sullivan: Frontiers of Science, from Microcosm to Macrocosm. The New York Times, February 24 (1976) 18.

Krolak, P., Felts, W. & G. Marble, [1971]: "A Man-Machine Approach Toward Solving the Traveling Salesman Problem," CACM 14 327-334.

Land, A., [1979]: "The Solution of 100-City Symmetric Traveling Salesman Problems," Research Report, London School of Economics, London.

Leontief, W. [1936]: "Quantitative Input and Output Relations in the Economic System of the United States," Review of Economic Systems 18 105-126.

Leontief, W., [1951]: The Structure of the American Economy 1919-1939 Oxford University Press, New York.

Leontief, W. [1963]: "The Structure of Development" Scientific American, 1963.

Leontief, W., [1966]: Input-Output Economics, New York, Oxford University Press.

Lin, S. & B. Kerningham, [1973]: "An Effective Heuristic Algorithm for the Traveling Salesman," Operations Research, 21 498-516.

Marsten, R.E. [1981]: "XMP: A Structured Library of Subroutines for Experimental Mathematical Programming," ACM Transactions on Mathematical Software 7 (1981) 481-497.

Marsten, R. E. [1983]: "User's Guide to IP83," Technical Report, Dept. of Management Information Systems, University of Arizona.

Marsten, R.E. & T.L. Morin, [1978]: "A Hybrid Approach to Discrete Mathematical Programming," Mathematical Programming 14 (1978) 21-40.

Martin, R.K. & L. Schrage [1983]: "Subset Coefficient Reduction Cuts for 0/1 Mixed Integer Programming," Technical Report. Graduate School of Business, University of Chicago, Chicago, Illinois.

Miliotis, T., [1976]: "Integer Programming Approaches to the Traveling Salesman Problem," Mathematical Programming, 10 367-368.

Miliotis, T., [1979]: "Using Cutting Planes to Solve the Symmetric Traveling Salesman Problem," Mathematical Programming, 15 177-188.

Padberg, M. [1973]: "On the Facial Structure of Set Packing Polyhedra," Mathematical Programming, 5 (1973) 199-215.

Padberg, M. [1974]: "Characterizations of Totally Unimodular, Balanced and Perfect Matrices," in B. Roy (ed.): Combinatorial Programming: Methods and Applications, D. Reidel Publishing Co., Dordrecht-Holland, 1975 275-284.

Padberg, M. [1975]: "A Note on Zero-One Programming," Operations Research 23(1975) 833-837.

Padberg, M. [1977]: "On the Complexity of Set Packing Polyhedra," Annals of Discrete Mathematics, 1 (1977) 421-434.

Padberg, M. [1979]: "Covering, Packing and Knapsack Problems," Annals of Discrete Mathematics, 4 (1979) 265-281.

Padberg, M. & S. Hong, [1980]: "On the Symmetric Traveling Salesman Problem: A Computational Study," Mathematical Programming Studies, 12(1980) 78-107.

Padberg, M. [1980]: "(1,k)-Configurations and Facets for Packing Problems," Mathematical Programming 18 (1980) 94-99.

Padberg, M. & M.R. Rao [1982]: "Odd Minimum Cut-Sets and b-Matchings", Mathematics of Operations Research, 7 (1982) 67-80.

Padberg, M. & T. Van Roy, & L. Wolsey, [1982]: "Valid Linear Inequalities for Fixed Charge Problems," CORE Report No. 8232, Université Catholique, Louvain-la-Neuve, Belgium.

Padberg, M. & M. Grötschel [1983]: "Polyhedral Aspects of the Traveling Salesman Problem II: Computation," to appear in Lawler, E. et al. (eds.): The Traveling Salesman Problem, North-Holland, 1985.

Singhal, J. [1982]: "Fixed Order Branch and Bound Methods for Mixed Integer Programming," Dissertation, Dept. of Management Information Systems, University of Arizona, January, 1982.

Thompson, G.L. & R.L. Karp [1964]: "A Heuristic Approach to Solving Traveling Salesman Problems," Management Science 10 (1964) 225-247.

Trotter, L. [1976]: "A Class of Facets for Vertex-Packing Polyhedra," Discrete Mathematics 12 (1975) 373-388.

Van Roy, T. & L. Wolsey, [1983]: "Valid Inequalities for Mixed 0-1 Programs," CORE Report No. 8316, Université Catholique, Louvain-la-Neuve, Belgium.

Van Roy, T. & L. Wolsey, [1984a]: "Valid Inequalities and Separation for Uncapacitated Fixed Charge Networks," CORE Report No. 8410, Université Catholique, Louvain-la-Neuve, Belgium.

Van Roy, T. & L. Wolsey, [1984b]: "Solving Mixed Integer Programs by Automatic Reformulation," CORE Research No. 8432, Université Catholique, Louvain-la-Neuve, Belgium.

Weyl, H. [1935]: "Elementare Theorie der konvexen Polyheder," Comm. Math. Helv., 7 (1935), 290-306 translated in Contributions to the Theory of Games, Vol. I, 3-18, Annals of Mathematics Studies, No. 24, Princeton, 1950.

Wolsey, L. [1975]: "Faces for a Linear Inequality in 0-1 Variables," Mathematical Programming 8(1975) 165-178.

Young, H.P. [1978]: "On Permutations and Permutation Polytopes," Mathematical Programming Studies 8 (1978) 128-140.

NETWORK OPTIMIZATION

R. R. Meyer
Computer Sciences Department and Mathematics Research Center
The University of Wisconsin-Madison
Madison, Wisconsin, USA 53706

ABSTRACT

Network optimization not only has a long and distinguished history within the field of mathematical programming, but also continues to be one of the principal ongoing areas of research in optimization . This is due in part to the abundance of network applications, and in part to the fortuitous confluence of mathematical theory and computer science research that makes possible the optimization of linear network problems involving millions of variables, while simultaneously posing the challenge of adapting these techniques to nonlinear networks with thousands of variables. This paper endeavors to survey recent theoretical and computational developments in network optimization, with emphasis on convex as opposed to nonconvex or combinatorial problems.

1. Introduction

The earliest work in what may be regarded as modern network optimization coincides with the first modern paper in mathematical programming, namely that of [Kantorovich 1939] dealing with railway networks. It is worthwhile noting that both Kantorovich and Koopmans, who also did early work on the transportation problem [Koopmans 1949], received the Nobel Prize in 1975 in recognition of their contributions in the area of methods for the allocation of resources.

Other notable early achievements include Hitchcock's work on transportation [Hitchcock 1941] and the implementation of the stepping stone method on the National Bureau of Standards Eastern Automatic Computer in 1952, perhaps the earliest example of "production" grade mathematical programming software. After the initial dominance of primal simplex method approaches to linear network optimization, work by Ford and Fulkerson [Ford and Fulkerson 1962] led to the popularity of primal-dual methods during the 1960's. The computational pendulum drifted back to primal simplex methods again in the 1970's as a result of the development and exploitation of suitable computer science data structures by Johnson, Glover, Karney, Klingman, and others ([Johnson 1966], [Glover, Karney, and Klingman 1974]). At present the primal simplex method (specialized to take advantage of the tree structure of the basis) remains the method of choice for the general linear minimum cost network flow problem, although very promising results (particularly for transportation and assignment problems) have recently been reported in [Bertsekas and Tseng 1983] with a method based on some concepts from monotropic programming [Rockafellar 1984].

In section 2 of this paper we discuss some of the key concepts that have been used to develop efficient specializations of the primal simplex method to linear network optimization problems. The so-called network simplex method based on these techniques has been used to solve problems of up to 65,000,000 variables and 50,000 constraints [Barr and Turner 81]. These problems are perhaps the largest optimization problems ever solved. Multicommodity network techniques are described in section 3, along with extensions to networks with side constraints and variables, and so-called generalized networks, also known as networks with gains and losses. Finally section 4 surveys algorithms for networks with nonlinear convex objective functions, and notes the interrelationship between the use of network algorithms for computer network design and the use of local area computer networks (including the Crystal multicomputer at the Computer Sciences Department of the University of Wisconsin-Madison) to solve network optimization problems via parallel algorithms.

NATO ASI Series, Vol. F15
Computational Mathematical Programming
Edited by K. Schittkowski

2. Linear Single-Commodity Network Optimization

The problem class to be considered in this section is

$$\min_x \ \mathbf{cx}$$
$$s.t.\ \mathbf{Ax} = \mathbf{b} \qquad \text{(NET)}$$
$$\mathbf{l} \leq \mathbf{x} \leq \mathbf{u}$$

where $\mathbf{x}$ is the n-vector $(x_1, \ldots, x_n)^T$ in which x_i represents the flow on arc i; $\mathbf{c}$ is the vector of objective function coefficients (the juxtaposition of two vectors will be used to denote their inner product, so that $\mathbf{cx} = \sum_{i=1}^{n} c_i x_i$); $\mathbf{l}$ is a vector of lower bounds; $\mathbf{u}$ is a vector of upper bounds; A is the so-called *node-arc incidence matrix* of a directed graph or digraph with m nodes and n arcs, where if the kth arc goes from node i to node j (this arc will be denoted as (i, j))), then kth column of A has a $+1$ in row i and a -1 in row j (note that some authors use the opposite sign convention for these entries) and $0's$ elsewhere in the column; and $\mathbf{b}$ is the right-hand side (RHS) vector. When it is necessary to refer to the variable associated with an arc (i, j), this variable will be denoted as x_{ij}, and a similar convention will be used for cost coefficients. Each node is assumed to have at least one incident arc (i.e., an arc with that node as origin or destination) and each arc is assumed to have distinct origin and destination nodes (i.e., no self-loops are allowed). (Note that since $(1, 1, \ldots 1)A = 0$, a necessary condition for the feasibility of the system $Ax = b$ is that $\sum_{i=1}^{m} b_i = 0$. Without loss of generality we assume that this condition is satisfied as it is easily checked at input time.) Although for the sake of simplicity in presenting the key ideas of the network simplex method we massume that all variables have upper and lower bounds, the algorithm below may be augmented in a straightforward manner to handle unbounded variables and to detect unbounded problems. Also, for notational convenience we will assume that the lower bound has been translated to 0. With the sign convention on the entries of A, it is easily seen that a given row, say row i, of the product Ax corresponds to the sum of the flow variables for the arcs leaving node i minus the sum of the flow variables for the arcs entering node i. This quantity is often call the *divergence* of row i, and may be thought of as the *net* amount of flow leaving node i (if the node is a *conservation* node or a *supply* node) or the negative of the net amount remaining at node i if the node is a *demand* node. Thus, the three fundamental types of constraints occurring in the system $Ax = b$ are supply, demand, and conservation of flow constraints. Many other types of constraints can be converted to this network format, including supply and demand inequalities, bounds on the amount flowing through a node, bounds on the total amount flowing out of a subset of nodes, etc. In cases in which the transformation to the standard network format is not obvious, there exist algorithms which determine if classes of linear transformation can be used to convert the problem to that format [Bixby 81].

Finally, the corresponding digraph is assumed to be *connected*, so that between every pair of nodes (i, j) there exists a *path* of ordered nodes and arcs such that node i is the initial node of the path, j is the last node of the path, and if i_ℓ and $i_{\ell+1}$ are consecutive nodes in the path, then the ℓth arc in the path is either $(i_\ell, i_{\ell+1})$ or $(i_{\ell+1}, i_\ell)$. (Thus, the orientation of the arcs in the path is arbitrary.) (If the digraph is not connected, it is easily seen that the problem (NET) may be decomposed into a set of separate optimization problems, each of which corresponds to a maximal connected component of the digraph).

It should be noted that in the (NET) problem format, the columns of the matrix A correspond to arcs of the network whereas the rows of A correspond to nodes. In stating optimality conditions, we will associate dual variables with the rows of A, so from a network viewpoint, these dual variables are associated with the corresponding nodes, while the primal variables are associated with the arcs.

Computationally it has been found efficient [Grigoriadis 82] to start the network simplex method with an all artificial basic feasible solution (BFS). Given a problem in the form (NET), the network structure may be maintained in spite of the addition of a full set of artificial variables by the following procedure: (1) to add the artificial variable x_{n+i}, add to the matrix A the ith unit vector if $b_i \geq 0$, and the negative of the ith unit vector otherwise; (2) form the negative of the sum of the rows of this augmented matrix, and then add the resulting row as an *additional row* of constraint coefficients with corresponding RHS value $b^{\bullet}_{m+1} = -\sum_{i=1}^{m} b_i = 0$. It is easily seen that this additional row introduces a second non-zero of opposite sign in every artificial column, while adding only an additional zero to each column corresponding to an original variable. This new row may be thought of as corresponding to an *artificial node* $m+1$, to which each original node is connected by an arc of the appropriate direction. Computational experience [Grigoriadis 82] indicates that the best approach for handling the artificials is a gradual-penalty method in which a relatively small penalty coefficient (of the order of the objective function coefficient largest in optimal value) is first used for the artificials, followed by an increase in the penalty each time the optimal solution contains artificials at positive levels. If positive values of artificials persist even after the penalty has reached a sufficiently high level, then the original problem is declared infeasible. (It may be shown that the problem (NET) has the nice property that the penalty coefficient does not have to be raised above $mC/2$, where C is $\max_{1 \leq i \leq n} |c_i|$, assuming this is positive.)

A property of the node-arc incidence matrix A that results to a minor technical difficulty is the linear dependence of the rows of A. Since each column of A contains exactly two non-zeros, one $+1$ and one -1, it follows that the sum of the rows is the 0 vector. In order to be able to deal with basis matrices in a notationally simple manner, we will assume A has been augmented not only by the additional row and columns discussed above, but also by the column consisting of the $(m+1)st$ unit vector $(0,0,...,0,1)^T$, and that the corresponding variable, x_{root}, has been assigned lower and upper bounds of 0 and a 0 cost coefficient. Clearly, this does not affect the solution of the original problem (NET) in any way, but it does result in a constraint coefficient matrix of fall rank (see, e.g., [Kennington and Helgason 80]. From a graphical standpoint, x_{root} is considered to be the flow on a so-called *root arc* that originates at node $m+1$, but has no terminating node. (Node $m+1$ is thus termed the *root node* or simply the root). Without loss of generality we will assume that the problem (NET) also contains the variable x_{root} and its corresponding data as described above. (This variable actually plays no role in computation, since its value is always 0.) Since the root arc is included in every basis, the dual variables π associated with a basis B by the equation $\pi B = c_B$ have the property that $\pi_{m+1} = 0$. Moreover, since every other column has one $+1$ and one -1, these equations have the form

$$\pi_i - \pi_j = c_{ij} \qquad (PI)$$

for every other basic arc. The simple form of these equations helps to make the update of the dual variables after a pivot relatively easy, as we will see below. Because the initial basic arcs all begin or end at node $m+1$, the initial $\pi's$ (other than π_{m+1}) are simply $\pm K$, where K is the penalty cost associated with the artificial arcs.

Another key mathematical property of the problem (NET) that leads to enormous computational speedups in the application of the primal simplex method is the fact that there is a one-to-one correspondence between the $(m+1) \times (m+1)$ basis matrices of A and the *rooted spanning trees* of the corresponding network, where a rooted spanning tree is a connected subgraph that contains the root arc and all of the nodes of the original graph, but no cycles, a cycle being a non-trivial path whose origin and terminal nodes coincide (for a proof of this correspondence see, e.g., [Kennington and Helgason 80]). (An intuitive justification of this correspondence may be obtained by noting that connectedness is related to the *existence* of basic solutions for arbitrary RHS's - in particular for those RHS's corresponding to the selection of any given node as the sole supply node and any other node as the sole demand node, thereby necessitating a path between those nodes to correspond to the solution flow - whereas acyclicity is related to *uniqueness* of solutions, since the presence of a cycle would allow a feasible solution to be transformed into another feasible solution by the superposition of flows on the cycle without any effect on the divergence equations. More will be said about cycle flows in the discussion of the pivot operation below). For simplicity, we will simply refer to rooted spanning trees as trees .

The *computational* utility of trees arises from the fact that the pivot operation of the network simplex method may be carried out by utilizing data structures (representing the tree) that make it relatively easy to identify the variable (i.e., arc) that leaves the basis and to carry out the updating of the primal and dual variables. In fact, it is the simplicity of the arithmetic of those operations that allows single-commodity network flow problems to be solved about 100 times faster than general LP's of the same size.

It is important to note that with this approach there are two aspects to the performance of the pivot operation: 1) the numerical update, in which changes to the values of the primal and dual variables are effected, and 2) the data structure update, in which the tree representation must be changed when the basis changes. As might be expected, there is an important trade-off between those two updates, since more complex tree representations reduce the effort required for numerical updates, but are themselves more difficult to store and update as the tree changes. Indeed, a major share of the research into computational aspects of network optimization has been devoted to the determination of data structures that reduce the total time for the two portions of the updating process. While there is no current agreement on optimal data structures, there are some key ingredients for superior performance , and we will provide a brief overview of these elements .

It is convenient for both notational and algorithmic purposes to assume that the nonbasics x_N have value 0. This is clearly the case in the initial iteration with the all-artificial basis, and at any iteration in which a nonbasic x_i reaches upper bound, a "reflection" of x_i is performed by replacing x_i by $x_i' = u_i - x_i$. Computationally, this requires only the reversal of the direction of the arc corresponding to x_i, a change in the sign of c_i, and a record that this reflection has been made. (Since the RHS b is not actually used in the algorithm after the initial basis, it need not be modified.) Under the assumption that $x_N = 0$, a sufficient optimality condition is that

$$c_{ij} - \pi_i + \pi_j \geq 0 \qquad (OPT)$$

for each nonbasic arc (i, j). Thus, the so-called "*pricing out*" operation for an arc requires only the subtraction and addition from the cost coefficient of the $\pi's$ for the nodes at the ends of the arc. (Just as with general linear programs, a variety of sophisticated strategies have been used wtih the goal of obtaining "good" entering nonbasics without having to price out a large number of nonbasics at each iteration. While so-called "candidate list" strategies are effective, a simple "random" strategy has also been used successfully [Grigoriadis 82].)

Let us consider the operations that must be performed to accomplish a pivot in the network simplex method, given a nonbasic arc entering the current basis at lower bound. We will assume that the pivot is *non-degenerate*, i.e., that a flow increase is possible in the entering arc. (Degeneracy is actually fairly common in network optimization, particularly for special classes of problems such as the assignment problem, and special techniques have been devised [Barr, et al 77] to reduce the number of degenerate pivots. However, a discussion of these devices is beyond the scope of this paper, and for the general single-commodity network problem, the RNET code [Grigoriadis 82] to be discussed below has been found to be one of the most efficient of currently available network codes, yet contains no sophisticated anti-degeneracy features.)

The Network Simplex Pivot

We assume that a BFS and a corresponding set of dual variables are available.

1. If the optimality conditions (OPT) are satisfied, then quit with an optimal solution; else select an entering nonbasic arc (r, s) .

2. Increase the flow on arc (r, s) until a) it reaches its bound or b) until one of the basic arcs is driven to a bound, whichever comes first.

3. In case 2a), the primal variables are changed by appropriate amounts and we return to step 1, leaving the dual variables and tree unchanged.

4. In case 2b), the primal and dual variables and the tree representation are updated, then return to step 1 .

In order to determine how the flows in the basic arcs change as the flow in the entering arc increases, we take advantage of the fact that the addition of an entering arc (r, s) to the basis tree creates a graph with a *unique cycle* that consists of the path in the tree from r to s plus the new arc (r, s) (see Figure 1, where the arc orientation is shown only for cycle arcs). Since a flow change on the entering arc must yield a unique corresponding change in the basic flows, we will fully account for all changes in the basics by showing that appropriate flow changes on (only) the cycle arcs preserve feasibility. The orientation of the arcs of the cycle is determined by the orientation of the entering nonbasic arc (r, s), so that the arcs with positive orientation (i.e., in the same direction as (r, s)) are $(r, s), (p, q)$ and (t, p), whereas the arcs with negative orientation (i.e., opposite to (r, s)) are (r, q) and (t, s)). In both the example and the general case it is easily verified that feasibility (w.r.t. the divergence equations) is maintained if the flows on all positively oriented arcs are increased by a given amount and the flows on all negatively oriented arcs are decreased by the same amount. Unit flow changes on the cycle of the example are illustrated by Figure 2. Thus, for each unit change in $x_{r,s}$, there is a unit change (of appropriate sign) in each cycle arc. It is therefore relatively easy to determine which of the cycle arcs will be the first to hit a bound as they increase or decrease in flow depending on their orientation, since the simplex ratio test thus reduces to a comparison of distances to bounds. If this arc is (r, s), then the solution update is fairly trivial since dual variable values remain the same, and only the flows on the cycle arcs need be modified. Let us assume then, that a basic arc is the first to reach a bound, so that a full update is required. In the example, suppose that (p, q) hits its bound first. Again the primal solution update is obtained by modifying the cycle flows, but now the duals and tree representation must also be updated. From a *visual* viewpoint, it is easy to see how to change the tree representation in the example since it is clear that what is required is the severing and removal of arc (p, q) and the "rehanging" of the subtree that lies below this arc from node s via the new arc (r, s) (see Figure 3). Computationally, however, tree updates require a data structure perspective, and this process has some subtleties that we will consider below. Finally, to accomplish an update of the dual variables, note from (PI) that the values of the dual variables may be computed by moving down the tree from node $m + 1$, where $\pi_{m+1} = 0$. Each tree arc corresponds to a dual equation, and following the paths down from $m + 1$ we can obtain values of successive duals by noting that only two duals are present in each equation and the one higher in the tree has already been determined. Since the only paths from node $m + 1$ that are modified by the basis change are those to the nodes in the subtree below the leaving arc, the only dual variables that change are those for the nodes of that subtree. Moreover, in the example, if Δ is the change in π_r, then all of the other dual variables of the rehung subtree also change by exactly Δ since the *difference* between the dual variables on opposite nodes of any subtree arc is unchanged (i.e., it remains the cost associated with the arc). This property holds for the dual update in general, keeping in mind that the rehung subtree will contain one of the nodes of the entering arc.

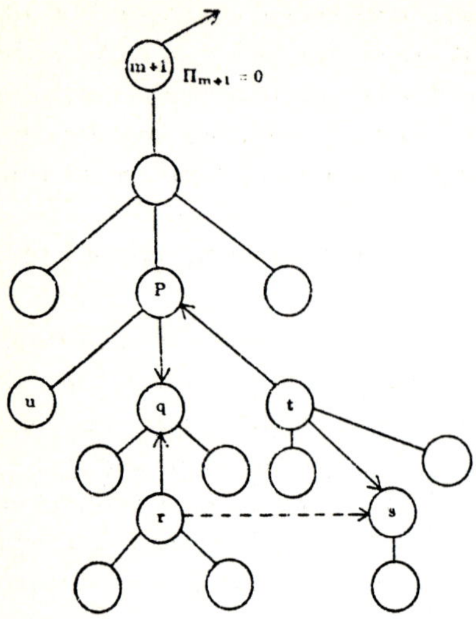

Figure 1. Tree augmented by entering arc (r,s)

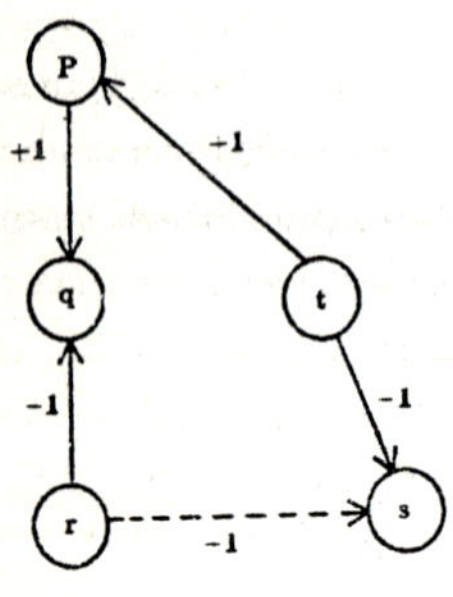

Figure 2. Flow changes on cycle

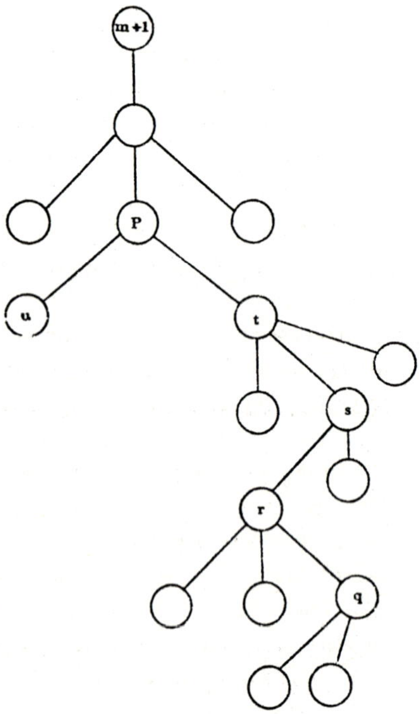

Figure 3. New basis tree

We will now define some concepts important for the description and update of basis trees. The unique path from a node i to the root node $m+1$ is called the *backpath* from i. The nodes on the backpath above i are the *predecessors* of i, and any node having i as predecessor is a *successor* of i. With this terminology, the determination of the cycle corresponding to an entering nonbasic (r, s) corresponds to generating the predecessors of both r and s and determining the first common predecessor (in the Figure 1 example, node p) . Conversely, the dual variables requiring update correspond to one of the nodes of the entering arc (the one on the rehung subtree) and its successors in the new tree. In summary, the primal update may be performed using predecessors, while the dual update involves successors.

Although the immediate predecessor of each node (other than the root) is uniquely determined, a node generally does not have a unique immediate successor and there are a variety of data structures suitable for representing the full set of successors of a node. The successor representation most commonly used in network optimization is the *preorder traversal* (or, simply, preorder) in which nodes are linearly ordered with the root being indexed first and the indexing proceeding to the leftmost immediate successor until this is no longer possible, then backtracking until a node with an unindexed successor is reached so that the indexing may be continued. The *initial* tree corresponding to the all-artificial basis has the nice property that no work is required to compute its corresponding immediate predecessor function p or the preorder traversal function s, since for $i \neq m+1$, $p(i) = m+1$ and $s(i) = i+1$ (for reasons to be noted below, it is useful to have node $m+1$ considered to be the "successor" of the "last" node, which in this case is node m).

Among contemporary network simplex algorithms, usage of only p and s would be considered to be a "minimal" collection of data structures for tree representation. Besides p and s, most algorithms use at least one additional node-length array to describe the tree in order to speed up the tree representation update and the determination of the intersection node of the backpaths of the nodes of the entering arc. For example, the set of successors of a node i may be generated by iterating the preorder function starting at i, and simultaneously checking the depth of the generated nodes, where the *depth* $d(i)$ of node i is the number of arcs in the backpath, since once a node j with $d(j) \leq d(i)$ is produced, the full set of successors of i is the set of previously generated nodes. Depth may also be used in cycle determination, since backtracking may be started from the node on the entering arc with greater depth and is continued simultaneously on both backpaths once the other depth is reached. Other additional data structures that have been used in the network simplex method include number of successors and "depth" with respect to the preorder index. Of course, all of the tree representation arrays must also be updated each time the basis changes, and the efficiency of a collection of data structures is also dependent on this factor. The inverse preorder function is used in RNET, for example, to reduce the cost of the preorder update. A good recent survey of some of the choices used in this area is given in [Kennington and Helgason 80].

There has been extensive computational testing of network flow codes on a standardized set of 35 medium to large problems produced by the NETGEN test problem generator described in [Klingman, Napier, and Stutz 74]. These problems have up to 1500 nodes and 5730 arcs. Direct comparison of codes with the same machine and compiler has been infrequent, but some idea of the relative performance of contemporary approaches is given by a survey in [Glover and Klingman 82], which concluded that the RNET code of [Grigoriadis 82] ran approximately as fast as the ARC-II code of [Barr, Glover, and Klingman 79], and that these two codes were faster than other publically availables codes. The reported total time using a CDC 6600 with the MNF Fortran compiler on the full set of problems was 102.85 seconds for RNET and 103.12 records for ARC-II. The longest solution time for RNET for an individual problem of this set was 8.11 seconds. Total storage for RNET is 7m + 4n, as compared to 7m + 3n for ARC-II. RNET uses the depth array and an array corresponding to the inverse of s in addition to predecessor and successor arrays. The study of [Glover and Klingman 82] also notes the preliminary testing of their new Code ARCNET, whose performance on the standard test problem set is slightly better than RNET.

Space does not permit a discussion of methods for important special classes of single-commodity network flow problems such as shortest path, maximum flow, transportation, and assignment problems. While it is possible to efficiently solve such problems using a general network flow code, still faster solutions can generally be obtained by further exploiting special structure. For example, a primal-dual method of F. Glover, R. Glover, and Klingman for the *assignment* problem was found to be nearly an order of magnitude faster [Glover and Klingman 82] than the network simplex method. Solution time for their method on a typical 400 node, 3000 arc assignment problem is 0.1 seconds on a dual Cyber 170/175. [Glover and Klingman 82] is also a good survey of recent research on special methods for shortest path and max flow problems.

3. Network-Related Linear Programs

There are a number of important classes of linear programs that are closely related to ordinary single-commodity networks. These include multicommodity flows, networks with additional general side constraints and variables (also known as embedded networks), and so-called generalized networks. In this section we provide a brief review of procedures for extending algorithms for ordinary networks to these more general problem classes.

Linear Multicommodity Network Flow

In a wide variety of applications several different commodities simultaneously flow through a network, and each commodity has its distinct set of supply, demand, and conservation constraints. Coupling between commodities occurs typically as a result of the sharing of arc capacities, and takes the form of "hard" constraints on the total flow of all commodities through individual arcs or "soft" constraints that impose nonlinear objective function penalties based on the total flow through arcs. In this section we will give a brief description of techniques for the case of linear coupling constraints, and then in following section consider generalizing this model to allow arbitrary additional rows and columns to be added to the basic network constraint format. Nonlinear networks are discussed in section 4.

We will assume the problem format

$$\begin{aligned} \min_{x} \; & \sum_{k} \mathbf{c_k x_k} \\ s.t. \; & \mathbf{A x_k} = \mathbf{b_k} \quad (k = 1, \ldots, K) \\ & \sum_{k} \mathbf{x_k} \leq \mathbf{b} \\ & \mathbf{0} \leq \mathbf{x_k} \leq \mathbf{u_k} \quad (k = 1, \ldots, K) \end{aligned} \qquad \text{(MUL)}$$

where A is a node-arc incidence matrix and $\mathbf{x_k}$ represents the set of flows corresponding to commodity k. (The techniques to be considered below may be extended in straightforward fashion to allow different networks constraints for each commodity as well as more general coupling constraints between commodities.)

Three fundamental approaches have been studied for this problem. These are Dantzig-Wolfe or price-directive decomposition, resource-directive decomposition, and primal partitioning. Dantzig-Wolfe decomposition uses an iterative process in which 1) the individual commodity constraints $\mathbf{Ax_k} = \mathbf{b_k}, \mathbf{0} \leq \mathbf{x_k} \leq \mathbf{u_k}$ are approximated by convex combinations of extreme points, 2) these approximations are substituted in the coupling constraints $\sum \mathbf{x_k} \leq \mathbf{b}$ to obtain an approximation (the "master" problem) that is then optimized, and 3) new extreme points of the single commodity problems are generated (assuming the optimality conditions are not satisfied) by optimizing via network techniques separate single commodity problems whose objective functions are determined by the optimal dual variables of the master problem. (In essence, this is the standard Dantzig-Wolfe method, except that network methods are used to obtain efficient solutions of the subproblems).

In resource-directive decomposition, the arc capacities are allocated among the commodities, the resulting single commodity subproblems are optimized, and the optimal dual variables are then used to determine a new allocation of capacities (assuming an optimal solution has not been obtained). [Kennington and Helgason 80] discuss tangential approximation and subgradient techniques for implementing resource-directive methods.

In primal partitioning, we take advantage of the fact that a basis for the multicommodity problem must contain bases for the single commodity problems. For notational purposes it is convenient to assume that slack variables y are added to the coupling constraints. Specifically, if $\bar{B}$ is a basis for (MUL), it may be partitioned as

$$\bar{B} = \begin{bmatrix} B & C \\ D & E \end{bmatrix}$$

where B is a collection of bases for the single commodity problems, and $[D\ E]$ is the portion of the basis corresponding to the coupling constraints. It may be verified that

$$\bar{B}^{-1} = \begin{bmatrix} B^{-1} + B^{-1}CQ^{-1}DB^{-1} & -B^{-1}CQ^{-1} \\ \\ -Q^{-1}DB^{-1} & Q^{-1} \end{bmatrix}$$

where $Q = E - DB^{-1}C$.

Note that in the limiting case in which all of the slack variables y are in the basis (so that nondegeneracy would imply that none of the joint capacity constraints were active), $C = 0$ and $E = I$, and the expression for $\bar{B}^{-1}$ reduces to

$$\bar{B}^{-1} = \begin{bmatrix} B^{-1} & 0 \\ -DB^{-1} & I \end{bmatrix}$$

More generally, it may be observed that relative simplicity of $\bar{B}^{-1}$ depends on the number of slack variables y in the basis, which in turn determine the complexity of E and the number of 0 columns of C. Of course, it is not necessary to explicitly compute $\bar{B}^{-1}$; instead we take advantage of the tree structure of the sub-bases of B in the computation of the solutions of the primal and dual systems of equations associated with $\bar{B}$. Details of the procedure may be found in [Kennington and Helgason 80]. Recent computational experience reported in [Kennington and Patty 84] indicates that the efficiency of primal partitioning is closely related to the number of active generalized upper bounding (GUB) rows in the optimal solution. In the applications considered in that paper, the number of active GUB rows was quite small (typically, fewer than 10), and primal partitioning was found to yield solutions about 6 times faster than linear programming approaches that did not take advantage of the network structure.

Networks with Side Constraints and Variables

From a theoretical standpoint, a straightforward generalization of multicommodity networks is the problem in which additional constraints and variables are added to standard network constraints. Algebraically, the structure of the bases for this class is similar to that of the multicommodity case, except that the B part of $\bar{B}$ is generally only a basis of a submatrix of the network portion. Computational experience with primal partitioning for this class has been rather limited, but again it is the case that efficiency relative to the plain simplex method is highly dependent on the significance in the solution of the non-network portion of the problem. In fact, this problem class reaches to (and sometimes past) the limit of generalization within which it is still computationally worthwhile to exploit network substructure. Results reported in [Glover and Klingman 81] showed primal partitioning to be about 4 times faster than

LP on some small problems and about 70 times faster on a problem with 1000 nodes, 5000 arcs, and 1 non-network row and column.

Generalized Networks

There are a great variety of additional applications that may be modelled by generalizing the network model to allow two arbitrary non-zero entries per column of the constraint matrix. (By using scaling, one of these may be assumed to be ± 1.) This class of problems is referred to as generalized networks or *"networks with gains and losses"*, since it can be thought of as allowing flows to increase or decrease as they traverse arcs. Physically, gains occur in such applications as financial models, in which interest is added as money flows from one time period to another, and losses occur in such phenomena as the transport of electricity and gas.

It may be shown that a basis for a network representation of the problem consists of a collection of one or more subgraphs, each of which consists of a rooted tree or a subgraph with exactly one cycle (such subgraphs are sometimes called one-trees). (Unlike ordinary networks, the presence of a cycle in a generalized network does not imply nonuniqueness of the solution.) By exploiting this basis structure using data structures that are extensions of those used for ordinary networks, it is possible to solve generalized networks about 50 times faster than by use of standard LP codes. In essence, cycles are dealt with as the roots of their components, and expressions are easily developed for the computation of primal and dual variables on the cycles. Details of the extension of the network simplex method to generalized networks may be found in [Kennington and Helgason 80]. A recent paper [Grigoriadis and Hsu 83] discusses some network analogs of partial pivoting that are useful for numerical stability during the pivot process, yet have very little computational overhead.

4. Nonlinear Networks

Nonlinear network optimization problems are of the form:

$$\begin{aligned} &\min f(\mathbf{x}) \\ &s.t.\ \mathbf{Ax} = \mathbf{b} \qquad\qquad \text{(NLP)} \\ &\mathbf{l} \leq \mathbf{x} \leq \mathbf{u} \end{aligned}$$

where f is a nonlinear function on the n-dimensional interval $[\mathbf{l},\mathbf{u}]$, and $\mathbf{Ax}=\mathbf{b}$ represents a system of network constraints. Significant applications that give rise to this format include computer network design [Cantor and Gerla 74], [Gavish and Hantler 82], [Magnanti and Wong 84], traffic assignment [Bertsekas and Gafni 82], [Dafermos 80], [Dantzig, et al, 79], [Feijoo and Meyer 83], [Lawphongpanich and Hearn 83], [Pang and Yu 82], hydroelectric power systems [Hanscom, et al, 80], [Rosenthal 81], [Escudero 84], telecommunications networks [McCallum 76], [Meyer 81], [Monma and Segal 82], and water supply systems [Beck, et al, 83], [Meyer 82,83]. These types of problems often have thousands of variables and are thus among the largest nonlinear programming problems that have ever been solved.

In the case in which f is *convex* and *separable*, i.e., of the form $f(\mathbf{x}) = \sum_{i=1}^{n} f_i(x_i)$, piecewise-linear approximation of the objective function has the nice property that the resulting approximating problems may be solved by linear network codes. [Kamesam and Meyer 82] showed that piecewise linear approximation of the f_i with iterative refinement of the grid yields a sequence of iterates converging to the optimal solution, and that Lipshitz continuity of the derivatives of the f_i implied a linear convergence rate of α^2 for the objective function, where α is the factor by which the grid size is multiplied at each iteration. Thus, for example, with $\alpha = 1/4$ (a typical value used in computation) the linear rate of convergence of the objective function will be $1/16$, in essence guaranteeing an additional figure of accurary with each iteration in the neighborhood of the optimal solution. The key issue in this regard is the speed with which the successive linear programs may be generated and solved. By using an implicit grid approach in which linear approximating segments are generated only as needed and at most two segments are stored for any variable at any given time, the efficiency of the problem generation phase is ensured. Major areas in which further research appears to be needed in the solution phase are the specification of the initial basis and the pricing out strategy. [Dembo and Klincewicz 81] and [Klincewicz 83] discuss the use of quadratic approximation for separable nonlinear networks.

In the case that the objective function f is convex and *non-separable* , it was commonly held that piecewise-linear approximation was impractical because of the "curse of dimensionality", i.e., the need to deal with k^n grid points in order to have a uniform grid of k points for each variable . The theory and computational results of [Feijoo and Meyer 83] showed that a convergent and efficient piecewise-linear approximation method could be achieved in the non-separable case by restricting the computation of approximating points to a small cluster situated on a set of axes translated to the current iterate. In essence, this is equivalent to using a locally-determined separable approximation at each major iteration, so that the techniques for the separable case are applicable. Moreover , this type of approximation applied to nonlinear networks has the advantage of preserving the network structure of the constraints.

Traffic equilibrium problems are an important class of nonlinear multi-commodity problems, i.e., they involve many different types of "commodities" flowing through the network, where depending on the application, a "commodity" will be associated with the traffic flowing out of a "source" node or between a designated origin-destination pair. Such problems arise in urban and regional transportation planning as well as computer network traffic models. Enormous problem sizes result from the fact that the number of variables and constraints is determined by the number of links and nodes multiplied by the number of commodities. These same features, however, also make these problems ideal candidates for solution via parallel algorithms, about which more will be said below. There are two fundamental types of traffic equilibrium problems: symmetric problems in which the congestion on a given link is determined by the total flow summed over all commodities in that link only, giving rise to a symmetric Jacobian matrix in the corresponding variational problem, and asymmetric problems in which the congestion on a link depends on the total flows in several links. It is well known [Steenbrink 74] that the former problem is equivalent to a convex optimization problem under relatively weak assumptions, whereas the latter gives rise to a variational inequality that has a more complex optimization formulation involving a non-differentiable objective function. In this paper we will concentrate on the former problem, although some of the decomposition ideas considered apply equally well to both problem classes and indeed to large-scale optimization problems that have non-network constraints. This problem is of the form (NLP), where the objective function represents total congestion over all of the links of the network, and the constraints represent the supply, demand, and conservation constraints for each of the individual commodities flowing through the network. (Alternatively, it is possible to transform the problem into a separable problem by including constraints that sum the flows on each arc, but this destroys the network structure of the constraints.) the The number of commodities may be quite large because, depending on the formulation, there can be a commodity corresponding to each node or to each origin-destination pair. However, in the urban traffic assignment problem it is typically the case that the only coupling between different commodities occurs in the objective function.(For other problems in which there are explicit joint capacity constraints, Lagrangian and penalty approaches can frequently be exploited to move these constraints into the objective function.) Because of this coupling property, the approximation of the objective function by a separable function leads to a decomposition of the problem into separate optimization problems, one for each commodity. These single commodity problems may then be optimized in parallel (many other algorithms for the symmetric and asymmetric problems also contain a phase in which the objective function is replaced by an approximation that allows decomposition to be used).

For simplicity we will assume that a feasible solution $\mathbf{x}^i$ (one satisfying all of the constraints) is available at the start of the ith iteration of the problem (NLP). (At the initial iteration an arbitrary starting point within $[\mathbf{l},\mathbf{u}]$ plays the same role.) To simplify notation below,consider the ***shifted function*** $f_{\mathbf{x}^i}(\mathbf{x}) := f(\mathbf{x}) - f(\mathbf{x}^i)$; it is clear that minimizing $f_{\mathbf{x}^i}$ is equivalent to minimizing f. A piecewise-linear separable approximation $\tilde{f}$ "centered" at $\mathbf{x}^i$ is then constructed as follows:

Consider the function

$$h : \mathbf{R}^n \longrightarrow \mathbf{R}$$
$$h(\mathbf{y}) = f_{\mathbf{x}^i}(\mathbf{y} + \mathbf{x}^i)$$

which is a translation of $f_{\mathbf{x}^i}$ to the origin, and let h_j be the restriction of h to the axis y_j, that is $h_j(y_j) = h(y_j\mathbf{e}^j)$ where $\mathbf{e}^j$ is the jth canonical unit vector. For a given ***grid size*** λ^i_j, generate a piecewise-linear approximation $\tilde{h}_j$ to h_j, with breakpoints at $\ldots, -2\lambda^i_j, -\lambda^i_j, 0, \lambda^i_j, 2\lambda^i_j, \ldots$ Defining $\tilde{f}_j(x_j) := \tilde{h}_j(x_j - x^i_j)$, the resulting piecewise-linear approximation to $f_{\mathbf{x}^i}$ is:

$$\tilde{f}(\mathbf{x}) = \sum_{j=1}^{n} \tilde{f}_j(x_j)$$

(Note that the computation of the $\tilde{f}_j$'s may be carried out ***in parallel***.) The original objective function is then replaced by the approximating function, and the resulting approximating problem is solved. This problem has the form:

$$\begin{aligned} &\min \tilde{f}(\mathbf{x}) \\ &s.t.\ \mathbf{A}\mathbf{x} = \mathbf{b} \\ &\mathbf{l} \le \mathbf{x} \le \mathbf{u} \end{aligned} \qquad (\mathrm{PLP}(i))$$

(Note that the function $\tilde{f}$ depends on the base point $\mathbf{x}^i$ and consequently varies from iteration to iteration.)

The problem (PLP(i)) has two key features: 1) because the objective function is now separable, the problem may be decomposed into K separate optimization problems, where K is the number of commodities in the original problem; and 2) because of the convexity and piecewise-linearity of their objective functions, each of these new problems is equivalent to a linear network optimization problem.

Specifically, assume that $\mathbf{x} = (\mathbf{x}_1, \ldots, \mathbf{x}_K)$, where $\mathbf{x}_k$ is the vector of variables corresponding to commodity k, and S_k is the correponding set of indices for those variables. Since there is no coupling between commodities in the constraints, the problem (PLP(i)) may be solved by solving ***in parallel*** (via very fast linear network codes) the set of problems:

$$\begin{aligned} &\min \sum_{j \in S_k} \tilde{f}_j(x_j) \\ &s.t.\ \mathbf{A}_k\mathbf{x}_k = \mathbf{b}_k \\ &\mathbf{l}_k \le \mathbf{x}_k \le \mathbf{u}_k \end{aligned} \qquad k = 1, \ldots, K$$

where $\mathbf{A}_k$, $\mathbf{b}_k$, $\mathbf{l}_k$ and $\mathbf{u}_k$ are for commodity k the corresponding components of $\mathbf{A}, \mathbf{b}, \mathbf{l}$ and $\mathbf{u}$. (The effect of this approach is therefore similar to Dantzig-Wolfe decomposition, in that it results in the decoupling of the subproblems; the difference is that the coupling is removed here from the objective function rather than the constraints. Clearly, the data communications requirements for most decomposition approaches will be similar.)

Given the set of solutions of these problems $\tilde{\mathbf{x}}_1, \ldots, \tilde{\mathbf{x}}_K$, if we let $\tilde{\mathbf{x}}^{i+1} = (\tilde{\mathbf{x}}_1, \ldots, \tilde{\mathbf{x}}_K)$, the next iterate $\mathbf{x}^{i+1}$ is obtained by solving the line search problem:

$$\begin{aligned} \min_{\theta} \quad & f(\mathbf{x}^i + \theta(\tilde{\mathbf{x}}^{i+1} - \mathbf{x}^i)) \\ s.t. \quad & \mathbf{l} \leq \mathbf{x}^i + \theta(\tilde{\mathbf{x}}^{i+1} - \mathbf{x}^i) \leq \mathbf{u} \\ & \theta \geq 0 \end{aligned}$$

Once the next iterate has been obtained, the procedure is repeated with a reduced grid size of $\lambda^{j+1} = \alpha\lambda^j$, where $\alpha \in (0, 1)$ ($\alpha = 0.25$ has worked well on our tests). It has been shown that this procedure will yield a sequence of iterates whose objective function values converge to the optimal value of the original problem (NLP).

Our preliminary computational results [Feijoo and Meyer 84] with this approach have been very promising. The method was implemented on the University of Wisconsin CRYSTAL multicomputer [DeWitt, et al 84], a set of VAX 11/750 computers (currently there are 20) with 2 megabytes of memory each, connected by a 10 megabit/sec Proteon ProNet token ring and accessed via VAX 11/780 "host" machines. For example, a 12-commodity, 432-variable traffic assignment problem was solved on a 12-machine partition of CRYSTAL in 23.5 seconds versus 112 seconds for solution via the sequential version of the algorithm on a single machine. Six seconds of the CRYSTAL time were used for communication by the master and an average of 0.3 seconds for communication by each slave . Similar computational results were obtained for related problems. Thus, communication between machines represents a relatively small percentage of the total computing time (summed over all of the machines), and this behavior should scale up favorably to yield nearly linear speedup for larger problems.

An important computational issue yet to be studied is the degree of optimization to be carried out in the optimization of the linear subproblems. In order to obtain a descent direction, these problems do not have to be fully solved, and it appears likely that, after a certain stage in the optimization, further pivots do not significantly improve the quality of the search direction. Research on an appropriate stopping rule for the individual subproblems should lead to significant efficiency gains, and also offers the possibility of improving the level of synchronization when problems are solved in parallel. Further study of this approach is needed in order to identify the convergence rate, which should be dependent on the relative magnitude of the off-diagonal terms of the Hessian matrix, and to estimate this rate in the case of problem classes such as traffic assignment optimization.

Other algorithms for convex nonseparable nonlinear networks include the Frank-Wolfe method [Dantzig, et al 79], simplicial decomposition ([Lawphongpanich and Hearn 83] and [Pang and Yu 82]), projection methods [Bersekas and Gafni 82], and the reduced gradient method [Beck, et al 83]. All of these techniques exploit in some manner the underlying network structure.

References

R. S. Barr and J. S. Turner, "Microdata File Merging through Large-Scale Network Technology", *Mathematical Programming Study 15* , 1-22 ,1981.

R. Barr, F. Glover and D. Klingman, "A New Alternating Basis Algorithm for Semi-Assignment Networks", in *Proceedings of the Bicentennial Conference on Mathematical Programming*, NBS, Gaithersburg, Maryland, USA ,1977.

R. Barr, F. Glover and D. Klingman, "Enhancements of spanning tree labelling procedures for network optimization", *INFOR*, 17, 16-34, 1979.

P. Beck, L. Lasdon and M. Engquist, "A Reduced Gradient Algorithm for Nonlinear Network Problems", *ACM Trans. on Math. Software*, 9, 57-70 ,1983..

D. P. Bertsekas and E. M. Gafni, "Projection methods for minimum cost network flow problems", *Mathematical Programming Study 17,* 1-22, 1981.

D. P. Bertsekas and P. Tseng, "Relaxation Methods for Minimum Cost Network Flow Problems", MIT Laboratory for Information and Decision Systems Report LIDS-P-1339 ,1983.

R.E.Bixby,"Hidden structure in linear programs", in *Computer- Assisted Analysis and Model Simplification* (H.J. Greenberg and J.S. Maybee , eds.), Academic Press, 1981.

D. G. Cantor and M. Gerla: "Optimal routing in packet switched computer networks", *IEEE Transactions on Computing* C-23, 1062-1068, 1974.

S. Dafermos,"Traffic equilibrium and variational inequalities", *Transportation Science*, 14, 42-54 ,1980.

G. Dantzig, R. Harvey, Z. Lansdowne, D. Robinson, and S. Maier: "Formulating and solving the network design problem by decomposition", *Transportation Research* 13B, 5-17, 1979.

R. Dembo and J. Klincewicz, "A Scaled Reduced Gradient Algorithm for Network Flow Problems with Convex Separable Costs", *Math. Prog. Study* 15, 125-147, 1981.

D. DeWitt, R. Finkel and M. Solomon, "The CRYSTAL multicomputer: design and implementation experience", Technical Report 553, Computer Sciences Department, The University of Wisconsin-Madison, September, 1984.

L.F.Escudero,"Using independent superbasic sets on nonlinear replicated networks with applications", paper presented at the NATO ASI on Computational Mathematical Programming , 1984.

B. Feijoo and R. R. Meyer, "Piecewise-Linear Approximation Methods for Nonseparable Convex Optimization", University of Wisconsin - Madison Computer Sciences Department Technical Report 521, 1983.

B. Feijoo and R. R. Meyer, "Optimization on the CRYSTAL Multicomputer", University of Wisconsin-Madison Computer Sciences Department Technical Report 562, 1984.

L.Ford and D.Fulkerson, *Flows in Networks*, Princeton University Press, Princeton, N.J., 1962.

B. Gavish and S. L. Hantler: "An algorithm for optimal route selection in SNA networks", Research Report RC 9549, IBM T. J. Watson Research Center, Yorktown Heights, N. Y., August, 1982.

F.Glover, D. Karney, and D. Klingman, "Implementation and computational comparisons of primal, dual, and primal-dual computer codes for minimum cost network flow problems," *Networks*, 4, 191-212, 1974.

F. Glover and D. Klingman, "The Simplex SON Algorithm for LP/Embedded Network Problems", *Math. Prog. Study* 15, 148-176 ,1981.

F. Glover and D. Klingman, "Recent developments in computer implementation technology for network flow algorithms", The University of Texas at Austin, Center for Cybernetic Studies Report CCS 377, 1982.

M. D. Grigoriadis, "Minimum-Cost Network Flows, Part I: An Implementation of the Network Simplex Method", Rutgers University Laboratory for Computer Science Research Report LCSR-TR-37 ,1982.

M. D. Grigoriadis and T. Hsu, "Numerical Methods for Basic Solutions of Generalized Flow Networks", Rutger University Laboratory for Computer Science Research Report LCSR-TR-43 ,1983.

M. A. Hanscom, L. Lafond, L. Lasdon and G. Pronovost, "Modeling and resolution of the medium term energy planning problem for a large hydro-electric system", *Management Science* 26, 659-668, 1980.

F.Hitchcock, "The distribution of a product from to several sources to numerous localities", *Journal of Mathematics and Physics*, 20, 224-230 ,1941.

E.L. Johnson,"Networks and basic solutions", *Operations Research*, 14, 619-623 ,1966.

P.V.Kamesam and R.R.Meyer, "Multipoint methods for nonlinear networks", to appear in *Mathematical Programming Studies 22* ,1984.

L.Kantorovich, *Mathematical Methods in the Organization and Planning of Production*, Publication House of the Leningrad State University, 1939. Translated in *Management Science*, 6, 366-422, 1958.

J. L. Kennington and B. W. Patty, "A Computational Comparison of Specialized versas General Codes for Multi-Commodity Network Problems", *COAL Newsletter*, 8-14 ,1984.

J. L. Kennington and R. V. Helgason, *Algorithms for Network Programming*, John Wiley , 1980.

J.G. Klincewicz, "A Newton method for convex separable network flow problems", *Networks*, 13, 427-442 ,1983.

D. Klingman, A. Napier and J. Stutz, "NETGEN - A program for generating large-scale (un)capacitated assignment, transportation, and minimum cost flow network problems", *Management Science*, 20, 814-821 ,1974.

T.C. Koopmans, "Optimum utilization of the transportation system", *Econometrica*, 17, 3-4 ,1949.

S. Lawphongpanich and D. W. Hearn, "Restricted Simplicial Decomposition with Application to the

Traffic Assignment Problem", Research Report 83-8, Department of Industrial and Systems Engineering, University of Florida, Gainesville, Fl., September, 1983.

T. L. Magnanti and R. T. Wong: "Network design and transportation planning: models and algorithms", *Transportation Science* 18, 1-55, 1984.

C. J. McCallum, "A generalized upper bounding approach to communications network planning problems", *Transportation Science*, 18, 1-55, 1984.

R.R. Meyer,"A Theoretical and computational comparison of 'equivalent' mixed-integer formulations", *Naval Research Logistics Quarterly*, 28, 115-131 ,1981.

R.R. Meyer, "Recursive piecewise-linear approximation methods for nonlinear networks", in *Evaluating Mathematical Programming Techniques*, Springer-Verlag,1982.

R.R.Meyer, "Computational aspects of two-segment separable programming", *Mathematical Programming*, 26, 21-39 ,1983.

C. L. Monma and M. Segal, "A primal algorithm for finding minimum-cost flows in capacitated networks with applications", *The Bell System Technical Journal* 61, 949-968, 1982.

J. S. Pang and C. S. Yu, "Linearized simplicial decomposition methods for computing traffic equilibria on networks", Technical Report, University of Texas at Dallas, Richardson, Texas, 1982.

R.T. Rockafellar, *Network Flows and Monotropic Programming*, Wiley, 1984.

R. E. Rosenthal, "A nonlinear network flow algorithm for maximization of benefits in a hydroelectric power system", *Operations Research* 29, 763-786, 1981.

P. A. Steenbrink, *Optimization of Transport Networks,* Wiley, London, 1974.

This research was supported in part by NSF grant MCS8200632 and ARO contract DAAG29-80-C-0041.

REFLECTIONS ON GEOMETRIC PROGRAMMING

M.J.RIJCKAERT AND E.J.C.WALRAVEN
INSTITUUT VOOR CHEMIE-INGENIEURSTECHNIEK
K.U.LEUVEN, LEUVEN, BELGIUM

INTRODUCTION.

Geometric Programming is a member of the family of non-linear programming techniques, that deals with problems whose functions are all generalized polynomials. Its development started in the sixties |10|. Since then close to 500 papers have appeared on the subject, dealing with computational or theoretical aspects and with applications |26, 30|.

Papers have appeared on applications of Geometric Programming in the following domains (the list cites only exemplary items, many of them appearing several times in the literature and is certainly not exhaustive) :

- chemical engineering (design of heat exchangers, cooling towers, condensers, membrane separation processes with special mentionning of the chemical equilibrium problem, which has a formulation similar to the Geometric Programming problem.)
- mechanical engineering (cutting tools, helical springs, tool engineering)
- civil engineering (design of cofferdam , wood beams, bridges and structures)
- metallurgical engineering (freezing of aluminium alloys)
- nuclear engineering (nuclear systems, tritium breeding)
- electrical engineering (transformer design, control systems, tolerances in circuits)
- hydraulics (hydraulic networks)
- environmental engineering (water resources, sewer networks, industrial waste treatment)
- location problems
- maximum likelihood estimation
- transportation planning
- portfolio selection
- reliability.

The above list shows that, at least for the published applications,

NATO ASI Series, Vol. F15
Computational Mathematical Programming
Edited by K. Schittkowski

there is a very heavy emphasis on engineering and design applications. The theoretical aspects of Geometric Programming are extensively covered in several books and review papers. (Readers interested in a bibliographical survey are referred once more to |26|.)
The third aspect of Geometric Programming, the computational one will be covered in greater detail in the present paper. However we will not give an overview of all computational aspects or of all algorithmic possibilities. (Many of these can be found in |25|).
Instead we will focus on 3 particular items and questions.

A. How do specific Geometric Programming codes behave compared to general codes with respect to mainly robustness and efficiency ? This question is of course of utmost relevancy. Besides there is no absolutely unanimous answer to that question in the literature. The usefulness of G.P. codes was advocated among others in comparative studies of Rijckaert and Martens |31|, Dembo |6|, Ratner et al. |24|. On the contrary Fatler et al. |11| and to some degree Lootsma |39| concluded the opposite. It should be noted that the study of Lootsma was based on a literature survey and took other features than efficiency and robustness into consideration. We hope to report in this paper on some new computational material, which should clarify the situation a little bit, without having the pretension of providing the ultimate answer.

B. We will present a hybrid code, obviously called HYBRID, formed by putting together in a sequential way two existing codes. Experience with such a code is reported and discussed.
We furthermore believe that the conclusions with respect to this particular code can be extrapolated outside the G.P.domain.

C. Finally in the process of describing the code GPKTC, we will report on recent improvements to this code and on some theoretical results with respect to numerical testing.

GEOMETRIC PROGRAMMING THEORY.

We'll start with some definitions and notations. A <u>monomial</u> u(t) is a function of a positive real m-dimensional vector variable t of the form:

$$u(t) = c \prod_{j=1}^{m} t_j^{a_j},$$

$$t_j > 0, \qquad j = 1, \ldots, m.$$

The coefficient c is a positive real constant, the exponents a_j are arbitrary real constants.

A posynomial $P(t)$ is a positive sum of monomials:

$$P(t) = \sum_{i=1}^{n} u_i(t) = \sum_{i=1}^{n} c_i \prod_{j=1}^{m} t_j^{a_{ji}},$$

$$t_j > 0, \quad j = 1, \ldots, m.$$

The constant n is the number of terms in the posynomial.

A signomial or generalized polynomial $g(t)$ is the difference of two posynomials :

$$g(t) = P^+(t) - P^-(t) = \sum_{i=1}^{n^+} u_i^+(t) - \sum_{i=1}^{n^-} u_i^-(t)$$

$$= \sum_{i=1}^{n^+} c_i^+ \prod_{j=1}^{m} t_j^{a_{ji}^+} - \sum_{i=1}^{n^-} c_i^- \prod_{j=1}^{m} t_j^{a_{ji}^-},$$

$$t_j > 0, \quad j = 1, \ldots, m.$$

The constants n^+ and n^- are the number of positive and negative terms respectively.

A generalized geometric programming problem or signomial problem is given by:

SP_o

minimize $g_o(t)$

subject to

$$g_k(t) \leq 0, \quad k = 1, \ldots, p,$$
$$t_j > 0, \quad j = 1, \ldots, m,$$

where

$$g_k(t) = \sum_{m_k^+}^{n_k^+} u_i^+(t) - \sum_{m_k^-}^{n_k^-} u_i^-(t), \qquad k = 0, \ldots, p.$$

The positive terms of all functions are numbered consecutively from 1 to n^+, and the negative terms from 1 to n^-. We have:

n_o^+ = number of positive terms in the objective function;

n_o^- = number of negative terms in the objective function:

$m_o^\sigma = 0$ if $n_o^\sigma = 0$, $\quad \sigma = +,-,$

$= 1$ if $n_o^\sigma > 0$, $\quad \sigma = +,-;$

$m_k^\sigma = n_{k-1}^\sigma + 1$, $\quad k = 1, \ldots, p$; $\sigma = +,-;$

$n_p^\sigma = n^\sigma$, $\quad \sigma = +,-.$

We define the following sets of indices:

$$I_k^\sigma = \{m_k^\sigma, \ldots, n_k^\sigma\}, \qquad k = 0, \ldots, p\ ;\ \sigma = +,-.$$

An equivalent formulation of SP_o is given by:

$$\text{minimize } P_o^+(t) - P_o^-(t)$$

$$\text{s. t.} \quad P_k^+(t) \lesseqgtr P_k^-(t), \qquad k = 1, \ldots, p,$$

$$t_j > 0, \qquad j = 1, \ldots, m.$$

If the objective function is a posynomial (i.e. if $P_o^-(t)$ vanishes) and if all $P_k^-(t)$ are monomials, one obtains a (prototype) geometric programming problem or <u>posynomial problem</u> PP_o.
With the primal signomial problem SP_o the following dual problem is associated :
DSP_o

find a stationary point of $v(\delta)$

$$\text{s. t.} \quad \delta_i^\sigma \geq 0, \qquad i = 1, \ldots, n^\sigma\ ;\quad \sigma = +,-,$$

$$\sum_{i\in I_o^+} \delta_i^+ - \sum_{i\in I_o^-} \delta_i^- = \sigma_o,$$

$$\sum_{i\in I_k^+} \delta_i^+ - \sum_{i\in I_k^-} \delta_i^- = 0, \qquad k = 1, \ldots, p,$$

$$\sum_{i=1}^{n^+} a_{ji}^+ \delta_i^+ - \sum_{i=1}^{n^-} a_{ji}^- \delta_i^- = 0, \qquad j = 1, \ldots, m,$$

where

$$v(\delta) = \sigma_o \prod_{i=1}^{n^+} \left|\frac{c_i^+}{\delta_i^+}\right|^{\delta_i^+\sigma_o} \prod_{i=1}^{n^-} \left|\frac{\delta_i^-}{c_i^-}\right|^{\delta_i^-\sigma_o},$$

$$\sigma_o = \frac{g_o(\overset{*}{t})}{|g_o(\overset{*}{t})|},$$

$\overset{*}{t}$ = a stationary point of SP_o.

For signomial problems only weak duality relations exist between so-called <u>equilibrium solutions</u> |9| of SP_o and DSP_o. A feasible solution $\overset{*}{t}$ to SP_o is termed a primal equilibrium solution if there is a feasible solution $\overset{*}{\delta}$ to DSP_o such that

$$\sigma_o\, g_o(\overset{*}{t}) \quad \overset{*}{\delta}{}_i^\sigma = u_i^\sigma(\overset{*}{t}), \qquad i\varepsilon I_o^\sigma \ ; \ \sigma = +,-;$$

$$\sigma_o\, g_o(\overset{*}{t}) \quad \overset{*}{\delta}{}_i^\sigma = \overset{*}{\mu}_k\, u_i^\sigma(\overset{*}{t}), \qquad i\varepsilon I_k^\sigma \ ; \ \sigma = +,-; \ k = 1, \ldots, p;$$

in which case $\overset{*}{\delta}$ is termed a dual equilibrium solution. The variables $\overset{*}{\mu}_k$ are the optimal Lagrange multipliers of SP_o. Equilibrium solutions satisfy the Kuhn and Tucker necessary conditions for optimality. Further, for corresponding equilibrium solutions $\overset{*}{t}$ and $\overset{*}{\delta}$, the relation

$$g_o(\overset{*}{t}) = v(\overset{*}{\delta})$$

holds.

The nature of the stationary point of the dual problem corresponding to a (local) minimum of the primal problem cannot be determined for a signomial problem. As a consequence, the duality theory can play only a minor computational role for the class of signomial problems.

For posynomial problems stronger duality relations exist because of the convex nature of the primal problem: the dual program is a maximization problem.

A much more detailed and profound treatment of the geometric programming theory can be found for the posynomial case in |10|, and for the signomial case in |9, 20, 21|.

<u>LOOSE CONSTRAINTS.</u>

It is worthwhile to note that to each primal constraint a Lagrange multiplier or dual variable μ_k is associated, and to each primal term $u_i^\sigma(t)$ a dual variable δ_i^σ, and vice versa. The multiplier of a loose constraint vanishes, and it then follows from the duality relations that all dual variables $\overset{*}{\delta}{}^{\sigma}_{i}$ associated with the terms of a loose constraint become zero in block. This phenomenon might cause serious computational difficulties.

Problems related to loose constraints can be avoided through the use of slack variables. Since geometric programming deals with positive variables only, the slacks can't be introduced in the usual way, e.g. as in linear programming. Kochenberger |15| introduced the slack variables in a penalty-function-like-manner, Duffin and Peterson |8| provided a theoretical basis for doing so, and Rijckaert and Martens |27| examined some numerical aspects.

CONDENSATION.

The geometric inequality has played an important role in the theoretical development of geometric programming, and it is the basis of a very powerful approximation technique. Consider two sets of positive numbers u_i and α_i such that

$$\sum_{i=1}^{n} \alpha_i = 1;$$

the geometric inequality states that |10| :

$$\sum_{i=1}^{n} u_i \geq \prod_{i=1}^{n} \left| \frac{u_i}{\alpha_i} \right|^{\alpha_i} .$$

The inequality becomes an equality if and only if

$$\sum_{i=1}^{n} u_i = \frac{u_i}{\alpha_i}, \qquad i = 1, \ldots, n.$$

The inequality can be used to approximate a posynomial P(t) by a monomial, an operation called condensation |7|. Consider thereto a set of function $\alpha_i(t)$ satisfying

$$\sum_{i=1}^{n} \alpha_i(t) = 1,$$

$$\alpha_i(t) > 0, \qquad i = 1, \ldots, n,$$

$$t_j > 0, \qquad j = 1, \ldots, m,$$

a posynomial P(t) and a positive point $\hat{t}$, then the in $\hat{t}$ condensed posynomial is defined as |7|:

$$\bar{P}(t,\hat{t}) = \prod_{i=1}^{n} \left|\frac{u_i(t)}{\alpha_i(\hat{t})}\right|^{\alpha_i(\hat{t})},$$

$$= \bar{c}(\hat{t}) \prod_{j=1}^{m} t_j^{\bar{a}_j(\hat{t})},$$

where

$$\bar{c}(\hat{t}) = \prod_{i=1}^{n} \left|\frac{c_i}{\alpha_i(\hat{t})}\right|^{\alpha_i(\hat{t})},$$

$$\bar{a}_j(\hat{t}) = \prod_{i=1}^{n} a_{ji}\, \alpha_i(\hat{t}), \qquad j = 1, \ldots, m.$$

It's clear that $\bar{P}(t,\hat{t})$ is a monomial. The numbers $\alpha_i(\hat{t})$ are called the condensation weights. From the geometric inequality it follows that

$$\bar{P}(t,\hat{t}) \leq P(t).$$

In all practical cases, the weights are specified by

$$\alpha_i(\hat{t}) = \frac{u_i(\hat{t})}{P(\hat{t})}, \qquad i = 1, \ldots, n,$$

and consequently the relation

$$\bar{P}(\hat{t},\hat{t}) = P(\hat{t})$$

holds. This technique is a powerful algoritmic tool, and it is used to linearize geometric programs |7|, to approximate a signomial problem by a posynomial problem |2|, to solve a square system of signomial equations |28|, and in estimating Lagrange multipliers |32|. The properties of the approximation were studied by Passy |19|.

ALGORITHMS AND SOFTWARE.

In our computational study we have limited our experiments to 5 codes (three specific and two general codes.).
One important reason for this limitation is that we could rely on previous computational studies. Comparative studies on GP |6, 31| indicated that codes solving the primal problem using the condensation technique performed the best, especially the codes GGP |4, 5| and GPKTC |18, 28, 29|. Recently the algorithm GPKTC has been updated implementing new algorithmic refinements |32, 36|.
For reasons mentionned earlier we added HYBRID to it. We are well aware of the fact that we have excluded several other possibilities (no dual code was tested). However it should be mentionned that for some of these codes, doubts can exist if they still can be classified as G.P.codes. For instance, Rajasekera J.R. and Peterson E.L. |40| report on a G.P.code, which relies to a very large extent on MINOS 4.0. Comparative studies concerning general NLP-codes |13, 35| showed that generalized reduced gradient algorithms and sequential quadractic programming algorithms performed among the best, specially the codes GRG2 |16, 17| and VFO2AD |22, 23|.
Finally, we would like to mention that in our comparative study attention was limited to codes for finding a local optimum in the signomial case (no concern about finding a global optimum).
We will now briefly describe the algorithms involved.

GGP

The algorithm GGP |1, 4| solves problems of the form:

SP_{2B}

$$\text{minimize } t_o$$

$$\text{s. t.} \quad P_k^+(t) \leq P_k^-(t), \qquad k = 1, \ldots, p,$$

$$t_j^L \leq t_j \leq t_j^U, \qquad j = 0, \ldots, m.$$

Notice that all variables have lower and upper bounds t_j^L and t_j^U. Each signomial problem can be formulated as SP_{2B}; if some variables are unbounded, artificial bounds should be provided. To solve this problem, GGP uses the algorithm of Avriel and Williams |2|. Condensing the 'negative' posynomials into monomials yields the condensed problem, which is a posynomial program:

$\overline{SP}_{2B}$

minimize t_o

s. t. $P_k^+(t) \leq \bar{P}_k^-(t,t^\ell)$, $k = 1, \ldots, p$,

$t_j^L \leq t_j \leq t_j^U$, $j = 0, \ldots, m$.

Starting from a primal feasible point of SP_{2B}, a series of condensed programs is constructed and under some weak conditions their solution is guaranteed to converge to an equilibrium point of the original problem SP_{2B} |2|. The algorithm GGP itself is used to generate a feasible point.
To solve the posynomial condensed problems GGP uses again the condensation and a cutting-plane technique |14|. The constraints of $\overline{SP}_{2B}$ are condensed in the current point, and next logarithmized to yield linear expressions. The resulting LP has a larger feasible region; hence, following the idea of Kelley's cutting-plane technique, a sequence of linear programs is solved.
The code has been tested in |4, 6, 11, 18, 31, 34, 36|.

GPKTC

The algorithm GPKTC |18, 28, 36| solves problems of the form :
SP_2

minimize t_o

s. t. $P_k^+(t) \leq P_k^-(t)$, $k = 1, \ldots, p$,

$t_j > 0$, $j = 0, \ldots, m$.

The basic algorithm solves problems with equality constraints only. Hence, if inequality constraints are present, slack terms are added |18, 27, 36| to get an augmented equality constrained problem :
ASP_2

$$\text{minimize } t_o + \sum_{k=1}^{p_o} \theta \gamma t_{m+k}^{-\beta}$$

$$\text{s. t.} \quad P_k^+(t) + P_k^-(t^o)\,\theta\, t_{m+k} \leq P_k^-(t), \qquad k = 1, \ldots, p_o,$$

$$P_k^+(t) \leq P_k^-(t), \qquad k = p_o+1, \ldots, p,$$

$$t_j > 0, \qquad j = 0, \ldots, m+p_o,$$

$$\theta, \gamma, \beta > 0,$$

where p_o is the number of inequality constraints, t^o is a starting point and $P_k^-(t^o)$ has to be seen as a constraint scale factor. The variables t_{m+k}, $k = 1, \ldots, p_o$, are slack variables. The parameter γ is in fact an objective function scale factor, and the parameter θ is used to detect inactive constraints. Advisable values for the parameters are :

$$\theta = 1,$$

$$\gamma = 0.01\, t_o^o,$$

$$\beta = 0.01$$

With these values, the solution of ASP_2 is a good approximation to the solution of SP_2. It is further possible to drop after some iterations the inactive constraints and the slack terms to get the solution of SP_2.

GPKTC solves the KT-equations. Since the problem is equality constrained, the complementary slack conditions need not to be considered, and one has to solve a square system of signomial equations of the form

$$P_k^+(t,\mu) = P_k^-(t,\mu),$$

where μ are the Lagrange multipliers. Condensation and logarithmization yield an approximating system of linear equations, which is easily solved. The solution is in turn used to condense the equations, and one obtains an iterative procedure.
This basic algorithm is extended with Lagrange multiplier estimation procedures and an active-set strategy |36|.
The code has been tested in |6, 11, 18, 31, 36|.

HYBRID.

The performance of GGP and GPKTC is complementary: GGP is robust, is fast in the beginning but slow at the end, it handles inequality constraints and bounds elegantly but can't force an equality constraint to be active, the computing time is sensitive to the starting point and the choice of the artificial bounds; GPKTC is not that robust but has a quadratic rate of convergence, it is very efficient for equality constrained problems but is bothered by inequalities and bounds, its computing time is more predictable, and it is extremely accurate. This complementarity of the properties of GGP and GPKTC was the incentive to link the two algorithms in a hybrid code |25|. In a first phase, GGP should find a point which is near to the solution; in a second phase, GPTKC will take over and locate the optimum very fastly and accurately.
In this way, most disadvantages of the two codes would be avoided, and the advantages exploited. The code HYBRID |33, 36| accomplishes these ideas efficiently and test results |36| are promising.

GRG2

The code GRG2 |16, 17| solves general nonlinear programming problems of the form:

NLP_{GRG}

$$\text{minimize } f_o(t)$$

$$\text{s. t.} \quad o \le f_k(t) \le t^U_{m+k}, \qquad k = 1, \ldots p_o,$$

$$f_k(t) = 0, \qquad k = p_o+1, \ldots, p,$$

$$t^L_j \le t_j \le t^U_j, \qquad j = 1, \ldots, m.$$

This problem is converted to the following equality constrained form by adding nonnegative slack variables $t_{m+1}, \ldots, t_{m+p}$:

NLP_E

$$\text{minimize } f_o(t)$$

$$\text{s. t.} \quad F_k(t) \equiv f_k(t) - t_{m+k} = 0, \qquad k = 1, \ldots, p,$$

$$t^L_j \le t_j \le t^U_j, \qquad j = 1, \ldots, m+p,$$

$$t^L_j = 0, \qquad j = m+1, \ldots, m+p$$

$$t^U_j = 0, \qquad j = m+p_o+1, \ldots, m+p.$$

Let t^i be any feasible point of NLP_E. The variable t can be partitioned into a p-dimensional so-called basic variable y and an m-dimensional nonbasic variable x. If the basic variables are chosen so that the Jacobian matrix

$$J = \nabla_y F$$

is nonsingular in t^i, then the equations

$$F(y,x) = 0$$

may be solved at least conceptually for y in terms of x, yielding a function y(x) valid for all (y,x) sufficiently near (y^i,x^i). Hence NLP_E can be transformed into a simpler reduced problem:

NLP_R

$$\text{minimize } F_o(x) \equiv f_o(y(x),x) = f_o(t)$$

$$\text{s. t.} \quad x^L \leq x \leq x^U.$$

The function $F_o(x)$ is called the reduced objective function, and its gradient $\nabla_x F_o(x)$ the reduced gradient.

The code GRG2 solves NLP_{GRG} by solving (perhaps only partially) a sequence of reduced problems. The problems NLP_R are solved by a gradient method. From $\nabla_x F_o(x^i)$ a search direction d^i (gradient, conjugate gradient or quasi-Newton direction) is formed, and a one-dimensional search is initiated to solve the problem

$$\text{minimize } F_o(x^i + \alpha d^i)$$

$$\text{s. t.} \quad \alpha > 0.$$

The line search is done only approximately. For each $\alpha^\ell, F_o(x^i+\alpha^\ell d^i)$ must be evaluated, which is equal to $f_o(y(x^i+\alpha^\ell d^i), x^i+\alpha^\ell d^i)$. Hence the basic variable $y(x^i+\alpha^\ell d^i)$ must be determined, and this is done by solving

$$F(y, x^i+\alpha^\ell i^i) = 0$$

by a variant of Newton's method. If during the line search a basic variable bound is violated, a new reduced problem is formed.
The code GRG2 has been tested in |13, 16, 24, 35, 36|.

VMCON

The code VMCON |3| or VFO2AD solves general nonlinear programming problems of the form
NLP

$$\text{minimize } f_o(t)$$

$$\text{s. t.} \quad f_k(t) \leq 0, \qquad k = 1, \ldots, p_o,$$

$$f_k(t) = 0, \qquad k = p_o+1, \ldots, p.$$

It accomplishes a sequential quadratic programming technique, and it is based on the algorithm of Powel |22, 23|. Powell's algorithm solves a sequence of positive definite quadratic programming subproblems. Each solution determines a direction in which a one-dimensional minimization is performed.

Let t^ℓ be the current point, then the quadratic problem is given by

QP

$$\text{minimize } f_o(t^\ell) + d^T \nabla f_o(t^\ell) + 0.5\, d^T B(t^\ell, \mu^\ell)\, d$$

$$\text{s. t.} \quad \nabla^T f_k(t^\ell)\, d + f_k(t^\ell) \le 0, \qquad k = 1, \ldots, p_o,$$

$$\nabla^T f_k(t^\ell)\, d + f_k(t^\ell) = 0, \qquad k = p_o+1, \ldots, p.$$

where d is the search direction and B is a positive definite approximation to the Hessian of the Lagrangian. To update B, a BFGS approach is used.

The one -dimensional minimization problem is given by

$$\text{minimize } f(\alpha) = f_o(t) + \sum_{k=1}^{p_o} w_k^\ell \max|0, f_k(t)| + \sum_{k=p_o+1}^{p} w_k^\ell\, |f_k(t)|,$$

where

$$t = t^\ell + \alpha\, d,$$

$$w_k^\ell \ge 0, \qquad k = 1, \ldots, p.$$

The weights w_k are given by |12, 22, 23| :

$$w_k^1 = |\mu_k^1|,$$

$$w_k^\ell = \max||\mu_k|,\ 0.5(w_k^{\ell-1} + |\mu_k|)|, \qquad \ell > 1.$$

The code VMCON (VF02AD) has been tested in |13, 35, 36|.

Estimation of Lagrange multipliers

In |32| we proposed a multiplier estimation procedure for generalized geometric programming that uses the condensation technique. Consider problem SP_o, and add slack terms to get an equality constrained problem:

ASP_o

$$\text{minimize } g_o(t) + \sum_{k=1}^{p_o} \theta\, \gamma\ t_{m+k}^{-\beta}$$

$$\text{s. t.} \quad P_k^+(t) + P_k^-(t^o)\,\theta\, t_{m+k} \leq P_k^-(t), \qquad k = 1, \ldots, p_o,$$

$$P_k^+(t) \leq P_k^-(t), \qquad k = p_o+1, \ldots, p,$$

$$t_j > 0, \qquad j = 1, \ldots, m+p_o,$$

$$\theta, \gamma, \beta > 0.$$

To get a posynomial objective function, we formulate this problem as:
ASP_1

$$\text{minimize } t_o$$

$$\text{s. t.} \quad P_o^+(t) + \psi + \sum_{k=1}^{p_o} \theta\,\gamma\, t_{m+k}^{-\beta} \leq P_o^-(t) + t_o,$$

$$P_k^+(t) + P_k^-(t^o)\,\theta\, t_{m+k} \leq P_k^-(t), \qquad k = 1, \ldots, p_o,$$

$$P_k^+(t) \leq P_k^-(t), \qquad k = p_o+1, \ldots, p.$$

$$t_j > 0, \qquad j = 0, \ldots, m+p_o,$$

$$\theta, \gamma, \beta > 0,$$

$$\psi \geq 0.$$

The constant ψ is introduced to insure the positivity of t_o. We redefine the posynomials so that all constraints are in the form

$$P_k^+(t) \leq P_k^-(t).$$

Condensing these posynomials and next logarithmizing the monomials yield a linear programming problem in ln t with Lagrange multipliers λ:
LP

$$\text{minimize } e_1^T \ln t$$

$$\text{s. t.} \quad F \ln t \leq d.$$

where $e_1 = (1, 0, \ldots, 0)^T$. The matrix F and the vector d depend on the condensation point, and so do the optimal Lagrange multipliers λ given by

$$F^T \lambda = - e_1.$$

If the condensation point is a minimizer $\overset{*}{t}$ of ASP_1, then the overdetermined set of linear equations

$$\overset{*}{F}^T \overset{*}{\lambda} = - e_1$$

is consistent |32|.
The relation between the optimal Lagrange multipliers $\overset{*}{\lambda}$ of LP^* and the optimal multipliers $\overset{*}{\mu}$ of ASP, is given by |32, 36|:

$$\overset{*}{\mu}_k = \frac{\overset{*}{\lambda}_k \overset{*}{t}_o}{P_k^-(\overset{*}{t})}, \qquad k = 0, \ldots, p.$$

These equations can be used to estimate λ and μ as an alternative to the procedure described in |37|. The multipliers of ASP_1 are good approximations for those of SP_o if reasonable slack parameters are chosen.

<u>Scaling and testing</u>

Any nonlinear programming problem can be written in a lot of different but equivalent forms. Experience shows that one formulation is more suitable than another to solve the problem. However, which formulation is the best can't be known a priori and is dependent on the code to be used. Given a specified formulation, one can generate other formulations through eliminations, transformations and scaling |38|. Here we consider the following relevant scale factors:

s_o = constant scale factor for the objective function,

$s_k = c_{sk} \prod_{j=1}^{m} t_j^{a_{jsk}}$ = monomial scale factor for constraint k,

s_j = constant scale factor for variable j.

Some algorithms are not scale dependent while others are more or less. The code GGP is scale invariant, the codes GPKTC and HYBRID are scale invariant if the constraint scale factors are constants |36|. The code VMCON is invariant for a constant constraint scale factor, but the iteration depends on the scaling of the objective function and the variables; the code GRG2 is sensitive to the three scale factors. Another important factor is whether formulation SP_o (standard form) or SP_2 (objective function is a variable) is used. GGP admits only the

form SP_2 and GPKTC is insensitive to it |36|; VMCON and GRG2 perform differently, and to our experience SP_o is preferable.
These reflections make clear that test results may be highly dependent on the problem formulation. Hence one should be careful in comparing test results and check whether identical formulations were used or not. To avoid some of these possible inconsistencies and to get an acceptable scaling, we propose a standard scaling for the constraints in signomial programming. Consider the following formulation of a constraint :

$$P^{+}(t) \leq P^{-}(t).$$

The proposed scale factor is |36|

$$s^{-1} = P^{-}(\overset{*}{t})$$

and the scaled constraint becomes

$$\frac{P^{+}(t)}{P^{-}(\overset{*}{t})} \leq \frac{P^{-}(t)}{P^{-}(\overset{*}{t})}.$$

This means that at optimality the right hand sides of all constraints have the value one, which is acceptable, and that the formulation is more standardized (the Lagrange multipliers are independent of the original formulation). In fact, this can be seen as a generalization of the original posynomial constraint form |10|

$$P(t) \leq 1$$

which has an implicit scaling and is unambiguous. Of course $\overset{*}{t}$ is unknown, but scale factors based on a good starting point have a comparable effect.

NUMERICAL RESULTS.

Technical Details of the Comparison

The codes involved were all tested under the same conditions on the same computer. To illustrate the importance of such standardization of the testing environment, we have executed the well-known Colville standard code on the IBM-3033 of the Computer Center at the Katholieke Universiteit Leuven and recorded the following Colville-times :

X - compiler, OPT(0) : 4.035 s.
X - compiler, OPT(2) : 2.459 s.
VS - compiler, OPT(0) : 4.030 s.
VS - compiler, OPT(3) : 1.170 s.

Based on these figures, codes treated with the X- or VS-compiler without optimization, should normally give the same computer times. Additional experiments confirmed this statement for GPKTC, but GGP became 30% slower if translated with the VS-compiler. In the present study all codes were translated with the VS-compiler (IBM Fortran 77), Level 1.2.0 (Sept.82) and OPT(1).

To judge the performance of the different codes only two simple straightforward criteria will be used :

i) the number of successes and failures on a series of test problems,

ii) the average CPU-time to solve the selected problems. The times recorded were execution times without input operations. Output operations were strictly limited.

Starting Points

In contrast with our earlier study |31| (where we considered a much larger class of algorithms including several dual codes), the problem of selecting starting points could be easily handled here since all codes solve the primal problem. Each problem was solved from 15 different starting points, randomly generated but based on the idea that, in practice, the user will always have a rough idea of the optimal solution. The starting points divided in 3 classes of difficulty were generated according to the following randomized procedure :

$$t_j = \overset{*}{t}_j \, |1 + (2 r_i - 1) P|,$$

where

t_j = initial value,

$\overset{*}{t}_j$ = optimal value,

r_i = random number : $0 \leq r_i \leq 1$,

P = level parameter.

If a component computed by the above formula, violates a bound on the variable, this bound itself then replaces the computed starting values. For P = 0.2 the maximal relative error on the initial value is 20%, and in this way the 20% level is defined. We used starting points on 3 levels: 20%, 50% and 90%. In each level 5 points were computed so that each problem was solved for 15 starting points, and each code had to run 375 times.

Stopping Rule

In order to obtain comparable results, serious efforts must be made to apply equivalent stopping rules for conceptually different algorithms. Although in practice the value of the objective function plays an utmost important role, we have generally neglected it in our stopping rules and made the accuracy of our optimal solution one of our prime concerns. The stopping criteria adopted were based on the relative value of the derivatives of the Lagrangian and on the feasibility of the solution.

Implementing these criteria for GPKTC and HYBRID was straightforward as well as for VMCON (although the high accuracy was once in a while a concern for the latter code.)

For GGP and GRG2, the accuracy of the solution depends on the feasibility requirements put forward. For GPKTC, VMCON and HYBRID the feasibility improves gradually during successful iterations. However for GRG2 and GGP, the user has to specify a level of feasibility, which remains constant during the evolution of the computation. This implies that the user himself has to adjust the feasibility requirements and the other stopping rules. The most practical solution to this problem is to select severe feasibility requirements, but GGP and GRG2 do not like very much such requirements.

For GRG2 we have taken the following procedure : the feasibility requirement is gradually enhanced until either the K.T.-conditions are satisfied or the feasibility requirements fullfilled.

The above approach cannot be used at all for GGP, which does not compute Lagrange multipliers. The only way out here was to select the change in objective function value together with a feasibility requirement and to tune the constraints in the stopping rules so as to produce "equivalent" results. As a consequence of the arguments of the previous paragraphs GRG2 and GGP will produce the least accurate results. (The code GRG2 was treated the most benevolently this way.)

To summarize : the criteria considered are :

$$P_k^+(t) - P_k^-(t) \leq \varepsilon_1, \qquad k = 1, \ldots, p, \qquad \text{C.1.}$$

$$\left|\frac{g_o(t^i) - g_o(t^{i+1})}{g_o(t^{i+1})}\right| \leq \varepsilon_3, \qquad \text{C.2.}$$

$$\left|\frac{P_k^+(t) - P_k^-(t)}{P_k^+(t) + P_k^-(t)}\right| \leq 0.5\ \varepsilon_4, \qquad k = 1, \ldots, p, \qquad \text{C.3.}$$

$$\left|\frac{P_k^+(t,\mu) - P_k^-(t,\mu)}{P_k^+(t,\mu) + P_k^-(t,\mu)}\right| \leq 0.5\ \varepsilon_4, \qquad k = p+1, \ldots, p+m, \qquad \text{C.4.}$$

with ε_1 = from 10^{-5} to 10^{-15} in 4 steps

$\varepsilon_3 = 10^{-4}$,

$\varepsilon_4 = 10^{-5}$.

GPKTC and HYBRID use C.2, C.3 and C.4 (although C.4 alone would be sufficient)

VMCON uses C.4.

GGP uses C.2 and C.3.

GRG2 uses C.1, C.2 and C.4 or $\varepsilon_1 \leq 10^{-15}$

Test problems

The algorithms were tested on a set of test problems with the following characteristics

a) both posy - and signomial problems,
b) most problems are the representation of a physical model,
c) number of test problems : 25,
d) size : number of variables ranging from 2 to 15; number of constraints ranging from 1 to 12. (Note : the largest problem has 15 variables and 12 constraints).
e) All problems have highly nonlinear functions as objective function and as constraints.

Basically the test problems are the same as mentionned in |31|. However because all codes used are primal ones, the problem formulaion could be hold rigoreously the same except for two aspects : bounds on variables and scaling.

Indeed GPKTC and VMCON cannot handle bounds on variables in an implicit way. They have to consider these bounds as explicit constraints. Hence it looked fair to us to keep only these bounds in the explicit problem

formulation, which have a physical meaning and necessity.
As far as the scaling of the objective function and the constraints is concerned : GP codes are insensitive to it (see above) and VMCON is invariant under scaling of the constraints.
An automatic scaling was applied to GRG2 for the constraints and to GRG2 and VMCON for the objective function. The scaling factors are:

$$s_o = 1/g_o(t^o)$$

$$s_k = 1/P_{\bar{k}}(t^o) \qquad k = 1, \ldots, p$$

Numerical Results

The following tables summarize our results.
Detailed results can be found in |36|.

	Successful runs				CPU-time			
	20%	50%	90%	tot	20%	50%	90%	tot
HYBRID	124	124	124	372	7805	7158	7601	22565
GGP	124	125	123	372	15360	15849	16989	48198
GPKTC	118	112	100	330	2487	3588	4874	10948
GRG2	114	111	111	336	22171	28394	32583	83149
VMCON	111	110	104	325	30561	38074	43590	112225

	% Successful runs				CPU-time ratios			
	20%	50%	90%	tot	20%	50%	90%	tot
HYBRID	99	99	99	99	3.14	2.00	1.56	2.06
GGP	99	100	98	99	6.18	4.42	3.49	4.40
GPKTC	94	90	80	88	1.00	1.00	1.00	1.00
GRG2	91	89	89	90	8.92	7.91	6.69	7.59
VMCON	89	88	83	87	12.29	10.61	8.94	10.25

The CPU-time is expressed in milliseconds.

note : the average computing time is based on the successful runs only.

COMMENTS AND CONCLUSIONS.

Successful runs :

A run was called successful if the predetermined stopping rules were satisfied. However a strict interpretation of this statement would lead to extremely bad results for GRG2, which could only rather rarely satisfy the criteria. In many cases however the final iterate could be "clearly" recognized as the optimal solution and was henceforth accepted as such. This means that the results for GRG2 are somewhat biased but only in a favorable direction.

Robustness :

The results show that HYBRID and GGP perform excellently with regard to this criterion. More than 99% of the runs were successful. The other codes perform about equally well among each other but with very much improving results for GPKTC if a better starting point is provided. The choice of a starting point has little influence on HYBRID, GGP and GRG2. The specific codes perform in general better although all codes are doing quite well. (For GRG2 one should however be aware of the above mentionned accuracy problem !)

If a code fails it is useful to know why and what conclusion can be drawn from the failure. In this respect GGP, GRG2 and HYBRID are able to find out, when applicable, that the set of feasible points is empty. GGP and GRG2 produce a sequence of feasible points, a useful property in case of failure. The iterate points generated with GPKTC contain no information until optimality is achieved.

Efficiency :

Here GPKTC outperforms the other codes although one should not overlook the fact that only successful runs were accounted for. This results in a handicap for the more robust codes HYBRID and GGP.

The computing time of HYBRID and GGP is very insensitive to the starting point implied.

Furthermore we would like to add a few little remarks, concretizing some of the experience gained during this study. GPKTC and HYBRID are very sensitive to the presence of very small components in the optimal solution and this in a negative sense.

The choice of the artifical bounds on the variables greatly effects the performance of GGP and GRG2.

Global Evaluation :

The code HYBRID performs as one could have hoped for on theoretical grounds. Its properties (efficiency and robustness) are relevantly better than the corresponding worst property of the pair GGP-GPKTC.

In this regard the results might perhaps be called exemplary and might form an incentive for stimulating research on hybrid codes as an alternative for the research on new algorithms and on improvements in existing algorithms.
The question if specific codes are useful for GP problems seems to be answered positively by our results. At least they performed better than the general codes on our specific set of test problems.

REFERENCES.

|1| AVRIEL M., DEMBO R., PASSY U., Solution of generalized geometric programs, Int. J. of Num. Meth. in Eng., 9 (1975), 149-169

|2| AVRIEL M., WILLIAMS A.C., Complementary geometric programming, SIAM J. Appli. Math., 19 (1970), 125-141

|3| CRANE R.L., HILLSTROM K.E., MINKOFF M., Solution of the general nonlinear programming problem with subrouting VMCON, Argonne National Laboratory, Report ANL-80-64, 1980

|4| DEMBO R.S., Solution of complementary geometric programming problems, M. Sc. thesis, Technion, Haifa, 1972

|5| DEMBO R.S., GGP - A program for solving generalized geometric programs, Users' manual, Dept. Chem. Eng., Technion, Haifa,1972

|6| DEMBO R.S., Current state of the art of algorithms and computer software for geometric programming, J.O.T.A., 26 (1978), 149-184

|7| DUFFIN R.J., Linearizing geometric programs, SIAM Review, 12 (1970), 211-227

|8| DUFFIN R.J., PETERSON E.L., Geometric programs treated with slack variables, J. Appl. Anal., 2 (1972), 255-267

|9| DUFFIN R.J., PETERSON E.L., Geometric programming with signomials, J.O.T.A., 11 (1973), 3-35

|10| DUFFIN R.J., PETERSON E.L., ZENER C.M., Geometric programming, J. Wiley, New York, London, Sydney, 1967

|11| FATTLER J.E., REKLAITIS G.V., SIN Y.T., ROOT R.R., RAGSDELL K.M., On the computational utility of posynomial geometric programming solution methods, Math. Prog., 22 (1982), 163-201

|12| HAN S-P., A globally convergent method for nonlinear programming, Cornell university, Department of computer science, Report TR 75-257, 1975

|13| HOCK W., SCHITTKOWSKI K., Test examples for nonlinear programming codes, Lecture notes in economics and mathematical systems, 187, Springer-Verlag, Berlin, Heidelberg, New York, 1980

|14| KELLEY J.E., The cutting plane method for solving convex programs, SIAM J. Appl. Math., 8(1960), 703-712

|15| KOCHENBERGER G.A., Geometric programming - Extensions to deal with degrees of difficulty and loose constraints, University of Colorado, Doctoral thesis, 1969

|16| LASDON L.S., WAREN A.D., JAIN A., RATNER M., Design and testing of a generalized reduced gradient code for nonlinear programming, ACM Trans. on Math. Softw., 4 (1978), 34-50

|17| LASDON L.S., WAREN A.D., RATNER M.W., GRG2 user's guide, 1980

|18| MARTENS X.M., Geometrische programmering en haar gebruik in de Chemie-ingenieurstechniek, Katholieke Universiteit Leuven, Doctoral thesis, 1976

|19| PASSY U., Condensing generalized polynomials, J.O.T.A., 9 (1972), 221-237

|20| PASSY U., WILDE D.J., Generalized polynomial optimization, SIAM J. Appl. Math., 15 (1967), 1344-1356

|21| PETERSON E.L., Geometric programming, SIAM Review, 18 (1976),1-51

|22| POWELL M.J.D., Algorithms for nonlinear constraints that use Lagrangian functions, Math.Prog., 14 (1978), 224-248

|23| POWELL M.J.D., A fast algorithm for nonlinearly constrained optimization calculations, Proc. of the Dundee conference on numerical analysis, Lecture notes in mathematics, 630, Springer-Verlag, Berlin, 1878, 144-157

|24| RATNER M., LASDON L.S., JAIN A., Solving geometric programs using GRG: results and comparisons, J.O.T.A., 26 (1978), 253-264

|25| RIJCKAERT M.J., Computational Aspects of Geometric Programming, in Design and Implementation of Optimization Software, H.J.Greenberg (ed.), Sythoff & Noordhoff, Alphen a/d Rijn, 1978, 481-505

|26| RIJCKAERT M.J., DEBROEY V., A bibliographical survey of geometric programming, Katholieke Universiteit Leuven, Instituut voor Chemie-ingenieurstechniek, report CE-RM-8205, 1983

|27| RIJCKAERT M.J., MARTENS X.M., Numerical aspects of the use of slack variables in geometric programming, Katholieke Universiteit Leuven, Instituut voor Chemie-ingenieurstechniek, report CE-RM-7501, 1975

|28| RIJCKAERT M.J., MARTENS X.M., A condensation method for generalized geometric programming, Math. Prog., 11 (1976), 89-93

|29| RIJCKAERT M.J., MARTENS X.M., GPKTC - A generalized geometric programming code, Users' manual, Katholieke Unversiteit Leuven, Instituut voor Chemie-ingenieurstechniek, report CE-RM-7601,1976

|30| RIJCKAERT M.J., MARTENS X.M., A bibliographical note on geometric programming, J.O.T.A., 26 (1978), 185-204

|31| RIJCKAERT M.J., MARTENS X.M., A comparison of generalized geometric programming algorithms, J.O.T.A., 26 (1978), 205-242

|32| RIJCKAERT M.J., WALRAVEN E.J.C., Estimation of Lagrange multipliers in geometric programming, Op. Res., to appear

|33| RIJCKAERT M.J., WALRAVEN E.J.C., HYBRID - A generalized geometric programming code, Users' manual, Katholieke Universiteit Leuven, Instituut voor Chemie-ingenieurstechneie, report CE-RM-8401,1984

|34| SARMA P.V.L.N., MARTENS X.M., REKLAITIS G.V., RIJCKAERT M.J., A comparison of computational strategies for geometric programs, J.O.T.A., 26 (1978), 185-204

|35| SCHITTKOWSKI K., Nonlinear programming codes - Information, tests, performance, Lecture notes in economics and mathematical systems, 183, Springer-Verlag, Berlin, Heidelberg, New York, 1980

|36| WALRAVEN E.J.C., Ontwikkeling en evaluatie van primaire algoritmen voor geometrische progammering, Katholieke Universiteit Leuven, Doctoral thesis, 1984

|37| GILL P.E., MURRAY W., The computation of Lagrange multiplier estimates for constained minimization, Math. Prog., 17 (1979) 32-60

|38| FLETCHER R., Practical methods of optimization - Vol. 2: Constrained optimization, J. Wiley, Chichester, New York, Brisbane, Toronto, 1981

|39| LOOTSMA F., Performance evaluation of Nonlinear Optimization Methods via Multi-Criteria Decision Analysis and via Linear Model Analysis, in Nonlinear Optimization 1981, M.J.D.Powell(ed), Academic Press, New York, 1982, 419-454

|40| RAJASEKERA J.R., PETERSON E.L., A posynomial geometric programming computer algorithm (PGP1), OR Report N° 194, North Carolina State University, Raleigh, 1983.

PRINCIPLES OF SEQUENTIAL QUADRATIC PROGRAMMING METHODS FOR SOLVING NONLINEAR PROGRAMS

J. Stoer
Institut für Angewandte Mathematik
und Statistik
Universität Würzburg
Am Hubland
D-8700 Würzburg

In this paper it is tried to describe to some extent the theoretical background and several practical aspects of sequential quadratic programming (SQP) methods for solving the following standard problem

$$(P) \quad \min f(x)$$
$$x \in R^n: \quad g_j(x) \leq 0 \;, \; j=1,2,\dots,mi$$
$$g_j(x) = 0 \;, \; j=mi+1,\dots,m,$$

where $f, g_j \in C^2(R^n)$.

Of course, a complete discussion of all aspects is neither possible nor intended. The reason for considering SQP-methods is that according to numerical evidence these methods seem to be the best to solve problems (P) of medium size.

We will use the following notation: $MI := \{1,2,\dots,mi\}$, $ME := \{mi+1,\dots,m\}$, $g(x) := (g_1(x),\dots,g_m(x))^T$. If $K = (k_1,\dots,k_r)$, $k_i \in \{1,2,\dots,m\}$, then $g_K(x)$ denotes the vector $(g_{k1}(x),\dots,g_{k_r}(x))^T$. $Df(x) = f'(x)$ is the derivative of f, which is a row vector and the transpose of its gradient, $\nabla f(x) = Df(x)^T$; $D^2f(x) = f''(x)$ denotes the Hessian of f; the Jacobian of $g(x)$ is $Dg(x)$. S is the set of feasible solutions of (P) , and for $x \in S$, $I(x) := \{i \in MI \mid g_i(x) = 0\}$ denotes the set of inequalities active at x, and $K(x) := I(x) \cup ME$.
A local optimal solution x^* of (P) is an $x^* \in S$, for which there

NATO ASI Series, Vol. F15
Computational Mathematical Programming
Edited by K. Schittkowski

is a neighbourhood $U(x^*)$ such that $f(x) \geq f(x^*)$ for all $x \in U(x^*) \cup S$. The Lagrangian of (P) is $L(x,u) := f(x) + u^T g(x)$, where $u \in R^m$. Formulas, theorems are numbered consecutively in each section; theorem (2) means theorem (2) of the same section, formula (3.2) means formula (2) of section 3.

1. Optimality criteria

We first state some well-known optimality criteria, which are basic for all methods to solve (P).

(1) Theorem of Kuhn-Tucker: Let x^* be a local optimal solution of (P) and suppose that

a) $\nabla g_j(x^*)$, $j \in ME$ are linearly independent, and

b) there is an $s \in R^n$ such that $Dg_i(x^*)s = 0$ for $i \in ME$ and

$$Dg_i(x^*)s < 0 \quad \text{for} \quad i \in I(x^*) .$$

Then there is an $u^* \in R^m$, such that

$$D_x L(x^*,u^*) = Df(x^*) + u^{*T} Dg(x^*) = 0 \tag{2}$$

$$u_i^* \, g_i(x^*) = 0 \; , \; u_i^* \geq 0 \quad \text{for} \quad i \in I(x^*) .$$

(2) are the well-known first-order necessary conditions for a local optimum of x^* of (P), and a pair (x^*,u^*) satisfying $x \in S$ and (2) is called a Kuhn-Tucker (K.-T.) pair of (P). Second order conditions are in terms of the matrix $W(x,u) := D_x^2 L(x,u)$. A convenient form of these conditions, not the most general one, is given by

(3) Theorem: 1) Let (x^*,u^*) be a K.-T.pair of (P), and suppose that

a) $u_i^* > 0$ for $i \in I(x^*)$ ("strict complementarity")

b) $\nabla g_k(x^*)$, $k \in K(x^*) := I(x^*) \cup ME$ are linearly independent

c) $s^T W(x^*,u^*)s > 0$ for all $s \neq 0$ with $Dg_k(x^*)s = 0$ for $k \in K(x^*)$.

Then x^* is a strict local minimum of (P), $f(x) > f(x^*)$ for all

$x \in U(x^*) \cap S$ <u>on some neighbourhood</u> $U(x^*)$.

2) <u>Conversely, if</u> x^* <u>is a local minimum of</u> (P) <u>and assumptions</u> <u>a) and b) are satisfied</u> (<u>note that b) implies the hypotheses of</u> (1), <u>then</u>

$$s^T W(x^*,u^*)s \geq 0 \text{ for all } s \text{ with } Dg_k(x^*)s = 0 \text{ for } k \in K(x^*).$$

Proofs of these (and more general) results can be found e.g. in Fiacco, McCormick (1968), Jittorntrum (1984).

A very convenient standard assumption, which combines the assumptions of (1) and (3) is therefore

(V)

1) $f, g_j \in C^2(R^n)$

2) x^* <u>is a local minimum of</u> (P).

3) $\nabla g_k(x^*)$, $k \in K(x^*)$, <u>are linearly independent</u>.

4) <u>For the vector</u> $u^* \in R^m$ <u>such that</u> (x^*,u^*) <u>is a K.-T. pair of</u> (P) (which exists by (1)) <u>it is true that</u> $u_i^* > 0$ <u>for</u> $i \in I(x^*)$ (<u>strict complementarity</u>)

5) $s^T W(x^*,u^*)s > 0$ <u>for all</u> $s \neq 0$ <u>with</u> $Dg_k(x^*)s = 0$, $k \in K(x^*)$.

One usually adds

6) <u>The Hessians</u> $f''(x)$, $g_j''(x)$ <u>are Lipschitz continuous at</u> x^*:

$$\|f''(x) - f''(x^*)\|, \|g_j''(x) - g_j''(x^*)\| \leq L \|x-x^*\|$$

<u>for</u> x <u>close to</u> x^*.

Basic for most minimization algorithms is that a K.-T. pair (x^*,u^*) of (P) solves the set of n+m nonlinear equations

$$F(x,u) = 0, \tag{4}$$

where

$$(5) \qquad F(x,u) = \begin{pmatrix} D_x L(x,u)^T \\ u_1 g_1(x) \\ \vdots \\ u_{mi} g_{mi}(x) \\ g_{mi+1}(x) \\ g_m(x) \end{pmatrix} .$$

The Jacobian $DF(x,u)$ of F has the well-known form

$$DF(x,u) = J(x,u,W(x,u))$$

where, more generally, with B a symmetric $n \times n$-matrix,

$$(6) \qquad J(x,u,B) := \begin{pmatrix} B & , & Dg_{MI}(x)^T & , & Dg_{ME}(x)^T \\ \Delta u_{MI} Dg_{MI}(x), & & \Delta g_{MI}(x) & , & 0 \\ Dg_{ME}(x) & , & 0 & & 0 \end{pmatrix} .$$

Here

$$\Delta u_{MI} := \mathrm{diag}(u_1,\ldots,u_{mi}),\ \Delta g_{MI}(x) = \mathrm{diag}(g_1(x),\ldots,g_{mi}(x)) .$$

Note that in the case of no inequality constraints $MI = \emptyset$, $J(x,u,B) = J(x,B)$ is a symmetric matrix independent of u

$$(7) \qquad J(x,B) = \begin{pmatrix} B , & Dg(x)^T \\ Dg(x), & 0 \end{pmatrix} .$$

For the convergence analysis of many algorithms for solving (P) via (4), which are patterned after Newton's method for solving (4), it is important that $DF(x^*,u^*)$ is nonsingular at a K.-T. pair (x^*,u^*) of (P) , so that the following well-known result underlines the central role of (V):

(8) <u>Theorem</u>: <u>Under assumption</u> (V), $F(x^*,u^*)$ <u>is nonsingular</u>.

The nonsingularity of $DF(x^*,u^*)$ is also used in many perturbation arguments, ensuring the existence of solutions $x(v)$, $u(v)$ differentiable for small v, $x(0) = x^*$, $u(0) = u^*$, for the perturbed equations

$$\phi(x,u,v) = 0 \; , \; F(x,u) \equiv \phi(x,u,0)$$

via the implicit function theorem.

2. Basics on solving nonlinear equations

As already noted, most methods for solving (P) essentially are methods for solving nonlinear equations (1.4). In order to understand the behaviour of these methods, some basic facts on solving nonlinear equations

$$\text{(P1)} \qquad F(x) = 0 \; , \quad F: \; R^n \to R^n$$

are needed. The following assumption is comparable to (V), and is standard

(V1)
1) $F \in C^1(R^n)$
2) x^* <u>solves</u> $F(x) = 0$
3) $F'(x^*) = DF(x^*)$ <u>is nonsingular</u>
4) $\|F'(x) - F'(x^*)\| \leq L\|x-x^*\|$ <u>for</u> x <u>close to</u> x^* .

An important special case is the unconstrained minimization problem

$$\text{(P1')} \qquad \min_{x \in R^n} \; f(x)$$

which reduces to (P1), if $F(x) := \nabla f(x)$. (V1) is then replaced by

(V1')
1) $f \in C^2(R^n)$
2) x^* <u>is a local minimum of</u> $f(x)$
3) $f''(x^*)$ <u>is positive definite</u>
4) $\|f''(x) - f''(x^*)\| \leq L\|x-x^*\|$ <u>for x close to</u> x^* .

Most methods for solving (P1) are variants of Newton's method

(1) $$x_{i+1} := x_i - F'(x_i)^{-1} F(x_i)$$

obtained by replacing

$$\text{either} \quad F'(x_i) \to B_i \quad \text{or} \quad F'(x_i)^{-1} \to H_i$$

by more easily computable nxn-matrices B_i (H_i), and incorporating a choice of step sizes λ_i in (1), resulting in a method of the type

(2) $$x_{i+1} = x_i - \lambda_i d_i \quad , \quad d_i := B_i^{-1} F(x_i) \quad \text{or} \quad d_i := H_i\, F(x_i) \; .$$

The step size λ_i is usually determined by some line search, ensuring e.g. in the minimization case (P1'), $f(x_{i+1}) \approx \min f(x_i - \lambda d_i)$. In the sequal, we use the standard abbreviations

$$s_i := x_{i+1} - x_i \quad , \quad y_i := F(x_{i+1}) - F(x_i) \; .$$

For the analysis of the convergence behaviour of (2) the following facts are important (for a proof see Ortega, Rheinboldt (1970))

(3)

1) Under assumption (V1), Newton's method (1) is locally Q-square convergent: there is an $\epsilon > 0$ such that for all x_o with $\|x_o - x^*\| \leq \epsilon$, (1) generates a sequence $\{x_i\}$, which is well-defined for all $i \geq 0$, such that

$$\|x_{i+1} - x^*\| \leq c\|x_i - x^*\|^2 \quad \text{for all} \quad i.$$

2) The same holds for method (2) with $\lambda_i \equiv 1$, if $B_i (H_i)$ are any matrices with $\|B_i - F'(x^*)\|, \|H_i - F'(x^*)^{-1}\| = O(\|x_i - x^*\|)$.

In most practically important cases where the matrices $B_i (H_i)$ in (2) are determined by update-techniques (see below), the assumptions of (3),2) do not hold. Here the following (easy to prove) basic result of Dennis and Moré (1974) is central for the investigation of the Q-superlinear convergence of the $\{x_i\}$ generated by (2) with $\lambda_i \equiv 1$.

(4) <u>Theorem</u>: <u>Suppose</u> (V1), <u>that all</u> B_i (resp. H_i) <u>are nonsingular and that</u> (2) <u>with</u> $\lambda_i \equiv 1$ <u>generates a sequence</u> $\{x_i\}$ <u>with</u>

$$\lim_i x_i = x^*, \quad x_i \neq x^* \quad \text{for all} \quad i.$$

<u>Then the following are equivalent</u>

a) $\lim_i \|x_{i+1} - x^*\| / \|x_i - x^*\| = 0$ (Q-<u>superlinear convergence</u>)

b) $\lim_i \|(B_i - F'(x^*))s_i\| / \|s_i\| = 0$ (resp. $\lim_i \|(H_i - F'(x^*)y_i\| / \|y_i\| = 0$)

c) $\lim_i \| B_i s_i - y_i\| / \|s_i\| = 0$ (resp. $\lim_i \|H_i y_i - s_i\| / \|y_i\| = 0$).

Henceforth, we will mainly concentrate on methods (2) where the matrices B_i are used. The corresponding results for the other methods using matrices H_i are usually obtained by interchanging B_i with H_i and s_i with y_i (see e.g. (4)c).

3. Standard matrix update methods

After the computation of x_{i+1} from x_i and B_i by (2.2), the updating problem is to find a suitable matrix B_{i+1} for the next iteration of (2.2). In order to simplify the notation we drop the index i and write x, B, s, y for x_i, B_i, s_i, y_i and $\bar{x}, \bar{B}$ for x_{i+1}, B_{i+1}. Since $\bar{B}$ is to approximate $F'(\bar{x})$ and, up to higher order terms,

$$y = F(\bar{x}) - F(x) \approx F'(\bar{x})(\bar{x}-x) = F'(\bar{x})s$$

it seems reasonable to require $\bar{B}$ to satisfy the <u>Quasi-Newton-equation</u>

(1) $$y = \bar{B}s$$

which forces $\bar{B}$ to approximate $F'(\bar{x})$ in the direction s, but deviate in the least possible way from B. This leads to the principle of "least change secant update methods", systematically investigated by Dennis and Schnabel (1979). If one denotes by

$$Q(y,s) = \{A \in L(R^n) \mid y = As\}$$

the set of n×n-matrices satisfying (1), and $\|A\|_F^2 = \operatorname{tr} A^T A$ is the Frobenius norm for matrices, in its simplest form this principle requires to determine $\bar{B}$ such that

(2) $$\bar{B} := \arg\min\{\|\hat{B} - B\|_F \mid \hat{B} \in Q(y,s)\},$$

which in this case leads to Broyden's rank-1 up-date formula

(3) $$\bar{B} = B + \frac{(y - Bs)s^T}{s^T s}.$$

More generally, it seems reasonable to take additional properties of the matrix $F'(\bar{x})$ to be approximated by $\bar{B}$ into account, e.g. that $F'(\bar{x})$ is symmetric, or $F'(\bar{x})$ has a known sparsity structure: This kind of knowledge of $F'(\bar{x})$ is described by an affine subspace $\mathcal{A} \subseteq L(R^n)$, such that $F'(\bar{x}) \in \mathcal{A}$ and $\mathcal{A} \cap Q(y,s) \neq \emptyset$ (e.g. $\mathcal{A} = \{A \in R^n \mid A^T = A\}$).
Likewise, one may replace $\|A\|_F$ by a wheighted Frobenius norm $\|MAM\|_F$, $M = M^T$ a nonsingular weight-matrix.
This leads to the problem of finding $\bar{B}$ such that

(4) $$\bar{B} = \arg\min\{\|M(\hat{B} - B)M\|_F \mid \hat{B} \in \mathcal{A} \cap Q(y,s)\}.$$

Explicit formulas for $\bar{B}$ are known for the following special cases of $M, \mathcal{A}$:

a) $M = I$, $\mathcal{A} = L(R^n)$: $\bar{B}$ is given by (3)

b) $M = M^T$ nonsingular, $\mathcal{A} = \{A \mid A = A^T\}$:

(5)
$$\bar{B} = B + \frac{(y-Bs)c^T + c(y-Bs)^T}{c^T s} - \frac{(y-Bs)^T s}{(c^T s)^2} c\, c^T$$

where $c := M^{-2}s$. In particular,

b1) For $M := I$, $c := s$, (5) is the PSB-update (PSB = Powell-symmetric-Broyden).

b2) If $s^T y > 0$, then there is a pos.definite M such $My = M^{-1}s$, $c = M^{-2}s = y$ and one obtains the DFP (Davidon-Fletcher-Powell) update formula

(6)
$$\bar{B} = B + \frac{(y-Bs)y^T + y(y-Bs)^T}{y^T s} - \frac{(y-Bs)^T s}{(y^T s)^2} yy^T .$$

Marwil (1978) and Toint (1977) have given explicit formulas for the so-called sparse updates $\bar{B}$ (4) where $M := I$ and $\mathcal{A}$ describes the sparsity pattern of $F'(x)$, namely for

$$\mathcal{A} := \{A \in L(R^n) \mid A_{ij} = 0 \text{ , if } F'(x)_{ij} = 0 \text{ for all } x\}$$

and, if $F'(x) = F'(x)^T$, also for

$$\mathcal{A} = \{A \in L(R^n) \mid A = A^T \text{ and } A_{ij} = 0, \text{ if } F'(x)_{ij} = 0 \text{ for all } x\}.$$

For H-type updates $\bar{H}$, the requirement analogous to (4) is $N = N^T$ nonsingular

(7)
$$\bar{H} = \arg\min\{\|N(\hat{H}-H)N\|_F \mid \hat{H} \in \mathcal{A} \cap Q(s,y)\} ,$$

and the most important special case is the well-known BFGS-update

(8)
$$\bar{H}_{BFGS} = H + \frac{(s-Hy)s^T + s(s-Hy)^T}{s^T y} - \frac{(s-Hy)^T y}{(s^T y)^2} ss^T ,$$

which is obtained from (7), if $\mathcal{A} = \{A \in L(R^n) \mid A = A^T\}$ and $s^T y > 0$ and $N = N^T$ is chosen such that $Ns = N^{-1}y$. In terms of the inverses

$B := H^{-1}$, $\bar{B}_{BFGS} := \bar{H}^{-1}_{BFGS}$, (7) gets the simpler form

$$\bar{B}_{BFGS} := B + \frac{yy^T}{s^T y} - \frac{Bss^T B}{s^T Bs} \quad .$$

Broyden's β-class of updates is obtained as a mixture of the DFP and BFGS-updates:

(9)
$$\bar{B}_\tau = (1-\tau)\bar{B}_{BFGS} + \tau\bar{B}_{DFP} =: U(\tau,B,s,y) \quad \text{where}$$
$$U(\tau,B,s,y) := B - \frac{Bss^T B}{s^T Bs} + \frac{yy^T}{y^T s} + \tau ww^T, \; w := \sqrt{s^T Bs}\left[\frac{y}{y^T s} - \frac{Bs}{s^T Bs}\right] .$$

It is well-known that $\bar{B}_\tau$ is positive definite for all $0 \leq \tau \leq 1$, if B is positive definite and $s^T y > 0$.

It should be noted, that according to numerical evidence the BFGS-method seems to be best among all updates (9) currently used for solving minimization problems (P1').

A convincing theoretical reason for the observed superiority is not known.

The convergence behaviour of iteration methods

(10) $$x_{k+1} := x_k - B_k^{-1} F(x_k)$$

for solving (P1), $F(x) = 0$, or (P1')($F(x) := \nabla f(x)$), where the new matrix B_{k+1} is determined from B_k, $s_k := x_{k+1} - x_k$, $y_k := F(x_{k+1}) - F(x_k)$ by the least change secant update principle, has first been analysed by Broyden, Dennis and Moré (1973), more generally in Dennis and Walker (1981), Stachurski (1981), Griewank, Toint (1982):

(11) <u>Theorem:</u> <u>Suppose that</u> F <u>satisfies</u> (V1), <u>if</u> (P1) <u>is to be solved by</u> (10) (<u>respectively that</u> f <u>satisfies</u> (V1'), <u>if</u> $F(x) \equiv \nabla f(x)$ <u>and</u> (P1') <u>is to be solved</u>). <u>Suppose further that the matrices</u> B_k

in (10) *are determined by one of the following update methods*

- (4) *with* $M = I$, *where* $\mathcal{A}$ *contains* $F'(x)$ *for all* $x \in R^n$
- *any method of the Broyden* β-*class* (9): $B_{k+1} := U(\tau_k, B_k, s_k, y_k)$, $0 \leq \tau_k \leq 1$ *in the case of* (P1').

Then there is an $\varepsilon > 0$ *such that* (10) *is well-defined for all starting values* x_o, B_o *with* $\|x_o - x^*\| \leq \varepsilon$, $\|B_o - F'(x^*)\| \leq \varepsilon$ *and generates sequences* $\{x_k\}$ and $\{B_k\}$ *where* $\{x_k\}$ *converges Q-superlinearly* to x^*

$$\lim_k \|x_{k+1} - x^*\| / \|x_k - x^*\| = 0 \text{ , \underline{if} } x_k \neq 0 \text{ \underline{for all} } k \geq 0 \text{ ,}$$

and $\sup_k \|B_k^{\pm 1}\| < \infty$.

The same conclusions hold for the Broyden β-class (9), $0 \leq \tau_k \leq 1$, in the case of unconstrained minimization problems (P1') satisfying (V1'), if instead of (10) one has the method

$$(12) \qquad x_{k+1} = x_k - \lambda_k d_k \text{ , } d_k := B_k^{-1} \nabla f(x_k) \text{ ,}$$

where λ_k is determined by a line search such that either

$$f(x_{k+1}) = \min_{\lambda > 0} f(x_k - \lambda d_k) \quad \text{(exact line search)}$$

or only $f(x_{k+1}) < f(x_k)$ and

$$|\nabla f(x_{k+1})^T d_k| \leq \mu_k \nabla f(x_k)^T d_k, \text{ where } \sup_k |\mu_k| < 1, |\mu_k| = 0(\|\nabla f(x_k)\|)$$

(asymptotically exact line search)(for a proof see e.g. Stoer (1983)).

An outline of the proof of (11) for the updates (4) with $M = I$ is worthwhile, because it reveals the essentially simple structure of such proofs and shows the role played by the least change secant update principle:

1) One first observes that the matrix

$$\bar{B}_k := \int_0^1 F'(x_k + ts_k)dt$$

satisfies $y_k = \bar{B}_k s_k$ and $\bar{B}_k \in \mathcal{A}$, hence $\bar{B}_k \in \mathcal{A} \cap Q(y_k, s_k)$.

2) The definition of B_{k+1} by (4) gives for all $B \in \mathcal{A} \cap Q(y_k, s_k)$

$$\|B_{k+1} - B\|_F^2 = \|B_k - B\|_F^2 - \|B_{k+1} - B_k\|_F^2 \text{ (theorem of Pythagoras),}$$

hence also for $B := \bar{B}_k$, which gives

$$\|B_{k+1} - \bar{B}_k\|_F^2 = \|B_k - \bar{B}_k\|_F^2 - \|B_{k+1} - B_k\|_F^2 \leqslant \|B_k - \bar{B}_k\|_F^2 \quad . \tag{13}$$

Using (V1) one gets the estimates

(14) a) $\|\bar{B}_k - F'(x_k)\|_F, \|\bar{B}_k - F'(x_{k+1})\|_F, \|\bar{B}_k - F'(x^*)\|_F = O(\sigma_k)$

b) $\|B_{k+1} - F'(x^*)\|_F \leqslant \|B_k - F'(x^*)\|_F + O(\sigma_k)$,

where $\sigma_k = \max(\|x_k - x^*\|, \|x_{k+1} - x^*\|)$. Using the last of these estimates one then shows by induction on k, that if $\|x_o - x^*\|$ and $\|B_o - F'(x^*)\|$ are sufficiently small (and therefore the method initially behaves like Newton's method which is quadratically convergent), then x_k converges at least linearly to x^* so fast, that $\sum_k \sigma_k < \infty$ is small enough for (14)b) to imply $\sup_k \|B_k^{\pm 1}\| < \infty$.

3) The Q-superlinear convergence then follows from Theorem (2.4), if

$$\lim_k \|B_k s_k - y_k\| / \|s_k\| = 0 \ . \tag{15}$$

But

$$\|B_k s_k - y_k\| / \|s_k\| = \|(B_k - B_{k+1}) s_k\| / \|s_k\| \leqslant \|B_k - B_{k+1}\|_F,$$

and by (13),(14), $\sup_k \|B_k\| < \infty$,

$$\|B_k - B_{k+1}\|_F^2 = \|B_k - \bar{B}_k\|_F^2 - \|B_{k+1} - \bar{B}_k\|_F^2$$

$$\leq \|B_k - F'(x^*)\|_F^2 - \|B_{k+1} - F'(x^*)\|_F^2 + O(\sigma_k)$$

implying $\sum_k \|B_k - B_{k+1}\|_F^2 < \infty$ because of $\sum_k \sigma_k < \infty$.

Hence $\lim_k \|B_k - B_{k+1}\|_F = 0$, implying (15).

It remains an intriguing problem to derive more quantitative results on the rates of convergence of $\{x_k\}$. The following is known for the update methods from Broyden's β-class (9), $0 \leq \tau_k \leq 1$, used for the minimization problem (P1') under assumption (V1'): For sufficiently good starting values x_o, B_o positive definite, $\|x_o - x^*\|$, $\|B_o - f''(x^*)\|$ small, the following holds

1) $\|x_{k+n} - x^*\| = O(\|x_k - x^*\|^2)$ for (12) with asymptotically exact line searches (see Stoer (1975, 1984)).

2) $\|x_{k+n} - x^*\| = O(\|x_{k+1} - x^*\| \; \|x_k - x^*\|)$ for (12) with exact line searches (Powell (1983a)).

3) There is no $\varepsilon > 0$ such that for (12) with exact line searches

$$\|x_{k+1} - x^*\| = O(\|x_k - x^*\|^{1+\varepsilon})$$

for all problems (P1') with (V1') (Yuan (1983)) .

4) The results of 1),2) presumably are too pessimistic. Numerical evidence shows that most probably

$$\|x_{k+n} - x^*\| = O(\|x_{k+n-1} - x^*\| \; \|x_k - x^*\|).$$

But, so far this has been proved only under additional hypotheses, which need not hold in general (see Schuller (1974), Ritter (1980)).

With respect to the convergence of the matrices B_k , the existence of $\lim_k B_k$ has been shown recently by Ge Ren-Pu and Powell (1983) for the DFP and BFGS-methods with step sizes $\lambda_k \equiv 1$ and by Stoer (1984) for all methods of Broyden's β-class, step sizes $\lambda_k \equiv 1$, and for λ_k determined by asymptotically exact line searches. But $\lim_k B_k$ depends on the course of the algorithm, and in general $\lim_k B_k \neq f''(x^*)$.

4. The local convergence of SQP-methods

SQP-methods, or as they are also called Wilson-Han-Powell-type methods (see Wilson (1963), Han (1976), Powell (1978b))for solving (P) are essentially Newton's method for finding Kuhn-Tucker-pairs of (P) by solving the nonlinear equations (1.4). The basic structure is as follows: Given current iterates $z_k = (x_k, u_k) \in R^{n+m}$, a symmetric $n \times n$-matrix B_k , one first determines a Kuhn-Tucker pair (s,v) of the quadratic program

$$(1)\ (P(B_k)): \qquad \min\ Df(x_k)s + \tfrac{1}{2} s^T B_k s$$
$$s:\ g_i(x_k) + Dg_i(x_k)s \leq 0 \quad \text{for} \quad i \in MI$$
$$g_j(x_k) + Dg_j(x_k)s = 0 \quad \text{for} \quad j \in ME\ ,$$

then sets $x_{k+1} := x_k + s$, $u_{k+1} := v$ and computes a new B_{k+1} by some updating method.

By writing down the Kuhn-Tucker conditions for (s,v) it is easily seen, that $s=0$ if and only if (x_k, v) is a Kuhn-Tucker-pair of (P), so that any Kuhn-Tucker pair of (P) is a fixed point of the iterative method, independently of the choice of B_k. Let us study the behaviour of the method near such fixed points.

We first investigate the convergence behaviour of the method close to a point x^*, which satisfies all conditions of assumption (V) with the exception of 5), which is replaced by the nondegeneracy of $W(x^*,u^*)$ on the nullspace of $N_K := \{s \mid Dg_k(x^*)s = 0 \text{ for } k \in K(x^*)\}$:

(V') 5') <u>if</u> $t \in N_K$ <u>and</u> $t^T W(x^*,u^*)s = 0$ <u>for all</u> $s \in N_K$ <u>then</u> $t=0$.

(i.e. (x^*,u^*) is a Kuhn-Tucker pair for (P), but x^* need not be a local minimum of (P)). If $K := K(x^*) = I(x^*) \cup ME$, then due to the strict complementarity (V)4), close to x^* the above method is equivalent to solving the equality constrained quadratic program

$$\text{(2)} \qquad \begin{aligned} &\min Df(x_k)s + \tfrac{1}{2}s^T B_k s \\ &s:\ g_i(x_k) + Dg_i(x_k)s = 0\ ,\ i \in K(x^*)\ , \end{aligned}$$

so that we may assume without loss of generality for our local analysis that there are only equality constraints in (P), $MI = \phi$, $K(x^*) = ME$. Clearly (s,v) is a Kuhn-Tucker pair for (2) iff (see (1.7)

$$\text{(3)} \qquad \begin{pmatrix} B_k, Dg(x_k)^T \\ Dg(x_k), 0 \end{pmatrix} \cdot \begin{pmatrix} x_{k+1} - x_k \\ u_{k+1} - u_k \end{pmatrix} = - \begin{pmatrix} D_x L(x_k,u_k)^T \\ g(x_k) \end{pmatrix} = - F(x_k,u_k)$$

holds for $x_{k+1} := x_k + s$, $u_{k+1} := v$, which would be Newton's method for solving $F(x,u) = 0$ if $B_k = W(x_k,u_k)$. Note that our modified assumption (V) implies the nonsingularity of the matrix

$$D^2L(x^*,u^*) = \begin{pmatrix} W(x^*,u^*) & , & Dg(x^*)^T \\ Dg(x^*) & , & 0 \end{pmatrix} .$$

The solution of (2),(3) is usually accomplished as follows: One finds an orthogonal nxn-matrix $Q(x_k) = (Y(x_k), Z(x_k))$ such that

$$Dg(x_k)(Y(x_k),Z(x_k)) = (R(x_k)^T\ ,\ 0)\ ,\quad R(x_k) = \triangledown$$

Omitting arguments and indices, (3) is then equivalent to

$$(4)\qquad \begin{pmatrix} Y^TBY, & Y^TBZ, & R \\ Z^TBY, & Z^TBZ, & 0 \\ R^T, & 0, & 0 \end{pmatrix} \cdot \begin{pmatrix} s_Y \\ s_Z \\ v \end{pmatrix} = - \begin{pmatrix} Y^T\nabla f(x_k) \\ Z^T\nabla f(x_k) \\ g(x_k) \end{pmatrix} .$$

$$s := Ys_Y + Zs_Z \ , \quad x_{k+1} = x_k + s$$

or

a) $R^Ts_Y = -g(x_k) \Rightarrow s_Y$

b) $Z^TBZs_Z = - Z^T\nabla f(x_k) - Z^TBYs_Y \Rightarrow s_Z$

(5) c) $s = Ys_Y + Zs_Z$

d) $x_{k+1} = x_k + s$

e) $Rv = - Y^T\nabla f(x_k) - Y^TBs \Rightarrow u_{k+1} := v$.

Note that under our assumptions (3) - (5) is solvable, if $Z(x_k)^TB_k\, Z(x_k)$ is nonsingular, and that u_k is a dummy parameter. (3) is an iteration of the form

$$(6)\qquad z_{k+1} = \phi_k(z_k) \ , \quad z_k = \begin{pmatrix} x_k \\ u_k \end{pmatrix}$$

with fixed point $z^* = (x^*,u^*)$, where

$$(7)\qquad \phi_k(z) = z - \begin{pmatrix} B_k & ,Dg(x)^T \\ Dg(x), & 0 \end{pmatrix}^{-1} \cdot F(z) .$$

Assuming that $B_k = B$ is fixed the method (3) - (6) will locally converge to z^*, if the spectral radius $\rho(D\phi(z^*)) < 1$, and z^* will be a repulsive fixed point, if $\rho(D\phi(z^*)) > 1$. Using (4) one finds

$$(8)\qquad \rho(D\phi(z^*)) = \rho(I - (Z^{*T}BZ^*)^{-1}Z^{*T}W^*Z^*) \ , \ Z^* = Z(x^*), \ W^* = W(z^*) .$$

(Note that $Q(x)$ can be chosen as a matrix function, which is differentiable for x close to x^*).

This explains why it is reasonable to choose B_k always such that B_k or at least $Z_k^T B_k Z_k$ is positive definite: The advantage is not only that (3) then has a well defined solution and the solution of the quadratic program (2) presents no major difficulties if B_k is positive definite. In addition, $\rho(D\phi(z^*)) > 1$ if z^* is such that $Z^{*T}W^*Z^*$ has negative eigenvalues, so that "bad" Kuhn-Tucker points x^* of (P) which do not belong to local minima of (P) are repulsive fixed points of the iteration (6) and are therefore automatically avoided.

On the other hand if x^* is a "good" Kuhn-Tucker point of (P), that is one for which $Z^{*T}W^*Z^*$ is positive definite and therefore belongs to a local minimum of (P), then for positive definite $Z^{*T}B_kZ^*$ at least a suitable damped iteration

$$z_{k+1} = z_k - \lambda_k \begin{pmatrix} B_k & Dg(x_k)^T \\ Dg(x_k) & 0 \end{pmatrix}^{-1} F(z_k) , \quad \lambda_k > 0 \quad \text{small}$$

will locally converge to z^*, and, **for** bounded B_k, **the convergence speed will be the better,** the smaller $\|Z^{*T}B_kZ^* - Z^{*T}W^*Z^*\|$ is.

More can be said on the speed of convergence in a neighbourhood of good Kuhn-Tucker points x^* of (P). Using the Broyden-Dennis-Moré theory one can establish the Q-superlinear convergence of $\{z_k\}$, $\{x_k\}$ for various update techniques for the B_k. So, we now assume that $z^* = (x^*,u^*)$ is a Kuhn-Tucker pair of (P) satisfying (V).

One of the first results along this line is due to Han (1976), who dealt with the Q-superlinear convergence of the vectors $z_k = (x_k, u_k)$ to z^*:

(9) <u>Theorem:</u> <u>Assume</u> (V) <u>and suppose that the matrices</u> B_k <u>are generated either by the PSB-method, or, if</u> $W^* = W(x^*,u^*)$ <u>is positive definite, by the DFP or BFGS-methods, such that</u>

$$B_k s_k = y_k, \quad s_k := x_{k+1} - x_k, \quad y_k := \nabla_x L(x_{k+1},u) - \nabla_x L(x_k,u) ,$$

$u := u_k$ <u>or</u> $u := u_{k+1}$. <u>Then the method</u> (3) - (5) <u>converges locally</u> (i.e. <u>if</u> $\|z_o - z^*\|$, $\|B_o - W^*\|$ <u>are small enough</u>) <u>and Q-superlinearly in</u> z_k:

$$\|z_{k+1} - z^*\| = o\,(\|z_k - z^*\|) \to 0 .$$

The positive definiteness of W^* is needed for the DFP/BFGS methods, because this guarantees $s_k^T y_k > 0$ for x_k close to x^*.
Powell (1978a)described an updating schema, which generates positive definite nxn-matrices B_k only: Given B_k compute B_{k+1} by the BFGS update formula (3.8) so that

$$B_{k+1}s = y ,$$

where $s = x_{k+1} - x_k$, but (compare (9))

$$y := \theta y_k + (1-\theta)B_k s \quad , \; y_k := \nabla_x L(x_{k+1}, u_{k+1}) - \nabla_x L(x_k, u_{k+1})$$

where

$$\theta := \begin{cases} 1 & \text{if } s^T y_k \geq 0.2\, s^T B_k s \\ 0.8\, s^T B_k s \,/\, (s^T B_k s - s^T y_k) & , \text{otherwise} \end{cases} .$$

This choice guarantees $s^T y \geq 0.2\, s^T B_k s > 0$ and therefore the positive definiteness of B_{k+1} . Under assumption V, Powell was even able to prove the local R-superlinear convergence of (3) - (5) (i.e. $\|x_k - x^*\|$ is bounded by a sequence α_k with $\lim \alpha_k = 0$, $\lim\limits_k \alpha_{k+1}/\alpha_k = 0$) , even if W^* is not positive definite, so that the B_k certainly do not approximate W^* .
Stronger results on the Q-superlinear convergence of the x_k are obtained, if one exploits that one can eliminate v from (5).
An elegant approach is due to Goodman (1982), which has been noted and exploited by Nocedal and Overton (1983): The elimination of v

from $D_xL(x,v)^T = 0$, $g(x) = 0$ by multiplying the first equation by $Z(x)^T$ shows that $x = x^*$ solves (note that v is arbitrary)

$$r(x) :\equiv \begin{pmatrix} Z(x)^T \nabla f(x) \\ g(x) \end{pmatrix} \equiv \begin{pmatrix} Z(x)^T \nabla_x L(x,v) \\ g(x) \end{pmatrix} = 0 .$$

Again noting that $Z(x)$ can be chosen as a differentiable function of x close to x^* , one finds for the Jacobian of $r(x)$

$$(10) \qquad Dr(x^*) = \begin{pmatrix} Z(x^*)^T W^* \\ Dg(x^*) \end{pmatrix} ,$$

which is nonsingular under assumption (V). This implies the local square convergence of the Newton-type method

$$(11) \qquad \begin{pmatrix} Z(x_k)^T W_k \\ Dg(x_k) \end{pmatrix} s_k = - \begin{pmatrix} Z(x_k)^T \nabla f(x_k) \\ g(x_k) \end{pmatrix} \Rightarrow s_k , \quad W_k := W(x_k, u_k)$$

$$x_{k+1} := x_k + s_k ,$$

where $u_k = u(x_k)$ is given by any function u with $\|u(x_k) - u(x^*)\| = O(\|x_k - x^*\|)$, e.g. by the customary least squares estimate

$$(12) \qquad u(x) := \arg\min_u \|Df(x) + u^T Dg(x)\|_2$$

$$= -R(x)^{-1} Y(x) \nabla f(x) .$$

If one replaces W_k by a nxn-matrix B_k , one obtains a method

$$\begin{pmatrix} Z(x_k)^T B_k \\ Dg(x_k) \end{pmatrix} \cdot s_k = - r(x_k) , \quad x_{k+1} := x_k + s_k ,$$

which is equivalent to (5)a)-d).

Because of (10) theorem (2.4) gives (if $\lim x_k = x^*$)

$$\lim_k \frac{x_{k+1} - x^*}{x_k - x^*} = 0 \iff \lim_k \| [\begin{pmatrix} Z(x_k)^T B_k \\ Dg(x_k) \end{pmatrix} - \begin{pmatrix} Z^{*T} W^* \\ Dg(x^*) \end{pmatrix}] s_k \| / \| s_k \| = 0$$

(13)
$$\iff \lim \| Z(x_k)^T (B_k - W^*) s_k \| / \| s_k \| = 0$$

$$\iff \lim \| Z(x_k) Z(x_k)^T (B_k - W^*) s_k \| / \| s_k \| = 0 ,$$

which is the criterion of Boggs, Tolle and Wang (1982) for (3) - (5) to generate a Q-superlinearly convergent sequence $\{x_k\}$. In particular, if W^* is positive definite, and the DFP or BFGS-method is used for updating B_k, $B_{k+1}s_k = y_k$, $y_k := \nabla_x L(x_{k+1},u) - \nabla_x L(x_k,u)$, $u = u_k$ or u_{k+1}, then the sequence $\{x_k\}$ converges Q-superlinearly (Boggs, Tolle, Wang). This follows easily since (see proof of (3.11)), $\lim \| B_{k+1} - B_k \| = 0$ implies $\| (B_k - W^*) s_k \| / \| s_k \| \to 0$.

If one replaces the matrix $Z(x_k)^T W_k$ by an approximating $m{\times}n$-matrix C_k, one obtains methods (analysed by Nocedal and Overton (1983)) of the form

$$\begin{pmatrix} C_k \\ Dg(x_k) \end{pmatrix} \cdot s_k = - \begin{pmatrix} Z(x_k)^T \nabla f(x_k) \\ g(x_k) \end{pmatrix} \Rightarrow s_k$$

$$x_{k+1} := x_k + s_k .$$

The sequence x_k will converge Q-superlinearly to x^*, iff

$$\lim_k \| (C_k - Z^{*T} W^*) s_k \| / \| s_k \| = 0 ,$$

which in turn is guaranteed, using Dennis-Moré arguments, for e.g. for Broyden's rank-1 update method (3.3) to compute a C_{k+1} with $C_{k+1}s_k = y_k$, where y_k is any vector satisfying

$$y_k = Z(x^*)^T W^* s_k + \alpha_k \| s_k \| , \quad \alpha_k \to 0 ,$$

e.g. $y_k := Z(x_{k+1})^T \nabla f(x_{k+1}) - Z(x_k)^T \nabla f(x_k)$.

Another class of update methods, again thoroughly analysed by Nocedal and Overton (1983) is obtained by trying to update the (n-m)×(n-m)-matrix $\tilde{B}_k = Z^T BZ = Z(x_k)^T B_k Z(x_k)$ in (5) directly instead of $B = B_k$ and more or less ignoring the (n-m)× m-matrix $Z^T BY$, for which we write simply $\tilde{A}_k$. Instead of (5) we get the method, where we again drop indices with the quantities R_k, $\tilde{A}_k$, $\tilde{B}_k$, Z_k, Y_k;

a) $R^T s_Y = -g(x_k) \Rightarrow s_Y$

b) $\tilde{B}\, s_Z = -Z^T \nabla f(x_k) - \tilde{A} s_Y \Rightarrow s_Z$

(14) c) $s = Ys_Y + Zs_Y$

d) $x_{k+1} = x_k + s$

e) $u_{k+1} := u(x_{k+1})$ (see (12)) .

The proposal of choosing $\tilde{A} = \tilde{A}_k := 0$ was suggested by Murray and Wright (1978). The following important result, which is essentially due to Powell (1978a) shows that ignoring $\tilde{A}_k$ does not necessarily impede the superlinear convergence, if $\tilde{B}_k$ is accurate enough.

(15) Theorem: Assume (V) and that the matrices $\tilde{A}_k$, $\tilde{B}_k = \tilde{B}_k^T$ in (14) satisfy $\sup_k \|B_k^{\pm 1}\| < \infty$, $\sup_k \|\tilde{A}_k\| < \infty$. Furthermore suppose that (14) generates a sequence x_k with $\lim_k x_k = x$. Then the sequence $\{x_k\}$ is 2-step Q-linearly convergent, i.e. $\lim \|x_{k+1} - x^*\| / \|x_{k-1} - x^*\| = 0$, if

$$\lim_k \|(\tilde{B}_k - Z_k^T W^* Z_k) Z_k^T s_k\| / \|s_k\| = 0 .$$

A simple proof is as follows: By definition of s_Y, s_Z

$$\begin{pmatrix} s_Y \\ s_Z \end{pmatrix} = - \begin{pmatrix} \tilde{A}_k, & \tilde{B}_k \\ R_k^{-T}, & 0 \end{pmatrix}^{-1} \cdot \begin{pmatrix} Z_k^T \nabla_x L(x_k, u) \\ g(x_k) \end{pmatrix} , \text{ u arbitrary, e.g. } u = u^* .$$

Therefore

$$x_{k+1} = x_k + Y_k s_Y + Z_k s_Z =: \phi_k(x_k) ,$$

where, setting $Q = Q(x) = (Y(x), Z(x))$, $Y = Y(x)$, $Z = Z(x)$, $R=R(x)$

$$\phi_k(x) := x - Q \begin{pmatrix} \tilde{A}_k , & \tilde{B}_k \\ R^T , & 0 \end{pmatrix}^{-1} \begin{pmatrix} Z^T \nabla_x L(x,u^*) \\ g(x) \end{pmatrix} .$$

Its Jacobian $\phi_k'(x^*)$ is easily computed

$$\phi_k'(x^*) = I - Q_* \cdot \begin{pmatrix} \tilde{A}_k , & \tilde{B}_k \\ R_*^T , & 0 \end{pmatrix}^{-1} \cdot \begin{pmatrix} Z_*^T \cdot W_* \\ Dg(x^*) \end{pmatrix}$$

$$= I - Q_* \cdot \begin{pmatrix} I & 0 \\ 0 & \tilde{B}_k^{-1} \end{pmatrix} \cdot \begin{pmatrix} I , & 0 \\ Z_*^T W_* Y_* - \tilde{A}_k , & Z_*^T W_* Z_* \end{pmatrix} \cdot Q_*^T$$

$$= Q_* \begin{pmatrix} I & 0 \\ 0 & \tilde{B}_k^{-1} \end{pmatrix} \cdot \begin{pmatrix} 0 , & 0 \\ \tilde{A}_k - Z_*^T W_* Y_* , & \tilde{B}_k - Z_*^T W_* Z_* \end{pmatrix} \cdot Q_*^T .$$

Under the assumptions of the theorem $\sup_k \|(I - \phi_k'(x^*))^{-1}\| < \infty$, so that for $e_k := x_k - x^*$, $s_k := x_{k+1} - x_k$, because of $\phi_k(x^*) = x^*$ and

$$e_{k+1} = \phi_k'(x^*)\, e_k + O(\|e_k\|^2)$$

$$= -\phi_k'(x^*)\, s_k + \phi_k'(x^*)\, e_{k+1} + O(\|e_k\|^2)$$

we get

$$\|e_{k+1}\| = O(\|\phi_k'(x^*) s_k\| + \|e_k\|^2) .$$

But using the expression found for $\phi'_k(x^*)$, $\sup_k \|\tilde{B}_k^{-1}\| < \infty$, $\|Q(x_k) - Q(x^*)\| = O(\|e_k\|)$

$$\|\phi_k'(x^*) s_k\| = O(\|\tilde{A}_k - Z_*^T W_* Y_*) s_Y\| + \|(\tilde{B}_k - Z_*^T W_* Z_*) s_Z\|),$$

and $Z(x_k) - Z(x^*) = O(\|e_k\|)$, we find

$$\|e_{k+1}\| = O(\|s_Y\| + \|(\tilde{B}_k - Z_k^T W_* Z_k) s_Z\| + \|e_k\|^2) .$$

On the other hand, the formulae given for s_Y, s_Z show

$$\|s_k\| \sim \|s_Y\|+\|s_Z\|\sim\|Z_k^T\nabla_x L(x_k,\bar{u})\|+\|g(x_k)\| \sim \|e_k\| = O(\|e_{k-1}\|),$$

where we write $\alpha_k \sim \beta_k$, if $\alpha_k = O(\beta_k)$ and $\beta_k = O(\alpha_k)$. Also by

$$g(x_k) = g(x_{k-1}) + Dg(x_{k-1})s_{k-1} + O(\|s_{k-1}\|^2) = O(\|s_{k-1}\|^2) ,$$

$$\|s_Y\| = \|R_k^{-T}g(x_k)\| \sim \|g(x_k)\| = O(\|s_{k-1}\|^2) = O(\|e_{k-1}\|^2)$$

we get

$$\|e_{k+1}\| = O(\|e_{k-1}\|^2 + \|(\tilde{B}_k - Z_k^T W_* Z_k)Z_k^T s_k\|),$$

so that by dividing by $\|e_{k-1}\|$, and using that $\|s_k\| = O(\|e_{k-1}\|)$, we obtain the final result

$$\|e_{k+1}\|/\|e_{k-1}\| = O(\|e_{k-1}\| + \|(\tilde{B}_k - Z_k^T W_* Z_k)Z_k^T s_k\|/\|s_k\|) ,$$

and $\lim\|e_{k+1}\|/\|e_{k-1}\| = 0$ by the assumptions of the theorem. □

On the basis of this result, Nocedal and Overton (1983) were able to devise an updating scheme, which generates positive definite matrices $\tilde{B}_k$, which, together with the choice $\tilde{A}_k := 0$, guarantees that the method (14) is locally 2-step Q-superlinearly convergent. Their method is as follows: let $\{\alpha_k\}$ be the sequence $\alpha_k := \eta/(1+k)^{1+\nu}$, $k=0,1,\ldots,$ where $\nu,\eta > 0$, so that $\sum \alpha_k < \infty$. Assume $\tilde{B}_o$ to be positive definite. Consider the vectors $s_Z := Z_k^T s_k$, $s_Y := Y_k^T s_k$, $s_k = x_{k+1} - x_k$ generated in step k of (14). If

$$(16) \qquad \|s_Y\| \leq \alpha_k \|s_Z\|$$

compute $\tilde{B}_{k+1}$ from $\tilde{B}_k$ by the DFP or BFGS-update formula, so that $\tilde{B}_{k+1}\tilde{s}_k = \tilde{y}_k$, where $\tilde{s}_k := s_Z = Z_k^T s_k$ and $\tilde{y}_k$ is chosen from

$$(17) \qquad \tilde{y}_k = \begin{cases} Z^T(\nabla_x L(x_{k+1},u) - \nabla_x L(x_k,u)) \ , \ u := u_{k+1} \text{ or } u_k \ , \ Z := Z_{k+1} \text{ or } Z_k \\ Z_{k+1}^T \nabla f(x_{k+1}) - Z_k^T \nabla f(x_k) \end{cases}$$

(if (16) holds, then all these choices guarantee $\tilde{s}_k^T \tilde{y}_k > 0$ for x_k close to x^*).

If (16) does not hold, simply set $\tilde{B}_{k+1} := \tilde{B}_k$.

5. Exact penalty functions and merit functions

Early penalty functions such as

$$P(x,r) = f(x) + r \sum_{j\in MI} g_j(x)_+^2 + r \sum_{j\in ME} g_j(x)^2 \ , \ g_i(x)_+ := \max(0, g_i(x))$$

were introduced in order to reduce the solution of the constrained problem (P) to the problem of finding the unconstrained minima $x(r_k)$ of $P(.,r_k)$ for a set of parameters r_k tending to $+\infty$ and using that in general the $x(r_k)$ converge to a solution x^* of (P). Since the numerical difficulties of finding $\min_x P(x,r)$ increase rapidly as $r \to \infty$, these penalty functions have soon be replaced by so-called exact penalty functions, which have x^* as a local unconstrained minimum for finite values of the penalty parameter r. One of the first functions of this type was given by Pietrzykowski (1969):

$$(1) \qquad \theta(x;r) := f(x) + r \sum_{j\in MI} g_j(x)_+ + r \sum_{j\in ME} |g_j(x)| \ .$$

He showed that under assumption (V), there is $\rho > 0$ such that for all $r \geq \rho, x^*$ is a strict local minimum of $\theta(.,r)$. The drawback of this function is that it is not differentiable. Differentiable exact penalty functions for equality constrained programming problems (P), $MI = \emptyset$, were first given by Hestenes (1969) and Powell (1969), which were later generalized by Rockafellar (1973) also for problems (P) with inequality constraints. He proposed the augmented Lagrangian

$$(2)\ \Lambda(x,u,r) := f(x) + \sum_{i \in MI} \tfrac{1}{2} r_i \left(g_i(x) + \frac{u_i}{r_i}\right)_+^2 + \sum_{i \in ME} \tfrac{1}{2} r_i \left(g_i(x) + \frac{u_i}{r_i}\right)^2 - \tfrac{1}{2} \sum_{i \in ME \cup MI} \frac{u_i^2}{r_i} ,$$

which is defined for a positive vector $r = (r_1,\ldots,r_m) > 0$ of penalty parameters. $\Lambda(.,.,r)$ is continuously differentiable, and also twice continuously differentiable except at points (x,u) where $g_i(x) + \frac{u_i}{r_i} = 0$ for some $i \in MI$.

Again Λ is an exact penalty function in the following sense: Under assumption (V) there is an $\epsilon > 0$ and a $\rho > 0$ such that for $r_i \geq \rho$, $i=1,\ldots,m$, and all u with $\|u-u^*\| \leq \epsilon$, the function $\Lambda(.,u,r)$ has a strict local minimum $x(u)$ which depends differentiably on u such that $x(u^*) = x^*$. Results of this type form the basis of augmented Lagrangian methods, which have the following basic structure

0) Choose u^o, r^o

For $k = 0,1,\ldots,$

1) Find x^k: $\Lambda(x^k,u^k,r^k) = \min\{\Lambda(x,u^k,r^k) \mid x \in R^n\}$

2) Compute a new u^{k+1} and r^{k+1}.

The computation of u^{k+1} should be such that $u^k \to u^*$ (if u^o was close to u^*) and r^{k+1} should be readjusted, if r^k was too small (which e.g. has the effect that $\min \Lambda(x,u^k,r^k) = -\infty$), or too large (then finding x^k is numerically difficult). Suitable formulae for

u^{k+1} ensuring the local linear convergence of $u^k \to u^*$ and $x^k \to x^*$, if $r^k \equiv r$ is large enough, are known, e.g.

$$u_i^{k+1} = (u_i^k + r_i g_i(x^k))_+ \qquad \text{for } i \in MI$$

$$u_i^{k+1} = u_i^k + r_i g_i(x^k) \qquad \text{for } i \in ME .$$

We will not describe these methods in more detail, in as much they are not competitive with the SQP-methods for medium size problems (their domain of application are large problems (P), where the storing of matrices and the solution of quadratic programs gets expensive), but consider the use of exact penalty functions as merit functions within SQP-methods. A typical case is the use of $\theta(x,r)$ (1) as proposed by Han (1977): Suppose the solution of $(P(B_k))$ (4.1), where B_k is chosen as a positive definite matrix, has given a Kuhn-Tucker pair (s,v) of $(P(B_k))$ with $s \neq 0$.
Then, according to Han, for sufficiently large r (namely $r \geq \|v\|_\infty$) and positive definite B_k, the vector s is a direction of decrease of $\theta(x,r)$ in x_k, that is

$$D_s\theta(x_k;r) := \lim_{t \downarrow 0} \frac{\theta(x_k + ts;r) - \theta(x_k,r)}{t} < 0 .$$

This property is used to define $x_{k+1} := x_k + \lambda_k s$ by means of a line search, such that

$$(3) \qquad \theta(x_{k+1};r) \approx \min_{0<\lambda\leq 1} \theta(x_k + \lambda s;r) ,$$

that is, θ is used as a merit function to judge the quality of a step-size λ. Han was able to prove under relatively strong assumptions (among others $ME = \emptyset$, (P) is a convex program,...)that the method is "globally convergent": For each starting point x_o, r large, the sequence $\{x_k\}$ either terminates with a Kuhn-Tucker point x^* of (P) or has only Kuhn-Tucker points of (P) as accumulation points.

One drawback of $\theta(x,r)$ and this approach is known under the name "Maratos-effect" (Maratos (1978)): It is possible that even for x_k close to x^*, the stepsizes λ_k found in (3) are less than 1, because of $\theta(x_k+s,r) > \theta(x_k,r)$, so that the step size strategy prevents the final superlinear convergence of the $x_k \to x^*$. Other, minor difficulties arise, because the line search (3) with the non-differentiable function θ is not simple.

Desirable properties of merit functions $\theta(x,r)$ therefore are:

(4)

1) The directions $s = s_k$ generated by solving $(P(B_k))$ are descent directions of $\theta(.,r)$.
2) $\theta(.,r)$ is differentiable.
3) x^* is a local minimum of $\theta(.,r)$ for suitable r.
4) If x_k is close to x^*, then $\theta(x_k+s_k,r) < \theta(x_k,r)$ for suitable r.

Differentiable exact penalty functions which are suitable as merit functions have been developed by Boggs and Tolle (1980) for equality constraint problems (P), $MI = \emptyset$, satisfying (V). They considered augmented Lagrangians of the form

(5) $M(x,u;r) = L(x,u) - 1/(2r)\, D_xL(x,u)\cdot Q(x)\cdot D_xL(x,u)^T$,

where $r > 0$ is a parameter and $Q(x)$ is positive semidefinite, and for $x = x^*$, $Dg(x^*)Q(x^*)Dg(x^*)^T$ is positive definite. Then $M(x,.;r)$ is quadratic and strictly concave in u. For x close to x^*, $x \in U(x^*)$,

$$u(x;r) := (Dg(x)\cdot Q(x)Dg(x)^T)^{-1}\,[r\, g(x) - Dg(x)Q(x)\,\nabla f(x)]$$

is differentiable in x, and satisfies

$$M(x,u(x;r);r) = \max_u M(x,u;r)\,, \quad \text{for } x \in U(x^*)$$
$$u(x^*;r) = u^*\,.$$

Moreoever, for r sufficiently large, the function

$$\theta(x;r) := M(x,u(x;r);r)$$

is a differentiable exact penalty function (Boggs, Tolle (1980)):

$$\theta(x^*;r) = \min_{x \in U(x^*)} \theta(x;r) .$$

Hence, these functions satisfy (4),2),3) .
For the choice $Q(x) := Dg(x)^T (Dg(x)Dg(x)^T)^{-1} Dg(x)$, the formulae simplify and $u(x;r)$, $\theta(x;r)$ are given by

$$u(x;r) = (Dg(x) \cdot Dg(x)^T)^{-1} [r\, g(x) - Dg(x) \nabla f(x)]$$

$$\theta(x;r) = r \cdot g(x)^T (Dg(x) \cdot Dg(x)^T)^{-1} g(x) + L(x,u(x)) ,$$

where (see (4.12)), $u(x) := -(Dg(x)Dg(x)^T)^{-1} Dg(x) \nabla f(x)$.

Moreover, it can be shown for this special case, that the remaining properties (4),1) and 4) are almost satisfied: 1) holds locally for x_k close to x^*, and 4) also holds for x_k close to x^*, if in addition the B_k are such that the undamped method (4.1), $\lambda_k \equiv 1$, would converge Q-superlinearly, $\lim_k \|x_{k+1} - x^*\| / \|x_k - x^*\| = 0$ (see (4.13) for criteria).
Similar results hold for a related class of functions

$$P(x,u;r) := L(x,u) + \frac{r}{2} g(x)^T g(x) + \tfrac{1}{2} D_x L(x,u) Q(x) D_x L(x,u)^T$$

introduced by Di Pillo and Grippo (1979) (and generalized in Di Pillo, Grippo, Lampariello (1980)), which is quadratic and strictly convex in u. There are again a simple formulae for $u(x,r)$ with

$$P(x,u(x;r);r) = \min_u P(x,u,r) ,$$

and $\theta(x,r) := P(x,u(x;r);r)$ has for large r the properties (4) required for nice merit functions (see Bertsekas (1982)).
Even when using merit functions satisfying (4) for finding suitable stepsizes λ_k for SQP-methods, one should not overlook some principal limitations of this approach.

First of all, merit functions only serve to reduce the natural step size $\lambda_k = 1$ of Newton's method, if such a large step seems too bald, they do not change the directions $s = s_k$ of line search. This auxiliary role of merit functions is justified, since the directions $s = s_k$ generated by solving $(P(B_k))$ for positive definite B_k are in principle good directions for the reasons explained in section 4:

a) If x_k is no Kuhn-Tucker point of (P), then $s = s_k \neq 0$: non-Kuhn-Tucker points are therefore no limit points of any iteration $x_k \rightarrow x_{k+1} + \lambda s_k$ with any arbitrary but fixed step size $\lambda > 0$ independent of k.

b) A Kuhn-Tucker point (x^*, u^*) satisfying (V) is a point of attraction of a suitable damped iteration $x_{k+1} = x_k + \lambda_k s_k$, $\lambda_k > 0$, even if B_k are arbitrary positive definite matrices.

c) Bad Kuhn-Tucker points (x^*, u^*) of (P) (i.e. those which do not satisfy the <u>necessary</u> 2nd order condition for a local minimum of (P)) are not points of attraction of any iteration $x_{k+1} = x_k + \lambda s_k$ with arbitrary small step sizes $\lambda_k = \lambda > 0$ and arbitrary positive definite matrices $B_k = B$, which are independent of k.

Now, as a rule the determination of step sizes λ_k by means of merit functions satisfying (4) forces the SQP-method to converge to a local minimum of $\theta(.,r)$, if convergence takes place at all. Unfortunately enough, any merit function if used with sufficiently nasty problems (P), may have local minima $x(r)$ for all choices of r, which are not local optimal solutions of (P) and in general not even feasible solutions of (P). One should therefore be aware that the algorithm is caught with such a local minimum of $\theta(.,r)$: It is an indication of this danger if the merit function decreases λ_k more and more with each iteration even though the s_k are not tending to zero and the iterates x_k, e.g. are far from feasibility.

In addition the proper choice of r presents problems: at x_k the direction $s = s_k$ is only a direction of decrease, if r is suitable, say is large enough. As one does not know beforehand, how large r has to be chosen, so that r is large enough for each iteration $k \geq 0$, r has to be readjusted in each step, leading to variable r_k. Another reason for changing r_k, is that, if, say, r is too large in step k, then though s_k is a direction of decrease of $\theta(.,r)$ at x_k, the step $\lambda_k = \lambda_k(r)$ determined by (3) may be too small (usually $\lambda_k(r) \to 0$, as $r \quad +\infty$). But changing the parameter $r = r_k$ with every step leads to very delicate questions concerning the "global convergence" of such methods, since the x_k generated do not lead to monotonically decreasing values of a fixed function $\theta(.,r)$

$$\theta(x_o,r) > \theta(x_1,r) > \theta(x_2,r) > \cdots$$

(this has a nice interpretation), but to relations like

$$\theta(x_o,r_o) > \theta(x_1,r_o) \ , \ \theta(x_1,r_1) > \theta(x_2,r_1) \ ,\ldots \ .$$

There are examples, for the merit function $\theta(x,r)$(1) (Chamberlain (1979)), which show that certain quite naturally looking strategies for changing r_k in each step, may lead to cycling sequences x_k. There are other types of merit functions, which are used with great success, which vary with the iteration number k by definition, not only by changing a parameter r at will.

One example is the function

$$\theta_k(x,r) := \Lambda(x,v(k),r) = f(x) + v(k)^T g(x) + \frac{r}{2} g(x)^T g(x) \ , \ r \geq 0 \ .$$

Here, $v = v(k)$ is the optimal Lagrange vector of $(P(B_k))$(4.1) found in iteration k: It is easily seen from (4.3) that the direction $s = s_k$ satisfies

$$\begin{pmatrix} B_k \ , & Dg(x_k)^T \\ Dg(x_k) \ , & 0 \end{pmatrix} \cdot \begin{pmatrix} s_k \\ 0 \end{pmatrix} = - \begin{pmatrix} D_x L(x_k,v)^T \\ g(x_k) \end{pmatrix}$$

so that for positive definite B_k

$$D_x L(x_k,u)\, s_k = -\, s_k^T B_k s_k < 0$$

$$D_x (g(x)^T g(x))_{x=x_k} s_k = 2g(x_k)^T Dg(x_k) s_k = -2g(x_k)^T g(x_k) \leqslant 0 \ ,$$

which shows (4),1) $D_x \theta_k(x,r)\, s_k < 0$.

Clearly (4) 2) also holds. Moreover, if the B_k are such that the undamped sequence $\{x_k\}$, $x_{k+1} = x_k + s_k$, converges Q-superlinearly to x^*, then also for sufficiently large r (4),4) is true. Still another use of merit functions is made by Schittkowski (1981): He considers the general problem (P) , and uses as merit function the augmented Lagrangian $\Lambda(x,u;r)$(2) in the following way:

If at step k , $(P(B_k))$(4.1) has the Kuhn-Tucker pair (s,v), then the new iterate (x_{k+1},u_{k+1}) is computed from the old iterate (x_k,u_k) by

$$\begin{pmatrix} x_{k+1} \\ u_{k+1} \end{pmatrix} := \begin{pmatrix} x_k \\ u_k \end{pmatrix} + \lambda_k \cdot \begin{pmatrix} s \\ v-u_k \end{pmatrix} \qquad ,$$

where the step $\lambda_k > 0$ is determined by a line search, such that for a suitable r_{k+1}

$$\Lambda(x_{k+1},u_{k+1};r_{k+1}) \approx \min_{0\leqslant\lambda\leqslant 1} \Lambda(x_k + \lambda s,\ u_k + \lambda(v-u_k);r_{k+1}) \quad .$$

Here, r_{k+1} has to be readjusted in every step, so that the direction $\binom{s}{v-u_k}$ is a direction of decrease of $\Lambda(.,.;r_{k+1})$. Note also, that (x^*,u^*) is a saddlepoint and not a minimum of $\Lambda(.,.;r)$, so that the interpretation of $\Lambda(x,u;r) - \Lambda(x^*,u^*;r)$ as a kind of distance between (x,u) and (x^*,u^*) is doubtful. Further precautions have to be taken that for (x_k,u_k) close to (x^*,u^*) the step size $\lambda_k = 1$ is not prevented. Yet the use of this technique is not only justified by its practical success: A detailed theoretical analysis is possible and was given by Schittkowski (1981,1983).

6. Some aspects of the practical realization of SQP-methods

The realization of SQP-methods presents numerous practical problems. Also there are several very different possibilities to realize such methods.

Suppose that at the beginning of step k we are given x_k and the symmetric matrix B_k. If one tries to solve $P(B_k)$(4.1), it may well happen that for large $\|x_k - x^*\|$ $(P(B_k))$ may have no feasible solutions s. One way to overcome this difficulty is to solve instead of (4.1) the modified problem (Powell(1978b))

$$
\begin{aligned}
P(B_k)' \quad & \min_{s\in R^n,\,\delta\in R:} \; Df(x_k)s + \tfrac{1}{2} s^T B_k s + \tfrac{1}{2}\rho_k \delta^2 \\
(1) \quad & Dg_i(x_k)s + (1-\sigma_i\delta)g_i(x_k) \leq 0 \text{ for } i \in MI \\
& Dg_i(x_k)s + (1-\delta)g_i(x_k) = 0 \text{ for } i \in ME \\
& 0 \leq \delta \leq 1
\end{aligned}
$$

where $\rho_k > 0$ is a penalty parameter and

$$
\sigma_i := \begin{cases} 0 & , \text{ if } g_i(x_k) < 0 \\ 1 & , \text{ otherwise} \end{cases} .
$$

Note that $\delta = 1$, $s = 0$ is always feasible for (1), and if $\delta=0$ within the optimal solution (s,δ) of (1), s is also an optimal solution of $P(B_k)$(4.1)(Schittkowski (1983) has discussed a strategy for the choice of δ_k).

Moreover, it is recommended to choose B_k always as a positive definite matrix: the advantages of this have been discussed in sections 4 and 5; not the least advantage of such a choice is that (1) then has a well-defined global minimum, which can be obtained by quadratic programming methods.

As x_k approaches x^*, in (1) only those constraints i will be active, which belong to $K(x^*) = I(x^*) \cup ME$. The costly computation of the gradients $Dg_i(x_k)$ for $i \notin K(x^*)$ then is unnecessary.

Therefore great savings result, if one combines (1) with an active set strategy: Suppose that in addition to x_k, B_k one has a guess K_k of those restrictions of (1) which will be active. Such a guess might be

$$K_k := ME \cup \{i \in MI: \ g_i(x_k) \geq -\varepsilon \text{ or } v_i^{k-1} > 0\} ,$$

where $\varepsilon > 0$ and v^{k-1} is the vector of optimal Lagrange multipliers of the previous quadratic program $P(B_{k-1})$. Instead of (1) it is therefore better to solve a problem of the form

$$\text{(2)} \quad P(B_k)'' \quad \begin{array}{ll} \min & Df(x_k)\, s + \tfrac{1}{2} s^T B_k s + \tfrac{1}{2} \rho_k \delta^2 \\ s,\delta: & Dg_i(x_k)s + (1-\delta)g_i(x_k) \leq 0 \text{ for } i \in MI \cap K_k \\ & Dg_i(x_k)s + (1-\delta)g_i(x_k) = 0 \text{ for } i \in ME \cap K_k \\ & Dg_i(x_{k(i)})s + g_i(x_k) \leq 0 \text{ for } i \notin K_k \\ & 0 \leq \delta \leq 1 . \end{array}$$

Here $k(i) \leq k$ corresponds to previous iterates.

Next arises the problem of how to solve the quadratic programs of the form (2) for positive definite B_k . The customary methods for solving quadratic programs are primal methods, which start with a feasible solution and generate a finite sequence of further feasible solutions, such that the objective function increases strictly and the final feasible solution is optimal for (2)(see e.g. Beale (1959), Fletcher (1971), Gill and Murray (1978)). Other primal algorithms reduce (2) to linearly constrained least squares problems (see Schittkowski (1981b)). The drawback of these methods is that they need a feasible point for their start, which is either not known (see e.g. $P(B_k)$) and must be computed in an extra "Phase I",or an "artifical" feasible point like $s=0,\delta=1$ for $P(B_k)''$)is far away from the optimum. This drawback is avoided by the dual methods for solving quadratic programs (see e.g. Lemke (1962), Van de Panne and Whinston (1964)). One efficient method of this type has recently been proposed by Goldfarb and

Idnani (1983)(see also Powell (1983b)). This method works for strictly convex quadratic programs of the form

$$(3)\qquad \begin{aligned} \min\ & q(x) \equiv d^T x + \tfrac{1}{2} x^T D x \\ x:\ & a_i^T x \leq b_i\ , \quad i \in MI = (1,2,\dots,mi) \\ & a_i^T x = b_i\ , \quad i \in ME = (mi+1,\dots,m)\ . \end{aligned}$$

The method generates a sequence of "S-pairs" (x,J): If $J \subset ME \cup MI$ is an index vector, A_J denotes the matrix with columns a_i , $i \in J$ and b_J the vector with components b_i , $i \in J$, then (x,J) is called an S-pair of (3), if $A_J^T x = b_J$ and

$$q(x) = \min_{\substack{y:\ a_i^T y \leq b_i \text{ for } i \in MI \cap J \\ \quad a_i^T y = b_i \text{ for } i \in ME \cap J}} q(y)\ .$$

The method starts with (x_o, J_o), $J_o = \emptyset$, $x_o := \arg\min \{q(y) | y \in R^n\}$, and, if (x_k, J_k) is an S-pair, such that x_k still is not feasible for (3), say constraint r is violated by x_k , then the method constructs a new S-pair (x_{k+1}, J_{k+1}) such that $r \in J_{k+1} \subset J_k \cup \{r\}$ and $q(x_{k+1}) > q(x_k)$.

The appealing feature of this method when applied to the solution of the successive quadratic programs $P(B_k)$"(2) is that i) one needs no feasible point for the start, and ii) one can use the information contained in K_k from the previous problem $P(B_{k-1})$" quite efficiently in the following way:

If K_k is a good guess for the set of active constraints of $P(B_k)$", then during the construction of S-pairs, one may first confine the search for constraints violated by the current S-pair to the constraints in K_k , and only after an S-pair is found, which satisfies all constraints of K_k , one checks the violation of the remaining constraints (see Goldfarb and Idnani (1983) for details) .

Since equality constrained quadratic programs (EQP) are much easier to solve, there is also another approach for the realization of SQP-methods rather than by recursively solving inequality constrained quadratic programs (IQP) like (4.1), $P(B_k)$. These methods (see Murray, Wright 1978), Bartholomew-Biggs (1972,1975,1982) use a guess K_k for the active set of $P(B_k)$(4.1) and replace (4.1) by an EQP-problem of the form (see (4.2))

$$(4) \qquad \min Df(x_k)s + \tfrac{1}{2} s^T B_k s$$
$$s: Dg_i(x_k)s + g_i(x_k) = 0 \ , \quad i \in K_k \ ,$$

(4) is the special case $(r_k = \infty)$ of a closely related method of Biggs (1972,1979):

$$(5) \qquad \min Df(x_k)s + \tfrac{1}{2} s^T B_k s$$
$$s: Dg_i(x_k)s + g_i(x_k) = + \frac{1}{2r_k} u_i \ , \ i \in K_k,$$

the optimal solution of which is given by

$$(6) \quad (1/(2r_k)I + G_k B_k^{-1} G_k^T)u = -G_k B_k^{-1} \nabla f(x_k) + g(x_k) \Rightarrow u$$
$$s = B_k^{-1}(-G_k^T u - \nabla f(x_k)) \ .$$

Here $g(x)$ is the vector with components $g_i(x)$, $i \in K_k$, $G_k := Dg(x_k)$. The method goes back to ideas of Murray (1969) and is obtained if one tries to find a vector s such that

$$P(x_k+s, r_k) = \min_y P(y, r_k) \ ,$$

for the penalty function $P(x,r) = f(x) + rg(x)^T g(x)$.
(5) and (6) form the basis of the REQP-method of Bartholomew-Biggs (1979). In Bartholomew-Biggs(1983) a similar method is proposed,which is based on the penalty function $\Lambda(x,u,r)$(5.2) instead of $P(x,r)$. Of course the solutions of (4),(5) in general depend on the set K_k, the proper choice of which is very important and is the subject of

several "active set strategies" for the determination of K_{k+1} from the information gathered when solving (4) for the index k (e.g. by taking into account the Lagrange multipliers u of (4) etc.).
The advantage of EQP-problem (4) is that, if due to a good choice K_k, there is no need to change K_k with k, one can use the simplified algorithm (4.14) and apply updates for the reduced matrices $\tilde{B}_k (\tilde{A}_k := 0)$, e.g. those proposed by Nocedal and Overton (4.16),(4.17).
Extensive numerical tests, not only of programs realizing SQP-methods, have been carried out by Schittkowski (1980), showing the superiority of SQP-methods for medium sized problems over other methods. The tests also show that the success of a program for solving (P) not only depends on the underlying method but to a large extent of the quality of the code. More details about the actual behaviour of the verious programs and a list of available programs can be found in Schittkowski (1980).

REFERENCES

Baptist,P.,Stoer,J.: On the relation between quadratic termination and convergence properties of minimization algorithms.
Part I, Numer.Math.28, 343-366 (1977).
Part II, Applications: Numer.Math.28, 367-391 (1977).

Bartholomew-Biggs,M.C.: An improved implementation of the recursive equality quadratic programming method for constrained minimization. Techn.Report No.105, Numerical Optimization Centre. The Hatfield Polytechnic, Hatfield, UK (1979).

Bartholomew-Biggs,M.C.: Equality and inequality constrained quadratic programming subproblems for constrained optimization. Techn.Report No.128, Numerical Optimization Centre. The Hatfield Polytechnic, Hatfield, UK (1982).

Bartholomew-Biggs,M.C.: A recursive quadratic programming algorithm based on the augmented Lagrangian function. Techn. Report No.139, Numerical Optimization Centre. The Hatfield Polytechnic, Hatfield. UK (1983).

Beale,E.M.L.: On quadratic programming. Naval Research Logistics Quarterly 6, 227-243 (1959).

Bertsekas,D.P.: Constrained optimization and Lagrange multipler methods. New York: Academic Press (1982).

Biggs,M.C.: Constrained minimization using recursive equality quadratic programming. In: Numerical Methods for Nonlinear Optimization (F.A. Lootsma,ed.). Academic Press, 1972.

Biggs,M.C.: Constrained minimization using recursive quadratic programming: some alternative subproblem formulations. In: Towards global optimization (L.C.W. Dixon,G.P. Szego, eds.), North-Holland 1975.

Boggs,P.T.,Tolle,J.W.: Augmented Lagrangian which are quadratic in the multiplier. JOTA, 31, 17-26 (1980).

Boggs,P.T.,Tolle,J.W.: Merit functions for nonlinear programming problems. Operations Research and Systems Analysis Report No.81-2, Univ. of North Carolina at Chapel Hill, N.C. (1981).

Boggs,P.T.,Tolle,J.W.,Wang,P.: On the local convergence of quasi-Newton methods for constrained optimization. SIAM J. Control and Opt., 20, 161-171 (1982).

Broyden,C.G.: Quasi-Newton methods and their application to function minimization. Math.Comp. 21, 368-381 (1967).

Broyden,C.G.,Dennis,J.E.,More,J.J.: On the local and superlinear convergence of quasi-Newton methods. J. Inst.Math.Appl. 12, 223-245 (1973).

Chamberlain, R.,M.: Some examples of cycling in variable metric methods for constrained minimization. Math. Programming 16, 378-384 (1979).

Dennis,J.E.,More,J.J.: A characterization of superlinear convergence and its applications to quasi-Newton methods. Math.Comput. 28, 549-560 (1974).

Dennis,J.E.,More,J.J.: Quasi-Newton methods, motivation and theory. SIAM Rev.19, 46-89 (1977).

Dennis,J.E.,Schnabel,R.B.: Least change secant updates for quasi-Newton methods. SIAM Rev.21, 443-459 (1979).

Dennis,J.E.,Walker,H.F.: Convergence theorems for least-change secant update methods. SIAM J.Numer.Anal.18, 949-987 (1981).

Di Pillo,G.Grippo,L.: A new class of augmented Lagrangians in nonlinear programming. SIAM J. Control Optim.17, 618-628 (1979).

Di Pillo,G.,Grippo,L.,Lampariello,F.: A method for solving equality constrained optimization problems by unconstrained minimization. In: Optimization Techniques Part 2, (K.Iracki, K.Malanowski, S. Walukiewicz, eds.), Lecture Notes in Control and Information Sciences, Vol.23, Berlin: Springer (1980).

Fiacco,A.V.,McCormick,G.P.: Nonlinear Programming. Sequential Unconstrained Minimization Techniques. New York: Wiley 1968.

Fletcher,R.: The calculation of feasible points for linearly constrained optimization problems. UKAEA Research Group Report, AERE R6354 (1970).

Fletcher,R.: A FORTRAN subroutine for quadratic programming. UKAEA Research Group Report, AERE R6370 (1970).

Fletcher,R.: A general quadratic programming algorithm. J.Inst. of Math. and its Appl. 76-91 (1971).

Fletcher,R.: Practical Methods of Optimization, Vol.1, Unconstrained Optimization. New York: Wiley (1980).

Fletcher,R.: Practical Methods of Optimization, Vol.2, Constrained Optimization. New York: Wiley (1981).

Ge Ren-Pu,Powell,M.J.D.: The convergence of variable metric matrices in unconstrained optimization. Math.Programming 27, 123-143 (1983).

Gill,P.E.,Murray,W.: Numerically stable methods for quadratic programming. Math.Programming 14, 349-372 (1978).

Goldfarb,D.,Idnani,A.: A numerically stable dual method for solving strictly convex quadratic programs. Math.Programming 27, 1-33 (1983).

Goodman,J.: Newton's method for constrained optimization. Courant Institute of Math.Sciences, New York,N.Y. (1982).

Griewank,A.,Toint,Ph.L.: Local convergence analysis for partitioned quasi-Newton updates. Numer.Math.39, 429-448 (1982).

Han,S.-P.: Superlinearly convergent variable metric algorithms for general nonlinear programming problems. Math.Progr. 11, 263-282 (1976).

Han,S.-P.: A globally convergent method for nonlinear programming. JOTA 22, 277-309 (1977).

Hestenes,M.R.: Multiplier and gradient methods. JOTA 4, 303-320 (1969).

Hock,W.,Schittkowski,K.: Test examples for nonlinear programming. Lecture Notes in Economics and Mathematical Systems,Vol.187, Berlin-Heidelberg-New York: Springer (1981).

Jittorntrum,K.: Solution point differentiability without strict complementarity in nonlinear programming. Math.Programming Study 21, 127-138 (1984).

Lemke,C.E.: A method of solution for quadratic programs. Management Sci.8, 442-453 (1962).

Maratos,N.: Exact penalty function algorithms for finite dimensional and control optimization problems. Ph.D.Thesis, Imperial College, London (1978).

Marwil,E.S.: Exploiting sparsity in Newton-like methods. Ph.D.Thesis, Cornell University, Ithaca, NY. (1978).

Murray,W.: An algorithm for constrained minimization. In: Optimization (R. Fletcher, ed.). Academic Press (1969).

Murray,W.,Wright,M.H.: Projected Lagrangian methods based on the trajectories of penalty and barrier functions. Report SOL 78-23, Rept.of Operations Research, Stanford University, Cal.(1978)

Nocedal,J.,Overton,M.: Projected hessian updating algorithms for nonlinearly constrained optimization. Computer Science Department Report No. 95, 1983. Courant Institute, New York Univ.,New York, N.Y.

Ortega,J.M.,Rheinboldt,W.C.: Iterative Solution of Non-linear Equations in Several Variables, New York: Academic Press 1970.

Pietrzykowski,T.: An exact potential method for constrained maxima. SIAM J. Numer.Anal.6, 299-304 (1969).

Pietrzykowski,T.: The potential method for conditional maxima in the locally convex space. Num.Math.14, 325-329 (1970).

Powell,M.J.D.: A method for nonlinear constraints in minimization problems. In: Optimization (R. Fletcher, ed.). London: Academic Press (1969).

Powell,M.J.D.: The convergence of variable metric methods for nonlinearly constrained optimization calculations. 27-63 in: Nonlinear Programming 3, O.L. Mangasarian, R.R. Meyer, S.M. Robinson, eds., Acad.Press (1978a).

Powell,M.J.D.: A fast algorithm for nonlinearly constrained optimization calculation, in: G.A. Watson (ed.): Numerical Analysis. Lecture Notes in Mathematics, Vol.630. Berlin-Heidelberg-New York: Springer (1978b).

Powell,M.J.D.: Algorithms for nonlinear constraints that use Lagrangian functions. Math.Programming 14,224-248 (1978c).

Powell,M.J.D.: On the rate of convergence of variable metric algorithms for unconstrained optimization. Report DAMTP

1983/NA7, Department of Applied Mathematics and Theoretical Physics, Univ.of Cambridge, England (1983a).

Powell,M.J.D.: ZQPCVX a Fortran subroutine for convex quadratic programming. Technical Report DAMTP/1983/NA17, Department of Applied Mathematics and Theoretical Physics, Cambridge University, England (1983b).

Ritter,K.: On the rate of superlinear convergence of a class of variable metric methods. Numer.Math. 35, 293-313 (1980).

Rockafellar,R.T.: A dual approach to solving nonlinear programming problems by unconstrained optimization.Math. Programming 5, 354-373 (1973).

Schittkowski,K.: Nonlinear proramming codes. Information, tests, performance. Lecture Notes in Economics and Mathematical Systems, Vol. 183, Berlin-Heidelberg-New York: Springer (1980).

Schittkowski,K.: The nonlinear programming method of Wilson, Han and Powell with an augmented Lagrangian type line search function. Part 1: Convergence Analysis. Numer.Math. 38, 83-114 (1983a), Part 2: An efficient implementation with linear least squares subproblems. Numer.Math.38, 115-127 (1983b).

Schittkowski,K.: On the convergence of a sequential quadratic programing method with an augmented Lagrangian line search function. Math.Operationsforsch.u.Statist., Ser. Optimization 14, 197-216 (1983c).

Schuller,G.: On the order of convergence of certain Quasi-Newton methods. Numer.Math.23, 181-192 (1974).

Stachurski,A.: Superlinear convergence of Broyden's bounded β-class of methods. Math.Progr.20, 196-212 (1981).

Stoer,J.: On the numerical solution of constrained least squares problems. SIAM J. Numer.Anal.8, 382-411 (1971).

Stoer,J.: On the convergence rate of imperfect minimization algorithms in Broyden's β-class. Math.Programming 9, 313-335 (1975).

Stoer,J.: The convergence of matrices generated by rank-2 methods from the restricted β-class of Broyden. Numer.Math.44, 37-52 (1984).

Tanabe,K.: Feasibility-improving gradient-acute-projection methods: a unified approach to nonlinear programming. Lecture Notes in Num.Appl.Anal.3, 57-76 (1981).

Tapia,R.A.: Diagonalized multiplier methods and quasi-Newton methods for constrained optimization. JOTA, 22, 135-194 (1977).

Toint,Ph.L.: On sparse and symmetric matrix updating subject to a linear equation. Math.Comp.31, 954-961 (1977).

Toint,Ph.L.: On the superlinear convergence of an algorithm for solving a sparse minimization problem. SIAM Numer.Anal.1063-1045 (1979).

Van de Panne,C.,Whinston,A.: The simplex and the dual method for quadratic programming. Operations Research Quarterly 15, 355-389 (1964).

Wilson,R.B.: A simplicial algorithm for concave programming. Ph.D.Thesis, Graduate School of Business Administration, Harvard University, Cambridge, Mass. (1963).

Yuan,Y.: On the least Q-order of convergence of variable metric algorithms. Report DAMTP 1983/NA10. Department of Applied Mathematics and Theoretical Physics, Univ.of Cambridge, England.

MODEL BUILDING AND PRACTICAL ASPECTS OF NONLINEAR PROGRAMMING

Philip E. Gill, Walter Murray,
Michael A. Saunders and Margaret H. Wright
Department of Operations Research
Stanford University
Stanford, California 94305

1. ASPECTS OF MODELLING THAT AFFECT OPTIMIZATION

1.1. Introduction. This survey paper has two main purposes: to summarize (briefly) certain aspects of modelling that influence the performance of optimization algorithms, and to describe recent advances in methods for nonlinear programming that influence the solution of practical problems. These two themes are not unconnected. A well constructed mathematical model should be such that the bad effects of ill-conditioning, degeneracy and inconsistent constraints are minimized. Ironically, the purpose of good software is to deal effectively with precisely these problems. Therefore it is not surprising that much of the insight necessary to construct a well-posed mathematical model is pertinent to the formulation of robust algorithms.

The principal problem of concern will be the *nonlinear programming problem*:

NP
$$\begin{array}{ll} \underset{x\in\Re^n}{\text{minimize}} & F(x) \\ \text{subject to} & \ell \le \left\{ \begin{array}{c} x \\ A_L x \\ c(x) \end{array} \right\} \le u, \end{array}$$

where $F(x)$ (the *objective function*) is a smooth nonlinear function, A_L is a constant matrix of constraints, and $c(x)$ is a vector of smooth nonlinear constraint functions. The objective function F and the constraint functions taken together comprise the *problem functions*. Unless otherwise stated, the problem functions will be assumed to be at least twice-continuously differentiable. (Methods that require this degree of smoothness will usually work if there are isolated discontinuities away from the solution). We shall use $g(x)$ to denote the gradient of $F(x)$, and $a_i(x)$ the gradient of $c_i(x)$. The solution of NP will be denoted by x^*.

NATO ASI Series, Vol. F15
Computational Mathematical Programming
Edited by K. Schittkowski

We shall begin with a statement of ten "modelling principles" that may help to make the resulting problem NP more susceptible to existing nonlinear programming software. This is followed by a review of the practical aspects of quasi-Newton sequential quadratic programming (SQP) methods for nonlinear programming. We conclude by presenting examples of the application of SQP methods to some illustrative optimization problems.

1.2. Some basic modelling principles. Our observations of practical optimization problems have indicated that, even with the best available software, the efficient optimization of a model can be critically dependent on certain properties of the formulation. It is often the case that the formulator of the model must make numerous decisions that do not affect the accuracy to which the model reflects the real world, yet are crucial to whether the model is amenable to solution by an optimization algorithm.

Our experience with the role of modelling in numerical optimization will be summarized by a list of ten "modelling principles". These principles may serve as a guide for those who have little knowledge of the intricacies and potential pitfalls of modern optimization codes. They have been derived from our own experiences with real problems.

Of course, the nature of possible models varies so much that it is impossible to treat all relevant aspects of modelling. The main thesis of these principles is that developers of models should consider in the initial stages the ultimate need to solve an optimization problem, since it is unlikely that optimization software will ever reach the state wherein a general routine can be used with impunity for all problems.

For additional material on aspects of modelling that influence the performance of optimization methods, the reader is referred to Gill, Murray and Wright (1981).

BASIC MODELLING PRINCIPLES

I. Formulate a simple model first and add features in conjunction with running the optimization.

A model to be optimized should be developed by striking a reasonable balance between the aims of improved accuracy in the model (which usually implies added complexity in the formulation) and increased ease of optimization. This might be achieved by invoking an optimization procedure on successively more complicated versions of the model, in a form of "stepwise" refinement. Thus, the effects of each refinement in the model on the optimization process can be monitored, and fundamental difficulties can be discovered much more quickly than if no optimization were applied until the model was essentially complete. This is especially important when dealing with models that contain many interconnected subsystems, each requiring extensive calculation.

II. Attempt to use smooth problem functions.

Probably the most fundamental property of the problem functions with respect to ease of optimization is *differentiability*, which is important because algorithms are based on using available information about a function at one point to deduce its behavior at other points. If the problem functions are twice-continuously differentiable, say, the ability of an algorithm to locate the solution is greatly enhanced compared to the case when the problem functions are non-differentiable. Therefore, most optimization software is designed to solve smooth problems, and there is a great incentive to formulate differentiable model functions. A useful feature of methods for smooth problems is that they tend to give more information concerning the *quality* of the solution. For example, some nonlinear programming methods can be shown to exhibit a superlinear rate of convergence in the neighborhood of a local minimum. If a method terminates at a point for which this rate of convergence is exhibited, the user will have some confidence that the final point is close to a local minimum. (When solving a problem on a digital computer, we need to define carefully what we mean by a "smooth" problem. In reality, all software minimizes a function $fl(F(x))$, which is the floating-point representation of $F(x)$. The function $fl(F(x))$ is piecewise constant at the round-off level. If we define ϵ_A to be the absolute precision of F, i.e.,

$$|F(x) - fl(F(x))| = \epsilon_A(x),$$

then algorithms for smooth problems will work whenever changes in the variables produce changes in F that are much greater than ϵ_A. Note that the vast majority of optimization software assumes that F is computed to full precision; i.e., it is assumed that ϵ_A is of the order of $\epsilon_M \|F(x)\|$, where ϵ_M is the relative machine precision.)

III. Avoid defining problem functions that are the result of some iterative procedure (such as the solution of a differential equation or the evaluation of an integral).

Problem functions defined by an iterative procedure are often the source of subtle discontinuities that may impede the progress of the optimization. The solution of these subproblems to full machine precision (even if possible) generally requires considerable computational effort, and thus tends to be regarded as unwarranted by the modeller, since the integral or differential equation (or whatever) is only an approximation to some more complicated real-world phenomenon.

The use of an iterative procedure to define a problem function most often occurs when the variables of the problem are *functions of a continuous parameter* (in an optimal control problem, for example). In many instances, an effective strategy for this type of problem is to discretize the problem *before* applying the optimization method. Accurate solutions to the continuous problem are then found by refining the discretization between optimizations. Such a strategy illustrates again that it is often worthwhile to interleave modelling and optimization, since the creation of an increasingly accurate discretization is in fact a modelling process.

IV. Think carefully about the nature of the constraints.

It is not always appreciated that substantial improvements in performance and robustness can result when methods exploit the different properties of simple bounds, linear constraints and nonlinear constraints. Whenever possible, the user should isolate the linear constraints from the nonlinear constraints and use software that differentiates between constraint types during the optimization. Unfortunately, some problem formats *guarantee* that a linear constraint will be treated as a nonlinear constraint. For example, in the class of *geometric programming problems* the objective and constraint functions are sums of functions of the form

$$\phi_i(x) = \alpha_i x_1^{a_{i1}} x_2^{a_{i2}} \cdots x_n^{a_{in}}, \tag{1.1}$$

where the x_j are the variables (constrained to be positive) and the a_{ij} are constants. The transformation of a linear constraint into a sum of functions of the form (1.1) unnecessarily increases the degree of difficulty of the problem.

The transformation of a problem from one form to another was often unavoidable in the past because less software was available. When algorithms for unconstrained optimization were more numerous and more effective than for constrained problems, it was common practice for simple bound constraints to be treated by a change of variable. Today, however, algorithms for problems with only simple bounds or linear constraints are comparable in efficiency to unconstrained algorithms. Therefore, it is virtually never worthwhile to transform bound-constrained problems (in fact, it is often beneficial to *add* bounds on the variables — see below), and it is rarely appropriate to alter linearly constrained problems.

Transformations can be used effectively to transform nonlinear constraints into simple bound constraints (for example, by using polar coordinates instead of cartesian coordinates to deal with range constraints of the form $l \le \sum x_i^2 \le u$). However, care should be taken to ensure that the transformation does not lead to a new problem that is more difficult to solve, or has additional (spurious) solutions.

V. Do not attempt to eliminate equality constraints from the problem.

Modellers often assume that since there may be no physical significance to a point at which nonlinear equality constraints are violated, such constraints should be satisfied exactly at all stages of the optimization. Accordingly, users often attempt to "eliminate" nonlinear equality constraints from the problem by the following method. The variables are partitioned into "independent" and "dependent" sets. The minimization is then performed only with respect to the independent variables, and the dependent variables are determined by "solving" the equality constraints. To be more precise, let x denote the vector of dependent variables and u the vector of independent variables. The constrained problem

$$\begin{array}{ll} \underset{u,x}{\text{minimize}} & F(x,u) \\ \text{subject to} & c(x,u) = 0, \end{array}$$

is solved by expressing $c(x,u) = 0$ as $x = T(u)$ for some transformation T, and then minimizing $F(T(u),u)$ with respect to u. This strategy is particularly common when there are special fast methods for solving the equations $c(x,u) = 0$.

We do not recommend this approach if the constraints have any significant degree of nonlinearity. Firstly, it is difficult to impose any simple bounds upon the dependent variables. Secondly, the resulting algorithm is of the "constraint-following" type, which will tend to be less efficient than other methods. Our experience is that the total computational effort required to solve for the dependent variables at every trial point is not usually worthwhile, compared to expending a similar amount of effort in the optimization without eliminating the variables. The user would be better advised to use a general nonlinear programming method.

VI. Distinguish between "hard" and "soft" constraints.

In many problems, the bound constraints may be classified as either "hard" or "soft". For example, hard bounds could specify the region in which the problem functions are well-defined. A soft constraint might represent a bound whose violation we are prepared to tolerate if this resulted in a large decrease in the objective function. In some cases, variables may have both a soft bound and a hard bound.

A good method of treating a soft bound of the form $x_i \geq b_i$ is to exclude it from the simple bound constraints and include it in the objective function in the form of a mild penalty term of the form $\frac{1}{2}\rho_i [x_i - b_i]_-^2$, where $[y]_-$ denotes the function $\min\{0, y\}$. Modest values of ρ_i do not lead to the severe ill-conditioning that occurs with traditional penalty methods. If the variable is also constrained by a hard bound, the constraint can be treated explicitly in the usual way.

As a general rule, the presence of simple bounds simplifies a problem by reducing dimensionality. However, if a bound has only a slight impact upon the problem (i.e., if the objective function does not change rapidly as the bound is perturbed), the degree of difficulty of the problem may be *increased* because the algorithm needs to resolve small Lagrange multipliers. If many of the bounds with small multipliers would be classified as "soft", the use of a penalty term tends to remove these constraints from the active set.

Similarly, a penalty function of the form $\frac{1}{2}\rho_i(x_i - d_i)^2$ can be used to produce a solution that is biased towards some "desired" value d.

VII. Avoid modelling near-dependent equality constraints

Constraints occur in problem formulations for a variety of reasons. Often the very nature of the variables imposes an equality constraint — for example, if the variables $\{x_i\}$ represent proportions or probabilities, this gives rise to the constraint $\sum_i x_i = 1$ (as well as non-negativity restrictions). Constraints of this type are "genuine" equalities, in the sense that the computed solution must satisfy them exactly (where "exactly" means "within working precision"). However, it is not unusual in modelling that constraints that might seem initially to be firm equality constraints should be treated instead as constraints that need not be satisfied with maximum possible accuracy. For example, this situation occurs when the underlying model is known to contain inaccuracies. In some problems, forcing constraints of this type to be equalities may cause there to be no feasible solution, or may distort the properties of the solution if the corresponding constraints are ill-conditioned.

We shall use a simple two-dimensional example to illustrate not only the the potential difficulties associated with dependent constraints, but also one simple method for dealing with them.

Suppose that we require the minimum of a function of two variables subject to two independent linear equality constraints. In this case, the solution lies at the point of intersection of the constraints (the objective function has no influence on the solution). However, suppose that one of the two constraints is really a copy of the other, but that due to small errors in the modelling process, the constraints are not *exactly* dependent. Now, the two constraints are nearly parallel and the point of intersection lies at a point that may be very different from the solution of the essentially equivalent problem posed with a *single* constraint.

A clue to one method of resolution of this difficulty lies in the observation that if one or the other of the constraints were ignored during the optimization, it would be violated by only a very small quantity. Suppose that each dependent equality constraint is replaced by an *inequality* constraint with a very narrow range. For example, the linear constraint $a^Tx = b$ would be replaced by

$$b - \delta \leq a^T x \leq b + \delta,$$

where δ is a small, but not negligible, positive quantity. (Exactly the same transformation can be made for a nonlinear constraint.) In the example above, the active-set strategy of the optimization procedure will decide that a lower value of the objective function is achieved if one of the equality constraints is not satisfied exactly at the solution.

In Section 3.5 we shall use this type of range constraint in methods that are robust on problems with dependent constraints, and show that the idea of allowing certain constraints to be violated by a small quantity recurs in many practical solution techniques. This is only to be expected, since it is never possible to satisfy constraints *exactly* in finite-precision arithmetic.

VIII. Use information about the problem to scale the variables and constraints.

It is well known that the "right" scaling of variables and constraints can dramatically improve the efficiency and accuracy of optimization methods. The *scale* of a problem is the measure of the relative importance of the variables and constraints — or, equivalently, the scale of a problem is a statement of what is "large" and what is "small" in a problem.

In the absence of any other information, an algorithm will generally assume that each variable or constraint has equal weight in the optimization. For example, if a unit change in a variable produces a small change in F compared to other variables, the algorithm will tend to make a larger change to that variable. Clearly, if the variables do have equal significance, this situation is quite satisfactory. However, if the small variation in F is due to the fact that the objective function is almost independent of the variable, the problem should be *rescaled*. (Note that in this case, fixing a variable or constraint is an acceptable form of problem scaling.)

The most common form of scaling is to define new variables and constraints using a linear transformation. (For simplicity, in the following discussion we shall assume that the nonlinear constraints are equalities of the form $c_i(x) = 0$.) Suppose we define new variables $\bar{x}$ and constraints $\bar{c}$ such that

$$x = D_1 \bar{x} \qquad \text{and} \qquad \bar{c} = D_2 c, \tag{1.2}$$

where D_1 and D_2 are nonsingular matrices (usually diagonal). With this scaling, the derivatives of the original and transformed objective function are related by $\nabla_{\bar{x}} F = D_1 \nabla F$ and $\nabla^2_{\bar{x}} F = D_1 \nabla^2 F D_1$. Similarly, for the constraints we have $\bar{A} = D_2 A$, where A and $\bar{A}$ are the Jacobian matrices of the original and transformed constraints. When D_1 and D_2 are diagonal, an interpretation of the scaling (1.2) is that D_1 and D_2 "rank" the elements of the gradient vector and constraints so that each is of equal importance.

The important thing to bear in mind is that a badly scaled problem is essentially an *ill-conditioned* problem. Badly scaled variables lead to ill-conditioned Hessian matrices; badly scaled constraints give rise to near-singular Jacobian matrices. Within optimization routines, convergence tolerances and other criteria are necessarily based upon an implicit definition of "small" and "large", and thus variables with widely varying orders of magnitude may cause difficulties for some algorithms.

Arguably the most important rule of scaling is that the variables of the scaled problem should be of similar magnitude and of order unity in the region of interest. If typical values of all the variables are known, a problem can be transformed so that the variables are all of the same order of magnitude. In this case, the transformation D_1 of (1.2) is a diagonal matrix whose elements are typical values of x.

An important property of a theoretical algorithm is that of invariance with respect to the scaling (1.2). A scale-invariant algorithm has the property that when it is applied to both the original problem and the transformed problem, the resulting sequences of iterates $\{x_k\}$ and $\{\bar{x}_k\}$ satisfy $x_k = D_1 \bar{x}_k$ for all k. This property does not imply that it is unnecessary to choose a sensible scaling for a problem when a scale-invariant algorithm is being used. *Scale-invariance cannot be achieved in a practical implementation of an algorithm.* Not only is computer arithmetic not scale-invariant, but also it is impossible to devise a scale-invariant test that an algorithm has converged to a point which satisfies the necessary or sufficient conditions. Many algorithms treat quantities that are "sufficiently small" in magnitude as "negligible" (in effect, as zero). However, since there is no universal definition of "small", it is impossible to formulate a scale-invariant procedure for distinguishing between quantities that *should* be treated as zero, and scale-dependent quantities that *should not* be neglected. To illustrate this point, consider the application of any quasi-Newton method that is scale-invariant in the sense just described. If the algorithm were applied to the scaled and unscaled problem, we may regard the two sequences $\{x_k\}$ and $\{\bar{x}_k\}$ as belonging to two different spaces. Each value x_k in the x-space is related to a value $\bar{x}_k$ in the $\bar{x}$-space by the formula $\bar{x}_k = D_1 x_k$. The optimality test of the algorithm must involve consideration of the magnitude of the gradient norm. However, since the gradient norm is tending to zero with a different value (either $\|g_k\|$ or $\|D_1 g_k\|$) in each space, this optimality test cannot be made scale-invariant. Thus, the *number* of elements in each sequence may be different. Furthermore, the initial estimate of the Hessian is almost always defined as a unit matrix. If this is done in both the x- and $\bar{x}$-spaces, the scale invariance is immediately lost.

Another part of an algorithm that is unavoidably scale-dependent is the criterion for deciding when a nonlinear constraint of the form $c_i(x) = 0$ is satisfied. Given a constraint value, say $c_i(x) = 10^{-4}$, it is necessary to know something about the natural scaling of the problem in order to make a sensible decision about whether the constraint is sufficiently close to zero. The only way of overcoming this problem is to let the user decide when a constraint is sufficiently satisfied. This leads to the idea of a user-defined *feasibility tolerance* δ_i that defines what is really "small" for each constraint. For example, the part of the termination criterion that concerns the test for feasibility would require that $|c_i(x)| \leq \delta_i$ for each constraint.

IX. Try to provide as much information about the function as possible.

As a general rule, algorithms tend to be more successful and robust when more information about the problem is provided. For example, if the problem functions are *smooth*, algorithms that use first and second derivatives perform much better than algorithms that use function values only.

It is generally perceived that a second-derivative algorithm is rarely useful because the cost of computing the second derivatives may be several orders of magnitude larger than that of calculating first derivatives. This is undoubtedly true in some situations, but there are a remarkably large number of problems for which the first and second derivatives of the problem functions may be obtained for about the same cost — for example, geometric programming problems, where the objective and constraint functions are of the form (1.1). Moreover, it is relatively straightforward to design software that will automatically differentiate the original problem functions.

Other examples of problems with cheap higher derivatives occur in exponential fitting and factorable programming (for example, see McCormick, 1983).

X. Take special care to check that the problem functions and their derivatives are programmed correctly.

Before embarking on a series of optimization runs, the user should verify that the code which defines the problem functions is correct. One obvious check is to evaluate the problem functions at a point where their values are known.

Errors in programming the function may be quite subtle in that the function value is "almost" correct. For example, the function may be accurate to less than full precision because of the inaccurate calculation of a subsidiary quantity, or the limited accuracy of data upon which the function depends. A common error on machines where numerical calculations are usually performed in double precision is to include even one single-precision constant in the calculation of the function; since some compilers do not convert such constants to double precision, half the correct figures may be lost by such a seemingly trivial error.

Incorrect calculation of derivatives is by far the most common user error. Such errors are almost never small, and thus no algorithm can perform correctly in their presence. This is why we recommend that some sort of consistency check on the derivatives be performed. The most straightforward means of checking for errors in the derivative involves comparing a finite-difference approximation with the supposedly exact value.

1.3. Some useful features of an implementation. In the following, we give some features of an "ideal" implementation that would help the user apply the modelling principles most effectively. This list is not intended to be exhaustive and we have included some items in the list that have not been referred to in the text. We do not claim that it is possible to implement every feature in all circumstances.

- The method should treat simple bounds, linear constraints and nonlinear constraints separately. Moreover, the method should effectively deal with nonlinear constraints, yet remain competitive with unconstrained or linearly constrained methods if constraint nonlinearities are not present.
- There should be the option of computing the problem functions only at points that satisfy the linear constraints.
- The software should print all the information necessary for the user to check how closely the final point satisfies the necessary and sufficient conditions for an optimum.
- The user should be able to specify a feasibility tolerance for each constraint.
- It should be possible to solve problems for which the precision of the problem functions is not necessarily close to the machine precision.
- The user should be able to scale the problem by specifying the scaling matrices D_1 and D_2 of (1.2).
- The user should have the option of checking the problem derivatives before starting the minimization.

2. QUASI-NEWTON METHODS FOR UNCONSTRAINED MINIMIZATION

Quasi-Newton methods are *iterative*, and generate a sequence $\{x_k\}$ that is intended to converge to x^*. At the k-th iteration, the new iterate is defined by

$$x_{k+1} = x_k + \alpha_k p_k, \tag{2.1}$$

where α_k is a non-negative scalar called the *step length*, and p_k is an n-vector called the *search direction*. These methods utilize the values of the problem functions and their gradients at trial iterates, but do not assume the availability of higher derivative information. (When explicit first derivatives are not available, quasi-Newton methods can be implemented using finite-difference approximations to the gradient — see, e.g., Gill, Murray and Wright, 1981.) A typical iteration of a quasi-Newton method comprises three related parts: computation of p_k, choice of α_k, and updating the necessary matrix factorizations.

The search direction p_k in unconstrained optimization is defined by the equation $H_k p_k = -g_k$, where H_k is an approximation to the Hessian matrix of F. Many computational benefits accrue from updating a factorized form of H_k. Suppose that $H_k = R_k^T R_k$, where R_k is an upper-triangular matrix (the Cholesky factor of H_k). The search direction is then obtained by solving two triangular systems:

$$R_k^T q = -g_k \quad \text{and} \quad R_k p_k = q. \tag{2.2}$$

The main purpose of the linesearch is to force steady progress by computing a step length α_k such that $F(x_k + \alpha_k p_k)$ is "sufficiently less" than $F(x_k)$; i.e., the decrease can go to zero only as the solution is approached.

It is widely accepted today that the best quasi-Newton update is given by the *BFGS formula*:

$$H_{k+1} = H_k - \frac{1}{u_k^T s_k} u_k u_k^T + \frac{1}{y_k^T s_k} y_k y_k^T, \tag{2.3}$$

where $s_k = x_{k+1} - x_k$, $u_k = H_k s_k$ and $y_k = g_{k+1} - g_k$. If H_k is positive definite, H_{k+1} as defined by (2.3) will be positive definite if and only if the approximate curvature $y_k^T s_k$ satisfies

$$y_k^T s_k > 0 \tag{2.4}$$

(see Dennis and Moré, 1977).

Dennis and Schnabel (1981) have shown that the BFGS update (2.3) may be expressed as a *rank-one* update to R_k. Let β and γ denote the scalars $(u_k^T s_k)^{\frac{1}{2}}$ and $(y_k^T s_k)^{\frac{1}{2}}$ respectively The BFGS update may then be written as $H_{k+1} = \bar{R}_{k+1}^T \bar{R}_{k+1}$, where

$$\bar{R}_{k+1} = R_k + vw^T, \quad \text{with} \quad v = \frac{1}{\beta} R_k s_k, \quad w = \frac{1}{\gamma} y_k - \frac{1}{\beta} u_k.$$

The matrix $\bar{R}_k$, which is not upper-triangular, may be restored to upper-triangular form by finding an orthogonal matrix P such that

$$P\bar{R}_{k+1} = R_{k+1},$$

where R_{k+1} is upper triangular. Then $H_{k+1} = \bar{R}_{k+1}^T \bar{R}_{k+1} = \bar{R}_{k+1}^T P^T P \bar{R}_{k+1} = R_{k+1}^T R_{k+1}$, as required. A suitable matrix P can be constructed from two sweeps of plane rotations; for more details, see Gill *et al.* (1974).

Since p_k satisfies $H_k p_k = -g_k$, two matrix-vector multiplications may be avoided in the implementation of the BFGS update (2.3). Substituting from (2.2), we obtain

$$v = \frac{1}{\sigma} q \quad \text{and} \quad w = \frac{1}{\gamma} y_k + \frac{1}{\sigma} g_k,$$

where $\sigma = |g_k^T p_k|^{\frac{1}{2}}$ and $\gamma = (y_k^T s_k)^{\frac{1}{2}}$.

Representation of H_k by its Cholesky factors avoids a serious problem that would otherwise arise in quasi-Newton methods: the loss of positive-definiteness through rounding errors. With exact arithmetic, satisfaction of (2.4) should ensure that the BFGS update generates strictly positive-definite Hessian approximations. However, in practice the formula (2.3) can lead to a singular or indefinite matrix H_{k+1}. When H_k is represented by its Cholesky factorization and updates are performed directly to the factorization, every H_k will be *numerically* positive definite.

Maintenance of positive-definiteness is considered to be a crucial element in the success of quasi-Newton methods in unconstrained optimization (see Dennis and Moré, 1977). In this context, the linesearch can also have the important function of guaranteeing that condition (2.4) holds at every iteration. In particular, methods of safeguarded polynomial interpolation (see, e.g., Brent, 1973; Gill, Murray and Wright, 1981) can find a value of α_k that satisfies

$$F(x_k) - F(x_k + \alpha_k p_k) \geq -\mu\alpha_k g_k^T p_k, \tag{2.5a}$$

$$|g(x_k + \alpha_k p_k)^T p_k| \leq -\eta g_k^T p_k, \tag{2.5b}$$

where $0 < \mu \leq \eta < 1$ and $\mu \leq \frac{1}{2}$. Condition (2.5a) ensures a "sufficient decrease" in F, and (2.5b) guarantees satisfaction of (2.4).

A popular alternative linesearch technique is known as *backtracking* (see, e.g., Dennis and Schnabel, 1983). Given a fixed $0 < \rho < 1$, a sequence $\{\beta_j\}$ is generated that satisfies $\beta_0 = 1$ and $\beta_j > \beta_{j+1} \geq \rho\beta_j$. The value of α_k is taken as the first element in the sequence $\{\beta_j\}$ satisfying the sufficient decrease criterion

$$F(x_k) - F(x_k + \beta_j p_k) \geq -\mu\beta_j g_k^T p_k,$$

where $0 < \mu < 1$. Since a backtracking method can never generate a value of α greater than unity, *(2.4) may not hold.* Most implementations of unconstrained quasi-Newton methods with a backtracking linesearch simply skip the update in this case. (As the iterates converge, (2.4) is satisfied for the initial step of unity, and hence the difficulty does not arise in the limit.) The simplicity of backtracking algorithms and their utility in convergence proofs have led to their frequent appearance in the literature.

3. METHODS FOR NONLINEAR EQUALITY CONSTRAINTS

3.1. Basic theory and notation. In this section, we consider methods for problems that contain only nonlinear *equality* constraints, i.e.

NEP
$$\begin{array}{ll} \underset{x\in\Re^n}{\text{minimize}} & F(x) \\ \text{subject to} & c_i(x) = 0, \quad i = 1,\ldots,m. \end{array}$$

We concentrate on this simplified problem in order to emphasize the treatment of nonlinear constraints.

The Kuhn-Tucker conditions for NEP state the existence of an m-vector λ^* (the *Lagrange multiplier vector*) such that

$$g(x^*) = A(x^*)^T\lambda^*. \tag{3.1}$$

(For a detailed discussion of first- and second-order Kuhn-Tucker conditions for optimality, see, for example, Fiacco and McCormick, 1968, and Powell, 1974.)

Let $Z(x)$ denote a matrix whose columns form a basis for the set of vectors orthogonal to the rows of $A(x)$; i.e., $A(x)Z(x) = 0$. An equivalent statement of (3.1) in terms of Z is

$$Z(x^*)^T g(x^*) = 0.$$

The vector $Z(x)^T g(x)$ is termed the *projected gradient* of F at x.

The *Lagrangian function*

$$L(x,\mu) = F(x) - \mu^T c(x),$$

where μ is an m-vector of Lagrange-multiplier estimates, plays an important role in understanding and solving constrained problems. Condition (3.1) is a statement that x^* is a *stationary point* (with respect to x) of the Lagrangian function when $\mu = \lambda^*$. One of the second-order sufficiency conditions for optimality is that the *projected Hessian of the Lagrangian function*,

$$Z(x)^T \nabla^2 L(x,\mu) Z(x) = Z(x)^T \Big(\nabla^2 F(x) - \sum_{i=1}^{m} \mu_i \nabla^2 c_i(x)\Big) Z(x),$$

is positive definite when $x = x^*$, $\mu = \lambda^*$.

In the following, we consider *sequential quadratic programming* (SQP) methods for NEP, in which the search direction is the solution of a *quadratic programming subproblem* and the steplength achieves a sufficient reduction in some "merit function". The purpose of the merit function is to enforce steady progress to the solution by balancing the (usually) conflicting aims of reducing the objective function and satisfying the nonlinear constraints.

The quadratic programming subproblem is of the form:

$$\underset{p \in \Re^n}{\text{minimize}} \quad g_k^T p + \frac{1}{2} p^T H_k p \tag{3.2a}$$

$$\text{subject to} \quad A_k p = -c_k, \tag{3.2b}$$

where c_k and A_k denote c and A evaluated at x_k. The so-called *linearized* constraints (3.2*b*) represent a first-order approximation to the nonlinear constraints of the original problem. The matrix H_k is an approximation to the Hessian of the Lagrangian function. The Lagrange multiplier vector of this subproblem (denoted by μ_k) satisfies

$$H_k p_k + g_k = A_k^T \mu_k,$$

and may be used as an estimate of λ^*.

SQP methods differ in their definitions of the matrix H_k, and, as we shall see later, formulation of the QP constraints (3.2*b*). In the next two sections, we shall see how the choice of matrix H_k is related to the method used to solve the equality-constraint QP (3.2).

3.2. Methods for equality-constraint QP. All methods for solving (3.2) may be viewed as alternative methods for solution of the *augmented system* of equations for p and μ

$$\begin{pmatrix} H & A^T \\ A & \end{pmatrix} \begin{pmatrix} p \\ -\mu \end{pmatrix} = -\begin{pmatrix} g \\ c \end{pmatrix}, \tag{3.3}$$

which expresses the optimality and feasibility conditions. (The subscript k has been suppressed for convenience.)

Methods for solving (3.3) are often based upon constructing an equivalent, but simpler, system. Let S be a nonsingular $(n+m)\times(n+m)$ matrix. The solution of (3.3) is equivalent to the solution of

$$S^T\begin{pmatrix} H & A^T \\ A & \end{pmatrix} S \begin{pmatrix} \bar{p} \\ -\bar{\mu} \end{pmatrix} = -S^T\begin{pmatrix} g \\ c \end{pmatrix} \quad \text{with} \quad \begin{pmatrix} p \\ -\mu \end{pmatrix} = S\begin{pmatrix} \bar{p} \\ -\bar{\mu} \end{pmatrix}. \tag{3.4}$$

We shall consider two commonly used choices for S derived from the LQ factorization of A: an $m\times m$ lower-triangular matrix L and an $n\times n$ matrix Q such that

$$AQ = (\, L \quad 0\,). \tag{3.5}$$

Assume that the columns of Q are partitioned so that

$$Q = (\, Y \quad Z\,),$$

where Y has m columns. Then let S be given by

$$S = \begin{pmatrix} Y & Z & \\ & & I \end{pmatrix}, \tag{3.6}$$

and let p_Y and p_Z denote the first m and last $n-m$ elements of $\bar{p}$, respectively. Substituting from (3.6) into (3.4), we obtain

$$\begin{pmatrix} Y^THY & Y^THZ & L^T \\ Z^THY & Z^THZ & \\ L & & \end{pmatrix}\begin{pmatrix} p_Y \\ p_Z \\ -\mu \end{pmatrix} = -\begin{pmatrix} Y^Tg \\ Z^Tg \\ c \end{pmatrix}.$$

Thus, p and μ may be found by solving the equations

$$Lp_Y = -c \tag{3.7a}$$

$$Z^THZp_Z = -Z^Tg - Z^THYp_Y \tag{3.7b}$$

$$p = Yp_Y + Zp_Z \tag{3.7c}$$

$$L^T\mu = Y^T(g+Hp). \tag{3.7d}$$

Note that the projected Hessian matrix Z^THZ appears explicitly in (3.7b). If (3.2) has a well-defined solution, this matrix is positive definite.

We consider two definitions of L and Q. In the methods of Gill and Murray (1974), Wright (1976), Murray and Wright (1978) and Gill *et al.* (1984c), L and Q are found by explicitly triangularizing A using Householder matrices or stabilized elementary matrices. In this paper, we consider only the use of Householder matrices, in which case the matrix Q is orthogonal.

Computation of the LQ factorization may be viewed as *updating* an existing factorization as new rows are added in the last position. Assume that the LQ factorization (3.5) of A is available, and consider the matrix $\bar{A}$, which is A augmented by the row a^T. Then

$$\bar{A}Q = \begin{pmatrix} A \\ a^T \end{pmatrix} Q \equiv \tilde{L} = \begin{pmatrix} L & 0 \\ t^T & s^T \end{pmatrix}, \tag{3.8}$$

where t and s are the relevant partitions of Q^Ta. Let $\tilde{Q}$ denote a Householder matrix of the form

$$\tilde{Q} = I - \frac{1}{\beta}uu^T,$$

where the vector u and scalar β are chosen to annihilate all but the first element of s, and to leave t unchanged. (For details of how these quantities are defined, see Stewart, 1973.) Then

$$\tilde{L}\tilde{Q} = \begin{pmatrix} L & 0 & 0 \\ t^T & \tau & 0 \end{pmatrix}$$
$$\text{or} \quad \bar{A}\bar{Q} \equiv \begin{pmatrix} \bar{L} & 0 \end{pmatrix}, \quad \text{where} \quad \bar{Q} = Q\tilde{Q}. \tag{3.9}$$

The so-called "standard" LQ factorization is a version of (3.8) and (3.9) in which the rows of A are added to the null matrix one by one. The initial Q matrix is taken as the identity, and the initial L is the null matrix. While computing the factorization, the sequence of Householder transformations is stored *in compact form* (i.e., Q is not stored explicitly); the vector a^TQ needed in (3.8) is obtained by applying the sequence of stored transformations. Once the initial factorization has been completed, the necessary explicit matrix Q is obtained by multiplying the compact Householder transformations together in reverse order.

A second choice for the matrices L and Q involves defining Q so that

$$Q^THQ = I. \tag{3.10}$$

In this case, $Z^THZ = I$ and $Z^THY = 0$, and the equations (3.7) for p and μ become

$$Lp_Y = -c \tag{3.11a}$$
$$p_Z = -Z^Tg \tag{3.11b}$$
$$p = Yp_Y + Zp_Z \tag{3.11c}$$
$$L^T\mu = Y^Tg + p_Y. \tag{3.11d}$$

The recurrence relations (3.8) and (3.9) may be adapted to compute Q satisfying (3.10) by defining the initial Q matrix to be R^{-1}.

Equations (3.11) are used to solve the augmented system in the QP method of Goldfarb and Idnani (1983). Similar techniques have been suggested previously for both positive-definite and indefinite quadratic programs. In the latter case, the relationship $Z^THZ = D$ is maintained instead of (3.10), where the matrix D may be diagonal (Murray, 1971) or block-diagonal (Bunch and Kaufman, 1978).

3.3. Properties of the SQP search direction. It is clear from (3.7*c*) that the search direction is the sum of two vectors: a range-space component p_R ($\equiv Yp_Y$), and a null-space component p_N ($\equiv Zp_Z$). The range-space vector satisfies the underdetermined equations (3.2*b*), and thus defines a step to the linearized versions of the nonlinear constraints. (If the columns of Y are orthogonal, p_R defines the step to the *nearest* point on the linearized constraints.) The null-space component p_N defines the step from $x_k + p_R$ to the minimum of the quadratic model of the Lagrangian function in the subspace orthogonal to the constraint normals. An explicit distinction between the aims of satisfying the constraints and minimizing the Lagrangian function is important because the properties of the equations that define the associated vectors are essentially different. The range-space vector is a *Newton step* in the sense that it is computed using exact derivative information. By contrast, the null-space component is a *quasi-Newton step* defined using approximate derivative information from H_k. The better accuracy of the range-space component implies that $\|c\|$ generally remains smaller than $\|Z^Tg\|$ as the solution is approached. During the final iterations, the behavior of SQP methods is characterized by the relationship

$$\|p_Y\|/\|p_Z\| \rightarrow 0; \tag{3.12}$$

i.e., the final search directions lie almost wholly in the null space of A.

3.4. The definition of H_k. Based on its success in the unconstrained case, the BFGS formula (2.3) seems a logical choice for updating an approximation to the Hessian of the Lagrangian function. However, the definition of the updating formula in the constrained case is complicated by the fact that there is some choice as to *which* matrix should be approximated.

An important feature of the BFGS update in unconstrained optimization is the maintenance of positive-definiteness. In the constrained case, the relevant positive-definite matrix is the *projected* Hessian Z^THZ. Accordingly, the first class of methods that we shall consider is based on maintaining a quasi-Newton approximation H_z to the projected Hessian. There are many closely-related variants of this approach. For example, for linearly constrained problems, see Gill and Murray (1974) and Murtagh and Saunders (1978). For nonlinearly constrained problems, see Murray and Wright (1978), Coleman and Conn (1982), Gabay (1979), and Nocedal and Overton (1982). A typical update for methods in this class is

$$\bar{H}_Z = H_Z - \frac{1}{u_z^Ts_z}u_zu_z^T + \frac{1}{y_z^Ts_z}y_zy_z^T, \tag{3.13}$$

where barred quantities refer to the updated values, $g_Z = Z^Tg$, $y_Z = \bar{g}_Z - g_Z$, $s_Z = Z^T(\bar{x} - x)$ and $u_Z = H_zs_z$. For these projected quasi-Newton methods, the matrix H that defines the QP subproblem (3.2) is ZH_zZ^T, which is positive semi-definite. A common feature of the projected quasi-Newton methods mentioned above is that the null-space component of the search direction is defined from the equations

$$H_Zp_Z = -g_Z$$

in order to avoid the necessity of recurring the matrix $Z^T HY$ (*cf.* (3.7*b*)).

In contrast to the unconstrained case, it is not always possible to choose a step length that guarantees the condition $y_Z^T s_Z > 0$. As a result, there may be iterations in which the quasi-Newton update cannot be performed because of loss of positive-definiteness. However, since (3.12) implies that the search directions will lie in the null space of A as the iterates converge, $y_Z^T s_Z$ is typically positive in the neighborhood of the solution.

An immediate consequence of storing only an approximation to the *projected* Hessian is that the QP multipliers μ cannot be computed, since the full matrix H is not available (*cf.* (3.7*d*)). However, if Q is orthogonal, the least-squares multipliers λ_L at a point x may be calculated from the equations

$$L^T \lambda_L = Y^T g(x).$$

The continuity properties of the associated Z are significant in projected quasi-Newton methods because Z defines the operation of projection. For example, in proving local convergence for algorithms that explicitly utilize Z, it is essential that small changes in x should lead to small changes in Z (see, e.g., Coleman and Conn, 1982a, b; and Nocedal and Overton, 1982). The standard method of computing the LQ factorization does not provide a continuous representation of $Z(x)$ (see Coleman and Sorensen, 1984). However, a continuous representation of Z may be defined using a trivial modification of the relations (3.8) and (3.9). Recall that, given the factorization (3.5) of the matrix A, we require the factors $\bar{L}$ and $\bar{Q}$ of $\bar{A}$ at the point $\bar{x} = x + \alpha p$. Exactly as in the standard factorization, (3.8) and (3.9) can be used to update the factors as the rows of $\bar{A}$ are added to the null matrix one by one, except that the explicit matrix Q *from the previous QP subproblem* is taken as the initial matrix Q in (3.8). Each Householder transformation is then multiplied into Q after the corresponding row has been transformed. With the standard procedure, the Householder transformations can be stored in compact form, but with this approach, each new row of $\bar{A}$ must be transformed by a full orthogonal matrix (which is somewhat more expensive unless some of the constraints are linear). The benefit is that Z is continuous at a point where A has full rank, and approaches a limit when computed at a sequence of points $\{x_k\}$ converging sufficiently fast to a suitable point x^* (see Gill *et al.*, 1985a).

The second major representation of H_k is as a quasi-Newton approximation to the *full* Hessian of the Lagrangian function. (This method is especially appropriate if the calculation of the search direction requires the full Hessian, e.g., the method defined by equations (3.11).) Consider a BFGS-type update of the form

$$\bar{H} = H - \frac{1}{u^T s} u u^T + \frac{1}{v^T s} v v^T, \tag{3.14}$$

where $s = \bar{x} - x$ and $u = Hs$. Since H is meant to approximate the Hessian of the Lagrangian function, a "natural" choice for v would be y_L, the difference in gradients of the Lagrangian function; i.e.,

$$y_L = \bar{g} - g - (\bar{A}^T - A^T)\lambda,$$

with λ taken as the best available multiplier estimate. However, since x^* is not an unconstrained minimum of the Lagrangian function, it may be *impossible*, with *any* linesearch, to find a step length for which $y_L^T s$ is positive. Hence, the update might be skipped at every iteration, which would not only destroy the local convergence properties, but also adversely affect the efficiency of the method away from the solution.

A popular method for dealing with this difficulty is to use y_L as v in (3.14) only when $y_L^T s$ is sufficiently positive; otherwise, v is taken as a perturbed vector $\bar{y}_L$ such that $\bar{y}_L^T s > 0$. A perturbation that we have found to be quite successful in practice is defined as follows. When $y_L^T s < 0$, compute the scalar

$$\bar{\omega} = \frac{-y_L^T s}{s^T(\bar{A}^T\bar{c} - A^T c)}.$$

The quantity in the denominator is an approximation to the curvature of $\|c\|_2^2$, which is positive at x^*. If $\bar{\omega}$ is negative, the update is skipped; otherwise,

$$v = y_L + \omega(\bar{A}^T\bar{c} - A^T c),$$

where ω is any positive scalar such that $\omega > \bar{\omega}$. The motivation for this choice of v is the result (due to Boggs, Tolle and Wang, 1982) that a necessary condition for q-superlinear convergence is that the approximate Hessian matrices must satisfy

$$\lim_{k\to\infty} \frac{\|Z_k Z_k^T(H_k - \nabla^2 L(x^*, \lambda^*))p_k\|}{\|p_k\|} = 0.$$

3.5. Inconsistent and ill-conditioned constraints. In the preceding discussion of SQP methods, we have assumed that the equations (3.2*b*) are *consistent*. When all the constraints of the original problem are linear, the subproblem constraints can be inconsistent only if the original problem has no solution. With nonlinear constraints, however, *the constraints of the subproblem may be inconsistent even when the original problem has a well-posed solution.* Techniques for dealing with inconsistent constraints fall into two major categories. In both cases, the search direction satisfies a *shifted* set of linear constraints

$$Ap = r - c. \tag{3.15}$$

The first approach generates a search direction designed to minimize a weighted combination of the quadratic approximation to the Lagrangian and the residual vector of the unperturbed constraints (the vector $c + Ap$). The compromise is achieved by choosing p as the solution of an unconstrained problem of the form

$$\underset{p}{\text{minimize}}\ \sigma\left(g^T p + \tfrac{1}{2}p^T H p\right) + \Phi(c + Ap), \tag{3.16}$$

where $\Phi(r)$ is a scalar-valued function that measures the "size" of r, and σ is a non-negative scalar. The theoretical basis of this class of method is that a minimum of (3.16) always exists, even when a solution of the QP subproblem (3.2) does not.

The algorithms of Biggs (1972a), Bartholomew-Biggs (1982) and Fletcher (1981) correspond to choosing Φ as the two-norm and one-norm, respectively. In either case, it can be shown that the unconstrained minimizer of (3.16) is the solution of a QP with objective function (3.2a) and constraints (3.15). The form of the vector r depends on the definition of Φ. If $\Phi(r)$ is defined as $\frac{1}{2}\|r\|_2^2$, all the components of r are generally non-zero, and p is defined by equations similar to (3.11) (see Bartholomew-Biggs, 1982, for more details). If $\Phi(r) = \|r\|_1$, p is the solution of an ℓ_1-QP (see Fletcher, 1981). At the solution of this problem, a subset of the original constraints (3.2b) will be satisfied exactly (i.e., the corresponding components of r will be zero). This approach therefore implicitly "discards" some of the violated constraints from the subproblem.

The methods of Bartholomew-Biggs and Fletcher are based on the properties of penalty functions (see Fletcher, 1983, for a survey). A feature of these treatments of inconsistent constraints is that σ is always non-zero in (3.16). An alternative approach is to define the composite function only when the constraints are found to be inconsistent. Other SQP methods with a strategy of this type have been suggested by Powell (1977), Schittkowski (1983), Tone (1983) and Gill *et al.* (1984b).

The second technique for the treatment of inconsistency is based on the observation that only the range-space portion of the QP search direction is ill-defined when the constraints are incompatible. With this approach, a well-defined procedure is used to compute p_R, and p_Z and p are obtained from (3.7b) and (3.7c). The most straightforward application of this approach is to define p_R as a solution of the linear least-squares problem

$$\underset{p_R}{\text{minimize}} \quad \|c + Ap_R\|_2, \tag{3.17}$$

which gives r as the *smallest* constraint shift (in the least-squares sense). This choice of p_R is equivalent to computing the first iterate of the Gauss-Newton method for minimizing the two-norm of the *nonlinear* constraint violations. Therefore, the strategy for dealing with the constraints has changed from the possibly unsolvable problem of finding a point such that $c(x) = 0$ to the always solvable problem of minimizing $\sum c_i^2(x)$.

Since A must be rank-deficient when the constraints are inconsistent, the solution of (3.17) is not unique. A suitable choice of p_R in this case is the minimum-length solution, which can be computed using the *complete orthogonal factorization:*

$$PAQ = \begin{pmatrix} L & 0 \\ 0 & 0 \end{pmatrix},$$

where P and Q are orthogonal matrices and L is a lower-triangular matrix whose dimension is equal to the rank of A.

Unfortunately, neither of these techniques resolves the difficulties caused by constraints that are *almost* inconsistent (i.e., ill-conditioned). Ill-conditioning in A will tend to cause p_R to be large in norm (see (3.7a)). In these situations it is necessary in practice to limit the norm of p. It might appear that the least-length solution of (3.17) would automatically be satisfactory. However, the computation of p using the complete orthogonal factorization involves serious practical difficulties — in particular, a strategy must be included for estimating the rank of A. It is well known that the definition of numerical "rank" is problem-dependent. The rank can never be determined without making an explicit judgment about scaling, i.e., a decision as to which quantities can be considered "negligible". The choice of rank is critical in the Gauss-Newton method because a slight alteration in the value of the tolerance used to estimate the rank may lead to completely different behavior.

If the composite function (3.16) is used, an *explicit* bound on the norm of p may be enforced by temporarily imposing additional constraints on the problem. (This type of procedure is used within trust-region algorithms for unconstrained optimization.) The effect of the trust-region constraints is to modify (implicitly or explicitly) the derivative information that defines the search direction. For example, if a temporary bound is placed on the two-norm of p, the search direction satisfies equations in which the second-derivative approximation is modified by a multiple of the identity matrix. Thus, $Z^T H Z$ and A are implicitly modified — an unfortunate result, since we would prefer the projected Hessian approximation to be independent of the conditioning of A.

The development of stable robust methods for dealing with ill-conditioned constraints is still an active area of research. One possible approach was suggested in Section 1.2, where small perturbations of constraints were used to resolve inconsistencies caused by modelling inaccuracies. By changing the constraints (3.2b) to suitable perturbed inequality constraints, (3.2) always has a bounded solution. For example, consider defining p_R as the solution of the *inequality-constrained* quadratic program

$$\begin{array}{ll} \underset{p_R \in \Re^n}{\text{minimize}} & \frac{1}{2} p_R^T p_R \\ \text{subject to} & -\delta \le A p_R + c \le \delta, \end{array} \qquad (3.18)$$

where δ is a vector of small quantities that are forced to approach the feasibility tolerances for the original constraints as x approaches x^*. The subset of constraints active at the solution of (3.18) may then be used to define Z and Y, from which p_Z and p can be computed using (3.7).

4. METHODS FOR NONLINEAR INEQUALITY CONSTRAINTS

4.1. Background. In the final problem to be considered, all the constraints are nonlinear *inequalities*:

NIP
$$\begin{array}{ll} \underset{x \in \Re^n}{\text{minimize}} & F(x) \\ \text{subject to} & c_i(x) \ge 0, \quad i = 1, \ldots, m. \end{array}$$

We consider this simplified form in order to concentrate on the treatment of nonlinear inequality constraints.

Let the matrix $A(x)$ denote the Jacobian of $c(x)$. The constraint c_i is said to be *active* at x if $c_i(x) = 0$, and *violated* if $c_i(x) < 0$. The Kuhn-Tucker conditions for NIP are similar to those for the equality-constraint case, except that they involve only constraints active at x^*, and impose a sign restriction on the Lagrange multipliers. The major difference between inequality- and equality-constrained problems is that the set of constraints active at the solution is *unknown* in the inequality case. Therefore, algorithms for NIP must include a procedure (termed an *active-set strategy*) that determines the correct active set — usually, by maintaining a *working set* that estimates the active set. In this section we discuss the additional algorithmic complexity in SQP methods that arises specifically from the presence of inequality constraints.

4.2. Formulation of the QP subproblem. Broadly speaking, two extreme types of QP subproblems can be posed when solving inequality-constrained problems. The first — called an *IQP strategy* — corresponds to representing *all* nonlinear inequality constraints as *inequalities* in the QP subproblem; this has been by far the most widely used formulation in published SQP methods. The *standard IQP search direction* is the solution of

$$\underset{p\in\Re^n}{\text{minimize}} \quad g^Tp + \tfrac{1}{2}p^THp \tag{4.1a}$$

$$\text{subject to} \quad Ap \geq -c, \tag{4.1b}$$

where H is an approximation to the Hessian of the Lagrangian function and A is the Jacobian of $c(x)$ evaluated at the current iterate. In general, the solution of (4.1) must be found by iteration. Thus, the structure of an SQP method with an IQP strategy involves *major* and *minor* iterations — the minor iterations being those of the quadratic programming method.

Because (4.1) includes all the constraints of NIP, it is convenient to take *the active set of the QP* as a prediction of the active set of the nonlinearly constrained problem. The theoretical justification for this strategy is that the QP (4.1) will make a correct prediction of the active set of the original problem in a neighborhood of x^* for *any* bounded positive-definite matrix H (Robinson, 1974). Furthermore, the multipliers of (4.1) approach the multipliers of NIP as the iterates converge to x^*, and hence it is common to take the QP multipliers as the next multiplier estimate.

The second extreme form of subproblem in SQP methods involves a QP with only *equality* constraints. In order to use an *EQP strategy*, some determination must be made before posing the QP as to which constraints are to be included. An EQP method should have the property that it will select the correct active set in some neighborhood of x^*. Therefore, such methods tend to choose constraints that satisfy properties of the active constraints in a neighborhood of the solution — e.g., are "small" in magnitude, or satisfy the sign requirements for the Lagrange multiplier estimates. A benefit of an EQP method is that, in general, the subproblem will be *easier to solve* than one with inequality constraints.

To a large extent, the active-set strategy will determine the choice of quasi-Newton update and Lagrange multiplier estimate. For example, if an IQP strategy is used, the method used to solve (4.1) will require specification of the full matrix H. On the other hand, an EQP strategy is usually implemented with an approximation to the projected Hessian. The following table summarizes the major features of the two active-set strategies.

EQP	IQP
• QP subproblem: $\underset{p\in\Re^n}{\text{minimize}} \quad g^Tp + \frac{1}{2}p^THp$ subject to $\hat{A}p = -\hat{c}$.	• QP subproblem: $\underset{p\in\Re^n}{\text{minimize}} \quad g^Tp + \frac{1}{2}p^THp$ subject to $Ap \geq -c$.
• Least-squares multipliers.	• QP multipliers.
• Projected Hessian approximation.	• Full Hessian approximation.

These two active-set strategies are the extremes of a whole range of possibilities. Other methods have been defined that use features from both approaches. For example, the method of Schittkowski (1981) solves a QP subproblem of the form (4.1), but uses a pre-assigned working set to specify which of the constraint gradients should be recomputed for the next major iteration. Similarly, it is possible to treat linear constraints with an EQP approach and nonlinear constraints with an IQP approach (see Gill *et al.*, 1984d).

It is important to note that IQP methods can be implemented so that, as the solution is approached, the amount of overhead per major iteration is the same as for an EQP method (i.e., solution of a single set of equations of the form (3.3)). This can be achieved by solving (4.1) with a QP method that allows the active set from one subproblem to be used to initialize the next. Since the active set of the subproblem eventually becomes the correct active set for the nonlinear problem, *QP subproblems near the solution reach optimality in only one minor iteration.*

4.3. Active-set strategies in quadratic programming methods. Quadratic programming methods for problems with inequality constraints solve a sequence of problems (3.3) in which the constraints in the working set are treated as *equalities*. The major differences among QP methods arise from the numerical procedures for solving the associated linear equations, and the strategies that control changes in the working set. (Modern QP methods are surveyed by Fletcher (1981) and Gill, Murray and Wright (1981).)

We briefly outline two methods of selecting constraints for the working set when solving (4.1). At each iteration, let p and λ denote the current estimates of the solution and optimal multiplier vector, and let r denote the residual vector $r \equiv Ap + c$. The "hat" notation indicates quantities associated with the working set. The vector δp is defined such that $p + \delta p$ is the solution of a QP with the original objective function, subject to the constraints of the working set held at equality, and $\delta\lambda$ denotes the corresponding change in the multiplier estimate. As indicated in Section 3.2, δp and $\delta\hat{\lambda}$ are the solution of the linear system

$$\begin{pmatrix} H & \hat{A}^T \\ \hat{A} & 0 \end{pmatrix} \begin{pmatrix} \delta p \\ -\delta\hat{\lambda} \end{pmatrix} = -\begin{pmatrix} g_Q - \hat{A}^T\hat{\lambda} \\ \hat{r} \end{pmatrix}, \tag{4.2}$$

where g_Q denotes $g + Hp$, the gradient of the quadratic function. The algorithms considered here always attempt to move from the minimum on one working set to the minimum on another by taking steps of the form $p+\delta p$ and $\lambda+\delta\lambda$. However, the maintenance of certain properties of the working set can cause a step α $(0 \le \alpha < 1)$ to be taken, where α depends upon the active-set strategy being used.

In an *active-set feasible-point QP method*, p is feasible ($r_i \ge 0$ for all i), but λ is not dual-feasible (i.e., $\lambda_i < 0$ for at least one i). Changes in the working set are designed to maintain feasibility of p, but to move interior to constraints that have negative Lagrange multipliers. At a typical iteration, the working set comprises the constraints satisfied exactly at p (i.e., $\hat{r} = 0$ in (4.2)). If $p+\delta p$ remains feasible (i.e., $r_i + a_i^T\delta p \ge 0$ for i not in the working set), then the full step of unity is permitted. A constraint with a negative multiplier (usually, the *most* negative) is then deleted from the working set. Otherwise, α is taken as the smallest step such that the residual of a constraint not in the working set becomes zero at α, and the corresponding constraint is *added* to the working set. For more details concerning the implementation of feasible-point quadratic programming methods, see Gill and Murray (1978) and Gill *et al.* (1984a, 1985b).

The second strategy is typical of *dual-feasible active-set methods*. In these methods, p is not feasible (i.e., some $r_i < 0$) but λ is always dual-feasible (all $\lambda_i \ge 0$). Changes in the working set are designed to maintain non-negative multipliers while moving to satisfy the violated constraints. At the beginning of a typical iteration, all the constraints in $\hat{A}$ are satisfied exactly except one (i.e., $\hat{r}$ is a multiple of a unit vector in (4.2)). The step length α is taken as one if $p+\delta p$ is dual-feasible (i.e., $\lambda_i + \delta\lambda_i \ge 0$). Otherwise, α is chosen as the largest step such that $\lambda_i + \alpha\delta\lambda_i = 0$ for an index i in the working set, and the corresponding constraint is deleted from the working set. After a unit step is taken, a constraint with a negative residual (usually, the most negative) is added to the working set. (Note that we have given a considerably simplified description of the dual-feasible iteration in order to emphasize the similarities between dual- and primal-feasible methods. In practice, δp and $\delta\lambda$ are not found directly from (4.2) because the new constraint may be dependent on the constraints already in the working set.) For further information concerning the implementation of dual-feasible quadratic programming methods, see Goldfarb and Idnani (1983) and Powell (1983b).

For both of these active-set strategies, each change in the working set leads to a simple change to $\hat{A}$, which in turn leads to a change in the factorizations used to solve (4.2).

Both of the active-set strategies described require an *initialization procedure* to obtain an initial primal- or dual-feasible point. As noted above, for efficiency within an SQP method, it is critical that this procedure should be able to utilize a pre-assigned working set.

The initialization procedure for the primal feasible-point method is equivalent to a linear programming problem. Consider the sum of infeasibilities

$$v(p) = -\sum_{a_i^T p + c_i < 0} (a_i^T p + c_i).$$

Note that $v(p)$ is a linear function that is zero at any feasible point, and positive at an infeasible point. Therefore, a feasible point can be found by minimizing $v(p)$, subject to continuing to satisfy the constraints with positive residuals at p. The function $v(p)$ may be minimized using an active-set strategy that is almost identical to that of the feasible-point active-set method. The principal differences are that the search direction is defined as $-ZZ^T\nabla v(p)$, and α is chosen as $\min(\alpha_1, \alpha_2)$, where α_1 is the maximum step that can be taken without violating one of the constraints that is currently satisfied, and α_2 reaches the furthest constraint along δp that is currently violated. (Several violated constraints may become satisfied during a single iteration.) For efficiency, the implementation of this procedure should reflect the similarity of the linear algebraic computations associated with iterations in both the feasibility and QP phases — in particular, each iteration involves an update of the same factorization of the working set. The computations in both phases may be performed by exactly the same program modules. The two-phase nature of the algorithm is reflected by changing the function being minimized from the sum of infeasibilities to the quadratic objective function. An important feature of this type of implementation is that if the pre-assigned working set is similar to the active set, just a few changes in working set are necessary to achieve feasibility. In particular, if the initial point is feasible, the procedure merely computes all the relevant factorizations (which are also needed for the QP iterations) and performs a feasibility check.

If a dual-feasible active-set strategy is used, the following initialization procedure may be employed. The procedure is based on finding a subset of the pre-assigned working set on which the multipliers are positive. First, the minimum of the quadratic on the pre-assigned working set is computed by solving (4.2) with $p = 0$ and $\lambda = 0$. If the $\delta\lambda_i$ are non-negative, the initial point can be taken as $p = \delta p$ and $\lambda = \delta\lambda$. Otherwise, a constraint with a negative multiplier is deleted, the factorizations are updated and (4.2) is solved again. This process is repeated until all the multipliers are non-negative or the working set is empty, in which case $p = \delta p$ and $\lambda = \delta\lambda$ define the required initial point. (Note that the unconstrained minimum is trivially dual feasible.)

An alternative initialization procedure is to start at the unconstrained minimum and give preference to adding the pre-assigned constraints. However, if the pre-assigned working set is similar to the active set, this scheme is likely to require more work than the procedure above. First, more operations are required to compute the factorizations by updating. Second, even if the pre-assigned working set defines the optimal feasible point, the number of QP iterations may not be equal to the dimension of the optimal working set, since it cannot be guaranteed that the multipliers will remain dual-feasible during the intermediate iterations.

4.4. Conditioning of the working set. One of the most important issues in the implementation of QP algorithms is *robustness*. During the solution of a nonlinear problem, quadratic subproblems of wildly varying degrees of difficulty are generated "automatically". Even if the original nonlinear problem is well-conditioned in the neighborhood of the solution, the QP subproblems of the early iterations may be very badly behaved. The most common difficulties include singular or nearly rank-deficient Jacobians, subproblems with very small feasible regions and severely ill-conditioned Hessian matrices. (For examples, see Section 5.)

One of the most critical features of a QP implementation is the strategy for maintaining a well-conditioned working set. The spectral condition number of $\hat{A}$ provides a measure of the degree of independence of the constraints in the working set. This number (the ratio of the largest to smallest singular values of $\hat{A}$) will decrease when a constraint is deleted from the working set and increase when a constraint is added. The worst case occurs when the new working set is singular, i.e., an attempt is made to add a constraint that is dependent on constraints already in the working set. However, if a near-dependent constraint is added to the working set, the condition number may increase substantially. Accordingly, it is important that the constraint-selection procedure should consider the condition number of the new working set.

With exact arithmetic and a non-singular initial working set, the active-set strategy described above for the primal-feasible method would never generate a singular working set. To see why, recall that $\hat{A}\delta p = 0$ at every iteration. Thus, δp will never intersect a constraint that is exactly linearly dependent upon $\hat{A}$. In practice, of course, the difficulty arises when the candidate constraints are *nearly* dependent. Determination of the condition number requires the singular values of $\hat{A}$, which would be too expensive to compute. Instead, a QP method can use an inexpensive *condition estimator* — for example, the ratio of the largest to smallest diagonals of the *LQ* factor L (see (3.5)), which is a *lower bound* on the condition number of $\hat{A}$.

Exercising control over this condition estimator turns out to particularly easy in a primal-feasible active-set method, if it is acceptable to violate constraints by a small tolerance. Suppose that each constraint has an associated user-defined tolerance that specifies the maximum permissible constraint violation. Let α_m denote the maximum step at which $p + \alpha_m \delta p$ does not violate any constraint by more than its feasibility tolerance. All constraints at distance α ($\alpha \leq \alpha_m$) from the current point are then viewed as acceptable candidates for inclusion in the working set. A criterion that we have found to be particularly successful in practice (due to Harris, 1973) is to add the constraint whose normal makes the largest angle with the search direction. In the case where the null space of $\hat{A}$ is of dimension one (for example, in the simplex method for linear programming), this choice gives the smallest condition estimator over the candidate set.

An unsatisfactory feature of the Harris scheme is that all the constraints active at the solution tend to be *violated* by their feasibility tolerances, even when the final active set is not ill-conditioned. However, this idea can be generalized so that constraint violations by δ are permitted when necessary to improve the conditioning of the working set, but an attempt is also made to minimize the constraint violations. With this strategy, the constraints active at the solution tend to be satisfied *exactly* rather than violated (see Gill *et al.*, 1985b). (An interesting result is that *negative* steps are sometimes necessary.)

5. SAMPLE RUNS

In this section we shall use several examples to illustrate the performance of nonlinear programming software on practical problems. All the problems were solved in double precision on an IBM 3081 using the VS Fortran compiler with optimization level 3.

5.1. A comparison of IQP and EQP methods. The purpose of the first set of runs is to illustrate the properties of methods based on the EQP and IQP active-set strategies. Two specific methods for *linearly* constrained optimization are considered. In order to aid the comparison, the methods have several features in common. Both methods recur an orthogonal factorization of the constraints in the working set and begin by computing a feasible point. Both methods treat simple bounds and linear constraints separately.

The first method is a standard IQP method for linearly constrained optimization. The Hessian of each QP subproblem is a positive-definite BFGS approximation to the full Hessian of the objective function. At each iteration, a steplength is computed that satisfies the linesearch conditions (2.5a) and (2.5b). Each QP subproblem is solved using a feasible-point active-set method with an orthogonal factorization of the constraints in the working set. The final working set from the QP of one iteration is used as the initial working set for the next. For additional details concerning the Fortran implementation, the reader is referred to Gill *et al.* (1984b).

In our discussion, we shall refer to the output printed during the run. A single line of output is printed at the end of each major iteration. The major iteration number is given in the first column (marked "ITN"). The next column "ITQP" gives the number of minor iterations needed to solve the QP subproblem. The "STEP" column gives the step α_k taken along the computed search direction. "NUMF" is the total number of evaluations of the problem functions. "OBJECTIVE" is the value of the objective function, $F(x_k)$. Columns "BND" and "LC" give the numbers of simple-bound constraints and general linear constraints in the working set. "NZ" is the dimension of the null space of the current matrix of constraints in the working set. The next five entries give information about the derivatives of the problem at the current point. "NORM GF" is the two-norm of the free components of the objective gradient g_k, and "NORM GZ" is the two-norm of $Z_k^T g_k$. "COND H", "COND HZ" and "COND T" are estimates of the condition numbers of the Hessian, projected Hessian and matrix of constraints in the working set. "CONV" is a set of four logical variables C_1, C_2, C_3 and C_4, used to inform the user of the quality of the current estimate of the solution, with the following meanings. C_1 is true if the projected-gradient norm is small; C_2 is true if constraints are satisfied to within the user-specified tolerance; C_3 is true if the signs of the multipliers indicate optimality; and C_4 is true if the last change in x was small. Finally, in some of the runs an "S" is printed as the last item of the iteration summary. This indicates that it was necessary to skip the BFGS update to the approximate Hessian.

The second implementation is based on an EQP active-set strategy used in conjunction with a BFGS approximation of the projected Hessian Z^THZ. The EQP is solved using the orthogonal LQ factorization and the Cholesky factorization of the projected Hessian (see (3.3) and (3.5)). If the objective function is decreasing at the step to the nearest satisfied constraint, the constraint is added to the working set. Lagrange multipliers are computed when the projected-gradient norm is less than some loose tolerance. If the smallest multiplier is negative, the corresponding constraint is deleted from the working set.

The iteration summary is the same as that for the IQP method except that only the condition estimate of the *projected* Hessian is printed (the full Hessian is not recurred), and additional information is printed about each change to the working set. When the status of a constraint changes, the index of the constraint is printed, along with the designation "L" (lower bound), "U" (upper bound) or "E" (equality). Indices 1 through "N" refer to the bounds on the variables, and the remaining indices refer to the general constraints. "KDEL" and "KADD" denote the indices of the constraints leaving and entering the working set. If an entry in one of these columns is zero, a constraint was not deleted or added. "MIN LM" is the multiplier associated with the constraint just deleted. If no constraint was deleted during the relevant iteration, the entry in this column is "--". The information printed in the "CONV" column is different from that given in the IQP method. C_1 is true if the projected-gradient norm is smaller than some loose tolerance; C_2 is true if the projected-gradient norm is smaller than some tighter tolerance; C_3 is true if the change in the objective function was sufficiently small; and C_4 is true if the change in x is small. Note that the loose tolerance on the projected-gradient norm must be satisfied before any multipliers are computed and any constraint is deleted.

Figure 1 gives the output from runs on a well behaved seven-variable linearly constrained problem with seven general linear constraints and upper and lower bounds on the variables. At the solution, two bounds and three general linear constraints are active. A comparison of the output from the two runs will illustrate one of the major differences between IQP and EQP active-set strategies — the different pattern of changes to the working set. In an EQP method, the problem functions are computed after every step along a search direction. Moreover, a constraint will be deleted from the working set only when the current point is considered to be sufficiently close to the minimum on the current working set (albeit to a very low accuracy). In other words, a constraint will be deleted only when the method has accumulated sufficient information about the *curvature* of the problem on the current working set. By constrast, many constraints may be added or deleted during a single major iteration of an IQP method. In the run given in Figure 1a, the IQP strategy finds a very close approximation to the correct active set during the *first* subproblem.

During the final few iterations, the methods are essentially identical, with a single minor iteration being performed in the IQP method. Note that the asymptotic superlinear convergence rate is evident from the unit steplengths in the "STEP" column and the sequence of converging "NORM GZ" entries. This column may be used to verify the convergence to a local minimum (see Gill, Murray and Wright, 1981).

5.2. Are IQP methods superior to EQP methods? It is a popular myth that the more "opportunistic" IQP active-set strategy will tend to find the correct active set faster than an EQP strategy. The next two problems are intended to demonstrate that this is not always the case. On some problems the IQP method will be faster, on others it will be slower.

In Figures 2a and 2b we give the IQP and EQP results for the minimization of the six-variable nonlinear test function "Exp 6" subject to simple-bound constraints upon the variables

ITN	ITQP	STEP	NUMF	OBJECTIVE	BND	LC	NZ	NORM GF	NORM GZ	COND HZ	COND H	COND T	CONV
0	5	0.0D-01	1	1.2127D 06	3	3	1	1.7D 03	8.29D 02	1.D 00	1.D 00	1.D 01	FFTT
1	1	5.2D-01	3	9.5919D 05	3	3	1	1.1D 03	1.14D 02	1.D 00	2.D 00	1.D 01	FFTF
2	2	1.0D 00	4	9.4923D 05	2	3	2	1.1D 03	2.50D 02	1.D 00	3.D 00	1.D 01	FTFF
3	1	1.0D 00	5	9.3120D 05	2	3	2	1.1D 03	7.19D 01	1.D 00	2.D 00	1.D 01	FTTF
4	1	1.0D 00	6	9.2960D 05	2	3	2	1.1D 03	1.55D 00	1.D 00	3.D 00	1.D 01	TTTF
5	1	1.0D 00	7	9.2960D 05	2	3	2	1.1D 03	5.15D-02	1.D 00	3.D 00	1.D 01	TTTF
6	1	1.0D 00	8	9.2960D 05	2	3	2	1.1D 03	1.23D-04	1.D 00	3.D 00	1.D 01	TTTF
7	1	1.0D 00	9	9.2960D 05	2	3	2	1.1D 03	5.97D-07	1.D 00	3.D 00	1.D 01	TTTT

EXIT NP PHASE. INFORM = 0 MAJITS = 7 NFEVAL = 9

Figure 1a. Results of an IQP method on a seven-variable linearly constrained problem.

ITN	JDEL	JADD	STEP	NUMF	OBJECTIVE	BND	LC	NZ	NORM GZ	MIN LM	COND HZ	COND T	CONV
0	0	0	0.0D-01	1	1.2127D 06	2	3	2	2.59D 02	--	1.D 00	7.D 01	T FFF
1	5L	14L	1.0D-02	2	1.1945D 06	1	4	2	9.78D 02	-1.5D 03	1.D 00	7.D 01	F FTFS
2	0	3L	1.6D-01	3	1.0755D 06	2	4	1	3.26D 02	--	1.D 00	7.D 01	T FTF
3	11U	0	4.9D-01	5	9.5223D 05	2	3	2	1.73D 02	-2.8D 03	1.D 00	1.D 01	T FTF
4	0	0	1.0D 00	6	9.4357D 05	2	3	2	2.21D 01	--	1.D 00	1.D 01	T FTF
5	3L	0	1.0D 00	7	9.3434D 05	1	3	3	1.62D 02	-1.4D 02	8.D 00	2.D 01	T FTF
6	0	4L	1.8D-01	8	9.2997D 05	2	3	2	3.79D 01	--	1.D 00	1.D 01	F FTFS
7	0	0	1.0D 00	9	9.2960D 05	2	3	2	3.90D 00	--	1.D 00	1.D 01	F FTF
8	0	0	1.0D 00	10	9.2960D 05	2	3	2	1.36D-03	--	1.D 00	1.D 01	T TTF
9	0	0	1.0D 00	11	9.2960D 05	2	3	2	2.19D-06	--	1.D 00	1.D 01	T TTT

EXIT LC PHASE. INFORM = 0 ITER = 9 NFEVAL = 11

Figure 1b. Results of an EQP method on a seven-variable linearly constrained problem.

(see Biggs, 1972b). The constraints are the vector of simple lower bounds $(0.5, 9, 0.9, 4, 3, 2)$ and simple upper bounds $(\infty, \infty, \infty, \infty, 4.9, \infty)$. The solution lies at the point $(1, 10, 1, 5, 4, 3)$, where no simple bounds are active. (For brevity, the output from some of the less important iterations has been omitted.)

Exp 6 is a problem for which the IQP method is substantially faster than the EQP method. Since no bounds are active at the solution, the ability of the quadratic subproblem to drop many of them during one major iteration allows the IQP method to identify the correct working set rapidly. By the seventh iteration, all the bounds have been deleted from the working set. An adequate solution is identified at iteration 32. By contrast, the requirement of some significant reduction in the projected gradient on each working set considerably slows down the rate at which the EQP method can delete constraints. It is not until iteration 55 that the correct working set is identified.

Cases where the more conservative strategy of the EQP method gives better performance are illustrated by the problem "Weapon" (see Bracken and McCormick, 1968). The objective function is

$$F(x) = \sum_{j=1}^{20} u_j \Big(\prod_{i=1}^{5} a_{ij}^{x_{ij}} - 1 \Big),$$

which is minimized subject to twelve general linear constraints and bounds on all the variables. At the solution, 75 bounds and seven linear constraints are active.

When applied to this problem, both methods require approximately the same number of

ITN	ITQP	STEP	NUMF	OBJECTIVE	BND	LC	NZ	NORM GF	NORM GZ	COND HZ	COND H	COND T	CONV
0	3	0.0D-01	1	1.8581D-01	2	0	4	1.1D 00	1.13D 00	1.D 00	1.D 00	1.D 00	FTTT
1	2	1.1D-01	3	1.1602D-01	1	0	5	6.4D-01	6.39D-01	9.D 00	4.D 00	1.D 00	FTTF
2	1	1.0D 00	4	6.1503D-02	1	0	5	3.2D-01	3.21D-01	8.D 00	4.D 00	1.D 00	FTFF
3	1	1.0D 00	5	1.6543D-02	1	0	5	1.2D-01	1.17D-01	1.D 01	4.D 00	1.D 00	FTTF
4	1	1.0D 00	6	1.2345D-02	1	0	5	4.6D-02	4.60D-02	1.D 01	4.D 00	1.D 00	FTTF
5	1	1.0D 00	7	1.1395D-02	1	0	5	3.2D-02	3.17D-02	9.D 01	2.D 01	1.D 00	FTTF
6	2	1.0D 00	8	5.6521D-03	0	0	6	1.3D-01	1.34D-01	1.D 02	1.D 01	1.D 00	FTTF
7	1	1.0D 00	9	4.8169D-03	0	0	6	4.0D-02	4.05D-02	9.D 01	1.D 01	1.D 00	FTTF
8	1	1.0D 00	10	4.5829D-03	0	0	6	1.4D-02	1.40D-02	8.D 01	9.D 00	1.D 00	FTTF
9	1	1.0D 00	11	4.4447D-03	0	0	6	1.5D-02	1.48D-02	1.D 02	2.D 01	1.D 00	FTTF
27	1	1.0D 00	29	1.1263D-07	0	0	6	6.2D-04	6.16D-04	7.D 03	7.D 02	1.D 00	FTTF
28	1	1.0D 00	30	7.8278D-09	0	0	6	8.1D-05	8.08D-05	6.D 03	7.D 02	1.D 00	FTTF
29	1	1.0D 00	31	1.8175D-09	0	0	6	3.7D-05	3.66D-05	6.D 03	8.D 02	1.D 00	FTTF
30	1	1.0D 00	32	1.1704D-09	0	0	6	1.8D-05	1.76D-05	7.D 03	7.D 02	1.D 00	FTTF
31	1	1.0D 00	33	1.1161D-09	0	0	6	4.3D-06	4.31D-06	7.D 03	7.D 02	1.D 00	TTTF
32	1	1.0D 00	34	1.1144D-09	0	0	6	4.0D-07	3.95D-07	7.D 03	7.D 02	1.D 00	TTTT

EXIT NP PHASE. INFORM = 0 MAJITS = 32 NFEVAL = 34

Figure 2a. Results of an IQP method on the six-variable problem Exp 6.

ITN	JDEL	JADD	STEP	NUMF	OBJECTIVE	BND	LC	NZ	NORM GZ	MIN LM	COND HZ	COND T	CONV
0	0	0	0.0D-01	1	1.8581D-01	4	0	2	8.17D-01	--	1.D 00	1.D 00	F FFF
1	0	0	1.3D-01	3	1.4147D-01	4	0	2	5.02D-01	--	1.D 00	1.D 00	T FTF
2	0	0	1.0D 00	4	2.3549D-02	4	0	2	4.81D-02	--	1.D 00	1.D 00	T FTF
3	6L	0	3.3D-01	6	2.3056D-02	3	0	3	2.56D-02	-5.2D-02	2.D 00	1.D 00	T FTF
4	5L	0	1.0D 00	7	2.1942D-02	2	0	4	7.82D-02	-4.0D-02	1.D 01	1.D 00	F FTF
5	0	0	1.0D 00	8	1.9597D-02	2	0	4	1.14D-01	--	1.D 02	1.D 00	F FTF
6	0	0	1.0D 00	9	1.2396D-02	2	0	4	3.46D-01	--	6.D 01	1.D 00	F FTF
7	0	0	1.0D 00	10	8.1084D-03	2	0	4	1.20D-01	--	4.D 01	1.D 00	F FTF
8	0	0	1.0D 00	11	6.7024D-03	2	0	4	6.57D-02	--	4.D 01	1.D 00	F FTF
9	0	0	1.0D 00	12	6.1975D-03	2	0	4	4.12D-03	--	3.D 01	1.D 00	T FTF
10	4L	0	1.0D 00	13	6.1445D-03	1	0	5	9.28D-03	-7.1D-03	2.D 01	1.D 00	F FTF
11	0	0	1.0D 00	14	5.9625D-03	1	0	5	1.25D-02	--	6.D 01	1.D 00	F FTF
12	0	0	1.0D 00	15	5.5139D-03	1	0	5	2.17D-02	--	2.D 02	1.D 00	F FTF
35	0	0	1.0D 00	39	1.1179D-05	1	0	5	3.10D-03	--	8.D 03	1.D 00	F FTF
36	0	0	1.0D 00	40	1.0212D-05	1	0	5	1.14D-03	--	9.D 03	1.D 00	F FTF
37	0	5U	8.3D-01	41	9.8389D-06	2	0	4	5.67D-04	--	4.D 02	1.D 00	F FTF
38	0	0	1.0D 00	42	9.8011D-06	2	0	4	4.54D-04	--	5.D 02	1.D 00	F FTF
39	0	0	1.0D 00	43	9.7912D-06	2	0	4	3.67D-05	--	5.D 02	1.D 00	F FTF
40	0	0	1.0D 00	44	9.7911D-06	2	0	4	1.02D-06	--	5.D 02	1.D 00	F FTT
41	0	0	1.0D 00	45	9.7911D-06	2	0	4	5.64D-07	--	6.D 02	1.D 00	F FTT
42	0	0	1.0D 00	46	9.7911D-06	2	0	4	1.21D-07	--	6.D 02	1.D 00	T FTT
43	2L	0	1.0D 00	47	9.7906D-06	1	0	5	3.64D-05	-2.4D-05	9.D 02	1.D 00	F FTT
44	0	0	1.0D 00	48	9.7878D-06	1	0	5	1.25D-04	--	1.D 03	1.D 00	F FTF
45	0	0	1.0D 00	49	9.7798D-06	1	0	5	2.90D-04	--	1.D 03	1.D 00	F FTF
53	0	0	1.0D 00	57	8.9619D-06	1	0	5	7.41D-05	--	1.D 04	1.D 00	F FTF
54	0	0	1.0D 00	58	8.9613D-06	1	0	5	6.24D-06	--	1.D 04	1.D 00	T FTF
55	5U	0	1.0D 00	59	8.9611D-06	0	0	6	3.61D-05	-1.4D-05	1.D 04	1.D 00	F FTT
116	0	0	1.0D 00	140	4.5034D-10	0	0	6	7.09D-05	--	1.D 05	1.D 00	F FTF
117	0	0	1.0D 00	141	5.7071D-11	0	0	6	5.01D-06	--	7.D 04	1.D 00	T TTF
118	0	0	1.0D 00	142	1.4296D-11	0	0	6	3.31D-06	--	1.D 05	1.D 00	T TTF
119	0	0	1.0D 00	143	1.5776D-12	0	0	6	2.44D-06	--	1.D 05	1.D 00	T TTF
120	0	0	1.0D 00	144	5.2595D-13	0	0	6	1.07D-06	--	8.D 04	1.D 00	T TTF
121	0	0	1.0D 00	145	1.1712D-14	0	0	6	1.52D-07	--	9.D 04	1.D 00	T TTT

EXIT LC PHASE. INFORM = 0 ITER = 121 NFEVAL = 145

Figure 2b. Results of an EQP method on the six-variable problem Exp 6.

ITN	ITQP	STEP	NUMF	OBJECTIVE	BND	LC	NZ	NORM GF	NORM GZ	COND HZ	COND H	COND T	CONV
0	58	0.0	1	-2.2839D+02	25	3	72	5.9D+01	5.41D+01	2.D+01	1.D+00	1.D+00	FFFT
1	10	1.0D+00	2	-1.1092D+03	16	3	81	1.9D+01	1.84D+01	3.D+01	2.D+00	1.D+00	FTFF
2	3	1.0D+00	3	-1.4382D+03	16	3	81	9.2D+00	9.16D+00	2.D+01	2.D+00	1.D+00	FFFF
3	15	1.0D+00	4	-1.5673D+03	29	4	67	5.7D+00	5.56D+00	2.D+01	2.D+00	1.D+00	FFFF
4	3	1.0D+00	5	-1.6401D+03	27	4	69	3.7D+00	3.43D+00	4.D+01	3.D+00	1.D+00	FTTF
5	4	1.0D+00	6	-1.6631D+03	29	5	66	2.6D+00	2.36D+00	3.D+0.1	4.D+00	5.D+00	FFFF
6	2	1.0D+00	7	-1.6796D+03	28	5	67	2.0D+00	1.77D+00	1.D+02	4.D+00	1.D+00	FTFF
7	8	1.0D+00	8	-1.6886D+03	34	6	60	1.5D+00	1.35D+00	2.D+02	4.D+00	1.D+00	FFFF
8	2	1.0D+00	9	-1.6915D+03	35	6	59	1.4D+00	1.22D+00	2.D+02	4.D+00	1.D+00	FFFFS
9	2	1.0D+00	10	-1.6931D+03	36	6	58	1.3D+00	1.12D+00	2.D+02	4.D+00	1.D+00	FFFFS
10	1	1.0D+00	11	-1.6945D+03	36	6	58	1.3D+00	1.06D+00	2.D+02	4.D+00	1.D+00	FTFFS
11	2	2.7D+00	13	-1.6978D+03	37	6	57	1.1D+00	9.54D-01	2.D+02	4.D+00	1.D+00	FTFFS
12	2	1.0D+00	15	-1.6989D+03	38	6	56	1.1D+00	9.21D-01	2.D+02	4.D+00	1.D+00	FTFFS
13	3	1.2D+00	17	-1.7001D+03	38	6	56	1.1D+00	8.92D-01	2.D+02	4.D+00	1.D+00	FTFFS
14	2	1.7D+00	19	-1.7016D+03	39	6	55	1.1D+00	8.43D-01	2.B+02	4.D+00	1.D+00	FTFFS
15	2	4.1D+00	21	-1.7049D+03	40	6	54	8.9D-01	6.61D-01	7.D+01	4.D+00	1.D+00	FTFFS
16	2	3.6D+00	23	-1.7071D+03	41	6	53	8.5D-01	6.26D-01	9.D+01	4.D+00	1.D+00	FTFFS
17	2	3.6D+00	25	-1.7088D+03	42	6	52	8.4D-01	6.25D-01	9.D+01	4.D+00	1.D+00	FTFFS
18	2	6.8D+00	28	-1.7113D+03	43	6	51	7.8D-01	5.77D-01	9.D+01	4.D+00	1.D+00	FTFFS
19	2	6.9D+00	31	-1.7133D+03	44	6	50	7.4D-01	5.49D-01	9.D+01	4.D+00	1.D+00	FTFFS
20	2	1.9D+00	33	-1.7137D+03	45	6	49	7.3D-01	5.37D-01	9.D+01	4.D+00	1.D+00	FTFFS
30	1	1.0D+00	53	-1.7281D+03	56	6	38	4.0D-01	2.42D-01	6.D+01	3.D+01	1.D+00	FTFFS
40	3	6.0D+00	73	-1.7323D+03	62	7	31	3.4D-01	1.37D-01	2.D+02	8.D+01	2.D+00	FFFF
50	2	1.0D+00	84	-1.7331D+03	63	7	30	3.4D-01	1.34D-01	1.D+03	1.D+02	2.D+00	FTFF
60	1	1.0D+00	94	-1.7337D+03	64	7	29	2.8D-01	7.20D-02	5.D+02	1.D+02	1.D+00	FTFF
70	5	1.0D+00	104	-1.7340D+03	65	7	28	2.9D-01	9.86D-02	6.D+02	1.D+02	1.D+00	FFFF
80	1	1.0D+00	114	-1.7345D+03	69	7	24	2.7D-01	6.03D-02	1.D+02	9.D+01	1.D+00	FTFF
90	1	1.0D+00	124	-1.7346D+03	69	7	24	2.6D-01	3.48D-02	6.D+02	1.D+02	1.D+00	FTFF
100	1	1.0D+00	134	-1.7346D+03	70	7	23	2.6D-01	2.03D-02	3.D+02	1.D+02	1.D+00	FTFF
110	1	1.0D+00	144	-1.7346D+03	71	7	22	2.6D-01	1.58D-02	2.D+02	1.D+02	1.D+00	TTFF
120	1	1.0D+00	154	-1.7347D+03	72	7	21	2.5D-01	2.48D-02	5.D+02	1.D+02	2.D+00	FTFF
130	1	1.0D+00	164	-1.7347D+03	72	7	21	2.5D-01	2.09D-02	9.D+02	1.D+02	2.D+00	FTFF
140	1	1.0D+00	174	-1.7347D+03	72	7	21	2.7D-01	2.87D-02	8.D+03	2.D+02	1.D+00	FTFF
150	1	1.0D+00	184	-1.7349D+03	73	7	20	2.8D-01	3.39D-02	2.D+03	2.D+03	1.D+00	FTFF
160	1	1.0D+00	194	-1.7349D+03	73	7	20	2.8D-01	3.59D-02	2.D+03	3.D+03	1.D+00	FTFF
170	2	1.0D+00	204	-1.7350D+03	75	7	18	2.5D-01	2.10D-02	8.D+02	5.D+03	2.D+00	FFFF
180	1	1.0D+00	214	-1.7350D+03	75	7	18	2.5D-01	9.33D-03	8.D+02	5.D+03	2.D+00	TTFF
190	1	1.0D+00	224	-1.7350D+03	75	7	18	2.5D-01	3.87D-03	6.D+02	5.D+03	2.D+00	TTFF
200	1	1.0D+00	234	-1.7350D+03	75	7	18	2.5D-01	1.02D-03	5.D+02	5.D+03	2.D+00	TTFF
210	1	1.0D+00	244	-1.7350D+03	75	7	18	2.5D-01	2.47D-04	5.D+02	5.D+03	2.D+00	TTFF
211	1	1.0D+00	245	-1.7350D+03	75	7	18	2.5D-01	1.31D-04	5.D+02	5.D+03	2.D+00	TTFF
212	1	1.0D+00	246	-1.7350D+03	75	7	18	2.5D-01	6.18D-05	5.D+02	5.D+03	2.D+00	TTFF
213	1	1.0D+00	247	-1.7350D+03	75	7	18	2.5D-01	4.90D-05	5.D+02	5.D+03	2.D+00	TTFF
214	1	1.0D+00	248	-1.7350D+03	75	7	18	2.5D-01	4.78D-05	5.D+02	5.D+03	2.D+00	TTFT

EXIT NP PHASE. INFORM = 0 MAJITS = 214 NFEVAL = 248

Figure 3a. Results of an IQP method on the Weapon problem.

function evaluations (see Figures 3a and 3b). However, the EQP method needs significantly less CPU time to obtain the solution — 3.42 seconds compared to 19.48 seconds for the IQP method. The reason for this discrepancy is that the EQP method maintains a much better approximation to the working set. While moving from the initial feasible point (a vertex) to the solution, the EQP method maintains a projected Hessian approximation that never becomes larger than 19 (the size of the projected Hessian at the solution is 18). This performance is to be contrasted with that of the IQP method, for which the dimension of Z^THZ increased to 81 during the first two major iterations. During the early QP subproblems the poor Hessian approximation causes the IQP method to delete too many constraints from the working set. Further minor iterations are then needed to add the constraints that were deleted unnecessarily. The large dimension of the projected Hessian also inhibits the ability of the method to recover quickly. Weapon has a Hessian matrix with many negative and zero eigenvalues. If there are not enough constraints

ITN	JDEL	JADD	STEP	NUMF	OBJECTIVE	BND	LC	NZ	NORM GZ	MIN LM	COND HZ	COND T	CONV
0	0	0	0.0D-01	1	-1.9539D 02	91	9	0	0.00D-01	--	1.D 00	4.D 00	T TFF
1	72L	0	1.0D 00	2	-3.1565D 02	90	9	1	5.28D 00	-2.7D 01	1.D 00	4.D 00	T FTF
2	101U	0	1.0D 00	3	-3.6372D 02	90	8	2	4.44D 00	-2.4D 01	1.D 00	1.D 00	T FTF
3	67L	0	1.0D 00	4	-5.2490D 02	89	8	3	5.47D 00	-2.2D 01	2.D 00	1.D 00	T FTF
4	97L	0	1.0D 00	5	-6.8450D 02	88	8	4	4.60D 00	-1.9D 01	2.D 00	2.D 00	T FTF
5	87L	0	1.0D 00	6	-8.1237D 02	87	8	5	3.43D 00	-1.3D 01	3.D 00	2.D 00	T FTF
6	77L	96L	2.1D-01	7	-8.4275D 02	87	8	5	7.90D 00	-1.3D 01	5.D 00	2.D 00	T FTF
7	64L	46L	1.4D-01	8	-8.7937D 02	87	8	5	1.21D 01	-1.2D 01	3.D 00	2.D 00	T FTF
8	0	0	1.0D 00	9	-9.9438D 02	87	8	5	5.28D 00	--	3.D 00	2.D 00	T FTF
9	82L	0	1.0D 00	10	-1.1043D 03	86	8	6	3.20D 00	-1.3D 01	3.D 00	3.D 00	T FTF
10	92L	0	1.0D 00	11	-1.2014D 03	85	8	7	2.53D 00	-1.3D 01	3.D 00	3.D 00	T FTF
20	22L	0	1.0D 00	21	-1.5083D 03	78	8	14	1.42D 00	-3.5D 00	6.D 00	3.D 00	T FTF
30	0	0	1.0D 00	31	-1.5892D 03	75	8	17	1.51D 00	--	3.D 01	3.D 00	T FTF
40	0	87L	5.5D-01	41	-1.6614D 03	78	7	15	1.11D 00	--	1.D 01	3.D 00	T FTF
50	0	0	1.0D 00	51	-1.6935D 03	77	6	17	1.08D 00	--	1.D 01	3.D 00	T FTF
60	0	0	1.0D 00	62	-1.7100D 03	79	7	14	4.36D-01	--	4.D 01	2.D 00	T FTF
70	0	0	1.0D 00	72	-1.7152D 03	78	8	14	3.36D-01	--	3.D 01	2.D 00	T FTF
80	0	0	1.0D 00	82	-1.7190D 03	78	8	14	2.49D-01	--	8.D 01	2.D 00	T FTF
90	0	0	1.0D 00	92	-1.7242D 03	78	8	14	2.05D-01	--	3.D 02	2.D 00	T FTF
100	0	0	1.0D 00	102	-1.7267D 03	77	8	15	2.74D-01	--	2.D 01	2.D 00	T FTF
110	0	0	1.0D 00	112	-1.7289D 03	77	6	17	2.18D-01	--	3.D 01	2.D 00	T FTF
120	0	0	1.0D 00	122	-1.7309D 03	78	6	16	1.09D-01	--	4.D 01	2.D 00	T FTF
130	0	0	1.0D 00	132	-1.7315D 03	77	6	17	1.09D-01	--	3.D 01	2.D 00	T FTF
140	0	0	1.0D 00	142	-1.7322D 03	79	6	15	6.96D-02	--	1.D 02	2.D 00	T FTF
150	0	0	1.0D 00	152	-1.7326D 03	78	6	16	5.98D-02	--	2.D 02	2.D 00	T FTF
160	0	0	1.0D 00	162	-1.7329D 03	78	6	16	3.99D-02	--	7.D 01	2.D 00	T FTF
170	0	0	1.0D 00	172	-1.7331D 03	77	6	17	3.22D-02	--	8.D 01	2.D 00	T FTF
180	0	0	1.0D 00	182	-1.7333D 03	76	6	18	6.32D-02	--	9.D 01	2.D 00	T FTF
190	73U	0	1.0D 00	192	-1.7335D 03	75	6	19	2.27D-02	-2.0D-02	2.D 02	2.D 00	T FTF
200	0	0	1.0D 00	202	-1.7337D 03	75	6	19	4.69D-02	--	2.D 02	2.D 00	T FTF
210	0	0	1.0D 00	212	-1.7344D 03	75	7	18	5.69D-02	--	2.D 01	2.D 00	T FTF
220	0	0	1.0D 00	222	-1.7347D 03	75	7	18	2.36D-02	--	3.D 01	2.D 00	T FTF
230	0	0	1.0D 00	232	-1.7348D 03	74	7	19	6.15D-03	--	3.D 01	2.D 00	T FTF
240	0	0	1.0D 00	242	-1.7350D 03	75	7	18	1.76D-04	--	4.D 01	2.D 00	T TTF
241	0	0	1.0D 00	243	-1.7350D 03	75	7	18	5.82D-05	--	4.D 01	2.D 00	T TTF
242	0	0	1.0D 00	244	-1.7350D 03	75	7	18	1.45D-05	--	4.D 01	2.D 00	T TTF
243	0	0	1.0D 00	245	-1.7350D 03	75	7	18	1.93D-06	--	4.D 01	2.D 00	T TTT

EXIT LC PHASE. INFORM = 0 ITER = 243 NFEVAL = 245

Figure 3b. Results of an EQP method on the Weapon problem.

in the working set, the linesearch function is likely to have negative curvature along the search direction and the quasi-Newton update will be skipped. If this occurs for several consecutive iterations, the original poor curvature information is left unaltered — as is the inability of the QP to predict the correct active set. This behavior was apparent during iterations 8–30 where no quasi-Newton update could be performed.

It is difficult to predict in advance whether a particular problem will be more suitable for an EQP method or an IQP method. IQP methods are likely to be less efficient on problems for which the QP multipliers change rapidly from one iteration to the next. Problems in this category tend to be highly nonlinear or to have many small Lagrange multipliers. In either case, the significant changes to the working set between iterations seriously impair the ability of the quasi-Newton update to build useful curvature information about the function. Conversely, IQP methods will tend to be efficient if the QP multiplier estimates are very accurate when computed at points that are far from the solution (for example, if the problem were close to being quadratic).

To a large extent, the relative efficiency of IQP and EQP methods depends upon the number of constraints active at the solution. EQP methods are usually implemented so that as many constraints as possible are included in the initial working set. It is therefore not surprising that they tend to be more efficient when more constraints are active at the solution. Finally, the relative efficiency of a method is critically dependent upon the ratio of the amount of work required to perform a single minor iteration compared to the work required to evaluate the problem functions. As this ratio increases (as it often does as the size of the problem increases), the advantage will swing towards the EQP method.

5.3. Typical performance of an SQP method for nonlinear constraints. The remaining runs were obtained from the program NPSOL (see Gill *et al.*, 1984b), an IQP quasi-Newton method for nonlinearly constrained optimization. The Hessian of each QP subproblem is a positive-definite BFGS approximation to the Hessian of the Lagrangian function. The QP subproblem is solved using a feasible-point active-set method with an orthogonal factorization of the constraints in the working set.

The merit function used in NPSOL is a smooth augmented Lagrangian function that utilizes the properties of *slack variables*. The inequality constraints of NIP can be reformulated as equality constraints by adding simply-bounded slack variables s_j. Estimates of the slack variables are used in the linesearch to give a smooth augmented Lagrangian function. At each major iteration, a vector triple $(p, \delta\lambda, \delta s)$ is computed that serves as a direction of search for the variables x, multiplier estimates λ, and slack variables s. (All the elements of the vector triple are available from the solution of the standard IQP subproblem considered in Section 4.2. The vector $\delta\lambda$ is defined as $\mu - \lambda$, where μ are the QP multipliers, and the vector δs is given by $Ap+c-s$. Note that the QP solver does *not* need to treat the elements of s as additional variables.) The steplength is required to produce a sufficient decrease in the augmented Lagrangian merit function

$$\mathcal{L}(x,\lambda,s) = F(x) - \sum_{i=1}^{m} \lambda_i\big(c_i(x) - s_i\big) + \frac{\rho}{2}\sum_{i=1}^{m}\big(c_i(x) - s_i\big)^2.$$

The value of ρ is initially set to zero, and is occasionally increased from its value in the previous iteration in order to ensure descent for the merit function. Thus the sequence of penalty parameters is generally non-decreasing, although NPSOL has the ability to reduce the value of the penalty parameter a limited number of times.

The iteration summary printed in each of Figures 4–6 is identical to that provided by the linearly constrained IQP method, except that the merit function value ("MERIT") is printed instead of the objective value, and the additional columns "NC", "NORM C" and "RHO" give the number of nonlinear constraints in the working set, the two-norm of the residuals of constraints in the working set and the penalty parameter used in the merit function. In all of the NPSOL runs, the feasibility tolerance for each nonlinear constraint was set at 10^{-6}.

Two runs were selected to illustrate the behavior of an IQP method when solving well behaved (but non-trivial) nonlinear problems. Figure 4 gives the results obtained on a version of the Hexagon problem. (For more details of this problem, see Wright, 1976. A slightly different formulation is given as Problem 108 by Hock and Schittkowski, 1981.) Hexagon is a popular test problem for nonlinear programming methods. All constraint types are included (bounds, linear, nonlinear), and the Hessian of the Lagrangian function is not positive definite at the solution. The problem has nine variables, finite bounds on six of the variables, four general linear constraints, and fifteen nonlinear constraints. Six nonlinear constraints are active at x^*.

The problem solved in Figure 5 is derived from a 30-bus optimal power flow (OPF) problem of optimizing the distribution of electrical power over a network. The problem has 67 variables, 60 nonlinear constraints, and upper and lower bounds on all of the variables. At the solution, 54 nonlinear constraints and three simple bounds are active.

ITN	ITQP	STEP	NUMF	MERIT	BND	LC	NC	NZ	NORM GF	NORM GZ	COND HZ	COND H	COND T	NORM C	RHO	CONV
0	6	0.0D-01	1	-1.4333D 00	0	0	5	4	2.0D 00	2.28D-01	1.D 00	1.D 00	3.D 00	1.15D 00	0.0D-01	FFTT
1	8	4.0D-01	3	-1.3952D 00	0	0	4	5	2.1D 00	9.21D-02	1.D 00	2.D 00	2.D 00	6.15D-01	5.3D-01	FFTF
2	1	1.0D 00	4	-1.2995D 00	0	0	4	5	2.0D 00	1.64D-01	2.D 00	2.D 00	1.D 00	6.60D-02	1.1D 00	FFTF
3	3	1.0D 00	5	-1.3222D 00	0	0	6	3	2.0D 00	1.06D-01	1.D 00	7.D 00	2.D 00	2.06D-01	1.1D 00	FFTF
4	1	1.0D 00	6	-1.3463D 00	0	0	6	3	2.1D 00	1.19D-01	2.D 00	4.D 02	2.D 00	1.40D-02	1.1D 00	FFTF
5	1	1.0D 00	7	-1.3496D 00	0	0	6	3	2.1D 00	2.74D-02	1.D 00	3.D 02	2.D 00	4.95D-03	1.1D 00	FFTF
6	1	1.0D 00	8	-1.3498D 00	0	0	6	3	2.1D 00	1.02D-02	2.D 00	2.D 02	2.D 00	1.12D-03	1.1D 00	FFTF
7	1	1.0D 00	9	-1.3499D 00	0	0	6	3	2.1D 00	8.10D-03	4.D 00	3.D 02	2.D 00	9.79D-05	1.1D 00	FFTF
8	1	1.0D 00	10	-1.3500D 00	0	0	6	3	2.1D 00	1.23D-03	4.D 00	2.D 02	2.D 00	5.00D-04	1.1D 00	FFTF
9	1	1.0D 00	11	-1.3500D 00	0	0	6	3	2.1D 00	9.35D-05	4.D 00	2.D 02	2.D 00	1.13D-06	1.4D 01	FTTF
10	1	1.0D 00	12	-1.3500D 00	0	0	6	3	2.1D 00	2.37D-05	4.D 00	2.D 02	2.D 00	5.90D-09	5.6D 00	FTTF
11	1	1.0D 00	13	-1.3500D 00	0	0	6	3	2.1D 00	2.35D-06	4.D 00	2.D 02	2.D 00	4.52D-10	5.6D 00	TTTT

EXIT NP PHASE. INFORM = 0 MAJITS = 11 NFEVAL = 13 NCEVAL = 13

VARIABLE	STATE	VALUE	LOWER BOUND	UPPER BOUND	LAGR MULTIPLIER	RESIDUAL
VARBL 1	FR	0.6094650D-01	0.0000000	NONE	0.0000000	0.6095D-01
VARBL 2	FR	0.5976502	NONE	NONE	0.0000000	0.1000D 11
VARBL 3	FR	1.000000	NONE	NONE	0.0000000	0.1000D 11
VARBL 4	FR	0.5976497	NONE	NONE	0.0000000	0.1000D 11
VARBL 5	FR	0.6094641D-01	0.0000000	NONE	0.0000000	0.6095D-01
VARBL 6	FR	0.3437710	0.0000000	NONE	0.0000000	0.3438
VARBL 7	FR	0.5000000	0.0000000	NONE	0.0000000	0.5000
VARBL 8	FR	-0.5000000	NONE	0.0000000	0.0000000	0.5000
VARBL 9	FR	-0.3437708	NONE	0.0000000	0.0000000	0.3438

LINEAR CONSTR	STATE	VALUE	LOWER BOUND	UPPER BOUND	LAGR MULTIPLIER	RESIDUAL
LNCON 1	FR	-0.5367037	NONE	0.0000000	0.0000000	0.5367
LNCON 2	FR	-0.4023498	NONE	0.0000000	0.0000000	0.4023
LNCON 3	FR	0.4023503	0.0000000	NONE	0.0000000	0.4024
LNCON 4	FR	0.5367033	0.0000000	NONE	0.0000000	0.5367

NONLNR CONSTR	STATE	VALUE	LOWER BOUND	UPPER BOUND	LAGR MULTIPLIER	RESIDUAL
NLCON 1	FR	0.8781070	0.0000000	NONE	0.0000000	0.8781
NLCON 2	FR	0.6875417	0.0000000	NONE	0.0000000	0.6875
NLCON 3	LL	-0.4990394D-10	0.0000000	NONE	0.8318393D-01	-0.4990D-10
NLCON 4	LL	-0.3452747D-09	0.0000000	NONE	0.3202626	-0.3453D-09
NLCON 5	FR	0.5272863	0.0000000	NONE	0.0000000	0.5273
NLCON 6	FR	0.3928143	0.0000000	NONE	0.0000000	0.3928
NLCON 7	FR	0.5881147	0.0000000	NONE	0.0000000	0.5881
NLCON 8	LL	-0.2877292D-09	0.0000000	NONE	0.1992985	-0.2877D-09
NLCON 9	LL	-0.1479264D-10	0.0000000	NONE	0.3202626	-0.1479D-10
NLCON 10	UL	0.0000000	NONE	0.0000000	-0.3437718	0.0000
NLCON 11	FR	0.5881142	0.0000000	NONE	0.0000000	0.5881
NLCON 12	LL	-0.1845829D-11	0.0000000	NONE	0.8318383D-01	-0.1846D-11
NLCON 13	FR	0.3928147	0.0000000	NONE	0.0000000	0.3928
NLCON 14	FR	0.6875419	0.0000000	NONE	0.0000000	0.6875
NLCON 15	FR	0.8781072	0.0000000	NONE	0.0000000	0.8781

Figure 4. Output from the solution of the well-behaved problem Hexagon.

On these two well behaved problems, the approximate Hessian and working set remain relatively well-conditioned. Similarly, the penalty parameters remain small and approximately constant. The two runs illustrate much of the numerical behavior of a quasi-Newton IQP method that is predicted from theoretical analysis. As x_k approaches the solution, just one minor iteration is performed per major iteration, and entries in the "NORM GZ" and "NORM C" columns exhibit the superlinear convergence rate discussed in Section 3.3. Note that the constraint violations converge earlier than the projected gradient. The final values of the projected gradient norm and constraint norm reflect the limiting accuracy of the two quantities. It is possible to achieve almost full precision in the constraint norm but only half precision in the projected-gradient norm.

ITN	ITQP	STEP	NUMF	MERIT	BND	LC	NC	NZ	NORM GF	NORM GZ	COND HZ	COND H	COND T	NORM C	RHO	CONV
0	20	0.0	1	-1.0911D+00	1	0	59	7	1.0D+00	7.73D-14	1.D+00	1.D+00	8.D+01	2.92D+00	0.0	TFFT
1	8	7.0D-01	3	1.0010D+00	1	0	55	11	1.0D+00	4.54D-01	2.D+00	3.D+00	9.D+01	1.67D+00	4.4D+00	FFTF
2	6	1.0D+00	4	1.0043D+00	1	0	56	10	1.0D+00	2.24D-01	7.D+00	6.D+00	8.D+01	6.31D-01	4.5D+00	FFTF
3	2	4.9D-01	6	9.9142D-01	1	0	56	10	1.0D+00	4.35D-02	1.D+01	1.D+01	9.D+01	7.12D-01	2.8D+00	FFTF
4	3	4.3D-01	8	9.8835D-01	1	0	54	12	1.0D+00	6.08D-02	9.D+00	1.D+01	9.D+01	1.85D-02	2.8D+00	FFTF
5	1	1.0D+00	9	9.8769D-01	1	0	54	12	1.0D+00	7.09D-02	2.D+01	3.D+01	9.D+01	1.09D-02	2.8D+00	FFTF
6	1	1.0D+00	10	9.8642D-01	1	0	54	12	1.0D+00	3.35D-02	3.D+01	3.D+01	9.D+01	7.49D-04	2.8D+00	FFTF
7	1	1.0D+00	11	9.8450D-01	1	0	54	12	1.0D+00	4.22D-02	6.D+01	3.D+01	9.D+01	3.84D-03	2.8D+00	FFTF
8	4	1.0D+00	12	9.8154D-01	3	0	55	9	1.0D+00	7.38D-02	8.D+01	4.D+01	9.D+01	1.18D-01	2.8D+00	FFFF
9	4	1.0D+00	13	9.7767D-01	5	0	54	8	1.0D+00	7.92D-02	8.D+00	9.D+01	8.D+01	4.29D-02	2.8D+00	FFFF
10	2	1.0D+00	14	9.7584D-01	6	0	54	7	1.0D+00	4.22D-02	8.D+00	9.D+01	9.D+01	4.59D-03	2.8D+00	FFTF
11	1	1.0D+00	15	9.7461D-01	6	0	54	7	1.0D+00	1.75D-02	2.D+01	9.D+01	9.D+01	6.48D-04	2.8D+00	FFTF
12	3	1.0D+00	16	9.7431D-01	4	0	54	9	1.0D+00	1.90D-02	2.D+01	9.D+01	9.D+01	1.78D-04	2.8D+00	FFFF
13	2	1.0D+00	17	9.7415D-01	3	0	54	10	1.0D+00	1.37D-02	3.D+01	9.D+01	9.D+01	9.88D-05	2.8D+00	FFTF
14	2	1.0D+00	18	9.7403D-01	2	0	54	11	1.0D+00	9.67D-03	5.D+01	8.D+01	9.D+01	1.36D-04	2.8D+00	FFTF
15	1	1.0D+00	19	9.7392D-01	2	0	54	11	1.0D+00	9.20D-03	9.D+01	8.D+01	9.D+01	1.99D-04	2.8D+00	FFTF
16	2	1.0D+00	20	9.7374D-01	2	0	55	10	1.0D+00	1.07D-02	7.D+01	1.D+02	9.D+01	3.30D-02	2.8D+00	FFFF
17	3	1.0D+00	21	9.7344D-01	3	0	54	10	1.0D+00	1.25D-02	3.D+01	3.D+02	9.D+01	8.56D-03	2.8D+00	FFFF
18	1	1.0D+00	22	9.7308D-01	3	0	54	10	1.0D+00	1.21D-02	1.D+02	7.D+02	1.D+02	1.99D-03	2.8D+00	FFFF
19	2	1.0D+00	23	9.7267D-01	4	0	54	9	1.0D+00	1.16D-02	3.D+01	9.D+02	1.D+02	5.01D-03	2.8D+00	FFFF
20	1	1.0D+00	24	9.7254D-01	4	0	54	9	1.0D+00	8.74D-03	5.D+01	1.D+03	1.D+02	8.28D-05	2.8D+00	FFFF
21	1	1.0D+00	25	9.7239D-01	4	0	54	9	1.0D+00	5.39D-03	7.D+01	1.D+03	1.D+02	7.99D-04	2.8D+00	FFTF
22	1	1.0D+00	26	9.7236D-01	4	0	54	9	1.0D+00	5.10D-03	2.D+02	1.D+03	1.D+02	8.60D-05	2.8D+00	FFTF
23	2	1.0D+00	27	9.7221D-01	3	0	54	10	1.0D+00	8.03D-03	3.D+02	1.D+03	1.D+02	4.21D-03	2.8D+00	FFTF
24	1	1.0D+00	28	9.7213D-01	3	0	54	10	1.0D+00	8.01D-03	4.D+02	1.D+03	1.D+02	2.33D-03	2.8D+00	FFTF
25	2	1.0D+00	29	9.7209D-01	4	0	54	9	1.0D+00	3.43D-03	1.D+02	9.D+02	1.D+02	1.48D-03	2.8D+00	FFTF
26	1	1.0D+00	30	9.7207D-01	4	0	54	9	1.0D+00	1.86D-03	1.D+02	8.D+02	1.D+02	8.32D-05	2.8D+00	FFTF
27	2	1.0D+00	31	9.7207D-01	3	0	54	10	1.0D+00	2.26D-03	1.D+02	8.D+02	1.D+02	2.99D-05	2.8D+00	FFTF
28	1	1.0D+00	32	9.7206D-01	3	0	54	10	1.0D+00	2.53D-03	3.D+02	8.D+02	1.D+02	5.88D-05	2.8D+00	FFTF
29	2	1.0D+00	33	9.7205D-01	4	0	54	9	1.0D+00	2.15D-03	2.D+02	8.D+02	1.D+02	1.08D-03	2.8D+00	FFTF
30	1	1.0D+00	34	9.7205D-01	4	0	54	9	1.0D+00	1.33D-03	3.D+02	9.D+02	1.D+02	1.11D-04	2.8D+00	FFFF
31	2	1.0D+00	35	9.7204D-01	3	0	54	10	1.0D+00	1.20D-03	2.D+02	9.D+02	1.D+02	5.17D-05	2.8D+00	FFTF
32	1	1.0D+00	36	9.7204D-01	3	0	54	10	1.0D+00	8.55D-04	3.D+02	8.D+02	1.D+02	1.24D-05	2.8D+00	FFTF
33	1	1.0D+00	37	9.7204D-01	3	0	54	10	1.0D+00	4.72D-04	3.D+02	6.D+02	1.D+02	2.68D-06	2.8D+00	FTTF
34	1	1.0D+00	38	9.7204D-01	3	0	54	10	1.0D+00	1.89D-04	3.D+02	5.D+02	1.D+02	2.57D-06	2.8D+00	FTTF
35	1	1.0D+00	39	9.7204D-01	3	0	54	10	1.0D+00	1.90D-04	3.D+02	5.D+02	1.D+02	3.25D-07	2.8D+00	FTTF
36	1	1.0D+00	40	9.7204D-01	3	0	54	10	1.0D+00	2.28D-04	3.D+02	7.D+02	1.D+02	2.36D-07	2.8D+00	FTTF
37	1	1.0D+00	41	9.7204D-01	3	0	54	10	1.0D+00	2.88D-04	4.D+02	9.D+02	1.D+02	1.14D-06	2.8D+00	FTTF
38	1	1.0D+00	42	9.7204D-01	3	0	54	10	1.0D+00	3.19D-04	5.D+02	8.D+02	1.D+02	4.37D-06	2.8D+00	FTTF
39	1	1.0D+00	43	9.7204D-01	3	0	54	10	1.0D+00	2.34D-04	5.D+02	7.D+02	1.D+02	6.05D-06	2.8D+00	FTTF
40	1	1.0D+00	44	9.7204D-01	3	0	54	10	1.0D+00	8.35D-05	5.D+02	8.D+02	1.D+02	1.35D-06	2.8D+00	FTTF
41	1	1.0D+00	45	9.7204D-01	3	0	54	10	1.0D+00	1.08D-05	5.D+02	8.D+02	1.D+02	6.87D-08	2.8D+00	TTTF
42	1	1.0D+00	46	9.7204D-01	3	0	54	10	1.0D+00	6.80D-07	5.D+02	7.D+02	1.D+02	1.66D-08	2.8D+00	TTTF
43	1	1.0D+00	47	9.7204D-01	3	0	54	10	1.0D+00	8.78D-08	5.D+02	7.D+02	1.D+02	3.60D-10	2.8D+00	TTTT

EXIT NP PHASE. INFORM = 0 MAJITS = 43 NFEVAL = 47 NCEVAL = 47

Figure 5. Output from the solution of the OPF problem.

The status and values of the variables and constraints at the final solution give useful information about the progress of a minimization and the degree of difficulty of the problem. Figure 4 includes the final solution output from NPSOL for Hexagon. The printout is divided into three sections, giving information about the final status of the variables, general linear constraints and nonlinear constraints, respectively. Within each section, "STATE" gives the status of the associated constraint in the predicted active set (FR if not included, EQ if a fixed value, LL if at its lower bound, and UL if at its upper bound). "VALUE" is the value of the constraint at the final iteration. "LOWER BOUND" and "UPPER BOUND" give the lower and upper bounds specified for the constraint ("NONE" indicates that no bound is enforced). "LAGR MULTIPLIER" is the value of the Lagrange multiplier. This will be zero if STATE is FR. The multiplier is non-negative if STATE is LL, and non-positive if STATE is UL. "RESIDUAL" gives the difference between the entry in the "VALUE" column and the nearer bound.

In the first section, "VARIABLE" is the name (VARBL) and index of a variable. In the linear constraints section, "LINEAR CONSTR" is the name (LNCON) and index of a linear constraint. "NONLNR CONSTR" is the name (NLCON) and index of a nonlinear constraint.

Note that, although the feasibility tolerance for the nonlinear constraints is of the order 10^{-6}, the final accuracy is considerably better than this. This is because the constraint violations are being refined during the last few iterations while the algorithm is working to reduce the projected-gradient norm. Another feature worth noting is that the the constraint values and Lagrange multipliers at the solution are "well balanced". For example, all the multipliers are approximately of the same order of magnitude. This behavior is typical of a well-scaled problem.

5.4. Performance on an ill-conditioned problem. Finally, we give the results of the IQP method on the problem Dembo 7. This problem is a geometric programming formulation developed by Dembo (1976) of a five-stage membrane separation process. The problem has sixteen variables, eight linear constraints, and eleven nonlinear constraints. All sixteen variables have simple upper and lower bound constraints. The problem causes many difficulties for a nonlinear programming algorithm because of bad scaling and linearly dependent constraints.

The results for Dembo 7 show a number of features that are common to badly behaved problems. First, note that the number of minor iterations does not decline quickly. Moreover, the presence of near-zero Lagrange multipliers sometimes causes the QP to require more than one iteration relatively close to the solution. A very common symptom of a badly behaved problem is the large value of the condition estimator of the full approximate Hessian, which is to be contrasted with the relatively modest value of the condition of the *projected* Hessian. This observation has some relevance to the choice of method for the QP subproblem. Clearly, special care must be taken when implementing any QP method that requires the factors of the full Hessian (as opposed to the projected Hessian). Starting the minor iterations at an unconstrained minimum of the QP subproblem will result in very large values of δp (see Section 4.3).

Note that the third bound constraint, the third linear constraint and eleventh nonlinear constraint all have very small residuals but are not in the working set. The values of the nonlinear constraints in the working set vary significantly in order of magnitude, indicating that the constraints are badly scaled. In constrast to the solution of Hexagon, in which the accuracy of the constraints was much better than required by the convergence tolerance, some of the nonlinear constraints are only just satisfied to the required feasibility tolerance.

Finally, we wish to emphasize that, despite severe ill-conditioning in the Hessian of the

ITN	ITQP	STEP	NUMF	MERIT	BND	LC	NC	NZ	NORM GF	NORM GZ	COND HZ	COND H	COND T	NORM C	RHO	CONV
0	30	0.0	1	2.8459D+02	1	3	9	3	1.1D+03	5.17D-01	1.D+00	1.D+00	7.D+03	2.19D-01	0.0	FFFT
1	13	9.6D-01	4	-2.1324D+02	3	2	5	6	8.1D+02	1.76D+00	4.D+01	6.D+06	2.D+02	1.50D+00	1.4D+06	FFFF
2	12	8.1D-03	6	3.7305D+02	2	2	7	5	8.1D+02	1.87D+00	1.D+00	1.D+08	1.D+05	1.49D+00	1.6D+03	FFFF
3	12	5.7D-02	8	4.4182D+02	2	1	7	6	8.1D+02	1.12D+02	5.D+01	3.D+08	5.D+03	1.41D+00	8.0D+02	FFFF
4	7	1.0D+00	9	4.0200D+02	2	3	5	6	8.2D+02	2.25D+00	1.D+00	1.D+08	4.D+02	1.86D-01	8.0D+02	FFFF
5	9	1.0D+00	10	3.9278D+02	2	1	7	6	8.1D+02	1.62D+00	2.D+01	4.D+08	7.D+01	4.18D-01	1.3D+03	FFFF
6	17	1.0D+00	11	3.0388D+02	2	3	5	6	1.7D+02	7.24D-01	3.D+08	2.D+10	4.D+03	3.03D-01	8.8D+01	FFFF
7	20	1.0D+00	12	1.3093D+02	4	2	8	2	1.5D+02	4.94D+00	8.D+01	7.D+11	1.D+05	9.00D-01	3.7D+01	FFFF
8	11	5.0D-01	14	-6.2190D+03	3	2	6	5	6.4D+02	3.12D+01	1.D+04	3.D+12	8.D+02	4.52D-01	6.4D+03	FFFF
9	2	1.0D+00	15	1.9003D+02	4	2	6	4	6.4D+02	3.21D+01	5.D+05	4.D+11	8.D+04	2.37D-02	3.8D+02	FFFF
10	1	1.0D+00	16	1.8996D+02	4	2	5	5	6.4D+02	3.19D+01	6.D+05	2.D+12	1.D+01	6.80D-06	3.6D+03	FTFF
11	3	1.0D+00	17	1.8987D+02	4	2	5	5	6.4D+02	3.17D+01	3.D+06	9.D+10	1.D+01	6.20D-07	6.2D+02	FTFF
12	6	1.0D+00	18	1.8883D+02	6	1	5	4	1.6D+02	2.99D+01	1.D+02	5.D+11	1.D+01	1.09D-01	6.2D+02	FFFF
13	2	1.0D+00	19	1.8809D+02	7	1	5	3	1.5D+02	2.29D-01	2.D+00	4.D+11	1.D+01	7.47D-03	6.2D+02	FFFF
14	4	1.0D+00	20	1.8357D+02	5	2	6	3	6.3D+02	2.35D-01	2.D+00	1.D+12	2.D+01	1.97D-03	6.2D+02	FFFF
15	3	1.0D+00	21	1.8352D+02	5	2	6	3	1.3D+02	2.36D-01	4.D+01	5.D+11	3.D+03	3.04D-06	1.1D+05	FTFF
16	4	1.0D+00	22	1.7857D+02	4	2	8	2	6.2D+02	2.41D-02	1.D+00	1.D+12	8.D+03	1.46D-02	6.5D+03	FFFF
17	1	1.0D+00	23	1.7509D+02	4	2	8	2	6.2D+02	6.76D-02	1.D+00	1.D+10	7.D+03	3.04D-03	2.9D+03	FFFF
18	2	1.0D+00	24	1.7506D+02	5	2	7	2	6.2D+02	6.25D-02	1.D+00	1.D+10	7.D+03	4.87D-05	2.9D+03	FFFF
19	8	1.0D+00	25	1.7505D+02	6	2	6	2	6.2D+02	1.05D+00	2.D+01	2.D+12	7.D+02	1.56D-02	2.9D+03	FFFF
20	2	1.0D+00	26	1.7479D+02	6	2	7	1	6.2D+02	8.91D-03	1.D+00	1.D+11	8.D+02	1.02D-03	2.9D+03	FFFF
21	1	1.0D+00	27	1.7480D+02	6	2	7	1	6.2D+02	8.54D-03	1.D+00	7.D+10	8.D+02	1.16D-06	3.3D+04	FTFF
22	3	1.0D+00	28	1.7480D+02	6	2	7	1	6.2D+02	1.11D+00	1.D+00	1.D+12	2.D+03	6.45D+00	6.4D+03	FFFF
23	10	2.5D-01	30	1.7479D+02	6	2	7	1	6.2D+02	1.38D-03	1.D+00	4.D+11	2.D+03	7.40D-05	6.4D+03	TFFF
24	1	1.0D+00	31	1.7479D+02	6	2	7	1	6.2D+02	7.93D-04	1.D+00	5.D+11	2.D+03	3.81D-06	3.8D+05	TTFF
25	3	1.0D+00	32	1.7479D+02	5	1	8	2	6.2D+02	5.51D-04	2.D+00	6.D+11	2.D+03	1.19D-07	4.7D+04	TTFF
26	2	1.0D+00	33	1.7479D+02	6	1	7	2	6.2D+02	6.89D-05	4.D+00	5.D+11	2.D+03	9.43D-08	3.7D+04	TTFF
27	2	1.0D+00	34	1.7479D+02	5	1	7	3	6.2D+02	1.30D-06	2.D+03	5.D+11	2.D+03	1.12D-07	3.7D+04	TTTF
28	1	1.0D+00	35	1.7479D+02	5	1	7	3	6.2D+02	1.23D-06	2.D+03	5.D+11	2.D+03	1.93D-07	3.7D+04	TTTT

EXIT NP PHASE. INFORM = 0 MAJITS = 28 NFEVAL = 35 NCEVAL = 35

VARIABLE	STATE	VALUE	LOWER BOUND	UPPER BOUND	LAGR MULTIPLIER	RESIDUAL
VARBL 1	FR	.8037732	.1000000	.9000000	.0	.9623D-01
VARBL 2	FR	.8161084	.1000000	.9000000	.0	.8389D-01
VARBL 3	FR	.9000000	.1000000	.9000000	.0	-.1388D-16
VARBL 4	UL	.9000000	.1000000	.9000000	-436.7986	.0
VARBL 5	FR	.9000001	.9000000	1.000000	.0	.1467D-06
VARBL 6	UL	.1000000	.1000000D-03	.1000000	-15.19769	.0
VARBL 7	FR	.1070316	.1000000	.9000000	.0	.7032D-02
VARBL 8	FR	.1908367	.1000000	.9000000	.0	.9084D-01
VARBL 9	FR	.1908367	.1000000	.9000000	.0	.9084D-01
VARBL 10	FR	.1908367	.1000000	.9000000	.0	.9084D-01
VARBL 11	FR	505.0452	1.000000	1000.000	.0	495.0
VARBL 12	FR	5.046043	.1000000D-05	500.0000	.0	5.046
VARBL 13	FR	72.63782	1.000000	500.0000	.0	71.64
VARBL 14	LL	500.0000	500.0000	1000.000	.6299044	.0
VARBL 15	LL	500.0000	500.0000	1000.000	.2499969	.0
VARBL 16	LL	.1000000D-05	.1000000D-05	500.0000	.1546715	.0

LINEAR CONSTR	STATE	VALUE	LOWER BOUND	UPPER BOUND	LAGR MULTIPLIER	RESIDUAL
LNCON 1	FR	-499.9991	NONE	.0	.0	500.0
LNCON 2	FR	-.1466659D-06	NONE	.0	.0	.1467D-06
LNCON 3	FR	.1387779D-16	NONE	.0	.0	-.1388D-16
LNCON 4	FR	-.8389163D-01	NONE	.0	.0	.8389D-01
LNCON 5	FR	-.1233521D-01	NONE	.0	.0	.1234D-01
LNCON 6	UL	-.5551115D-16	NONE	.0	-67.20963	.5551D-16
LNCON 7	FR	.5551115D-16	NONE	.0	.0	-.5551D-16
LNCON 8	FR	.9999983	NONE	1.000000	.0	.1744D-05

NONLNR CONSTR	STATE	VALUE	LOWER BOUND	UPPER BOUND	LAGR MULTIPLIER	RESIDUAL
NLCON 1	LL	.0	.0	NONE	7.025522	.0
NLCON 2	LL	-.6277010D-13	.0	NONE	103.7262	-.6277D-13
NLCON 3	LL	.1127570D-16	.0	NONE	263.3147	.1128D-16
NLCON 4	LL	-.2168404D-16	.0	NONE	104.7026	-.2168D-16
NLCON 5	LL	-.1562181D-06	.0	NONE	.0	-.1562D-06
NLCON 6	LL	-.1134245D-06	.0	NONE	.0	-.1134D-06
NLCON 7	LL	-.1848435D-13	.0	NONE	37.40537	-.1848D-13
NLCON 8	LL	-.2595146D-14	.0	NONE	402.2754	-.2595D-14
NLCON 9	LL	-.1387779D-16	.0	NONE	60.48867	-.1388D-16
NLCON 10	LL	-.2081668D-15	.0	NONE	.0	-.2082D-15
NLCON 11	FR	.4924576D-15	.0	NONE	.0	.4925D-15

Figure 6. Output from the solution of Dembo 7.

Lagrangian and serious dependencies among the constraints, Dembo 7 is solved in a relatively routine manner. Dependent constraints are successfully omitted from the working set in such a way that its condition estimator never gets much larger than 10^5. Moreover, the final convergence rate, although not superlinear, is quite rapid.

ACKNOWLEDGEMENTS

The authors would like to thank Robert Burchett of the General Electric Company for providing the 30-bus OPF problem. We are also indebted to Chris Fraley for help in obtaining the results for Dembo 7. This research was supported by the U.S. Department of Energy Contract DE-AM03-76SF00326, PA No. DE-AT03-76ER72018; National Science Foundation Grants MCS-7926009 and ECS-8312142; the Office of Naval Research Contract N00014-75-C-0267; and the U.S. Army Research Office Contract DAAG29-84-K-0156.

REFERENCES

Bartholomew-Biggs, M. C. (1982). "Recursive quadratic programming methods for nonlinear constraints", in *Nonlinear Optimization, 1981*, (M. J. D. Powell, ed.), pp. 213–221, Academic Press, London.

Biggs, M. C. (1972a). "Constrained minimization using recursive equality quadratic programming", in *Numerical Methods for Non-Linear Optimization* (F. A. Lootsma, ed.), pp. 411–428, Academic Press, London and New York.

Biggs, M. C. (1972b). Minimization algorithms making use of non-quadratic properties of the objective function, *J. Inst. Maths. Applics.* **8**, pp. 315–327.

Boggs, P. T., Tolle, J. W. and Wang, P. (1982). On the local convergence of quasi-Newton methods for constrained optimization, *SIAM J. Control and Optimization* **20**, pp. 161–171.

Bracken, J. and McCormick, G. P. (1968). *Selected Applications of Nonlinear Programming*, John Wiley and Sons, New York and Toronto.

Brent, R. P. (1973). *Algorithms for Minimization without Derivatives*, Prentice-Hall, Inc., Englewood Cliffs, New Jersey.

Bunch, J. R. and Kaufman, L. C. (1980). A computational method for the indefinite quadratic programming problem, *Linear Algebra and its Applics.* **34**, pp. 341–370.

Coleman, T. F. and Conn, A. R. (1982). Nonlinear programming via an exact penalty function, *Math. Prog.* **24**, pp. 123–161.

Coleman, T. F. and Sorensen, D. C. (1984). A note on the computation of an orthogonal basis for the null space of a matrix, *Math. Prog.* **29**, pp. 234–242.

Dembo, R. S. (1976). A set of geometric test problems and their solutions, *Math. Prog.* **10**, pp. 192–213.

Dennis, J. E., Jr. and Moré, J. J. (1977). Quasi-Newton methods, motivation and theory, *SIAM Review* **19**, pp. 46–89.

Dennis, J. E., Jr. and Schnabel, R. E. (1981). "A new derivation of symmetric positive definite secant updates", *Nonlinear Programming 4* (O. L. Mangasarian, R. R. Meyer and S. M. Robinson, eds.), pp. 167–199, Academic Press, London and New York.

Dennis, J. E., Jr. and Schnabel, R. B. (1983). *Numerical Methods for Unconstrained Optimization and Nonlinear Equations*, Prentice-Hall, Inc., Englewood Cliffs, New Jersey.

Fiacco, A. V. and McCormick, G. P. (1968). *Nonlinear Programming: Sequential Unconstrained Minimization Techniques*, John Wiley and Sons, New York and Toronto.

Fletcher, R. (1981). *Practical Methods of Optimization, Volume 2, Constrained Optimization*, John Wiley and Sons, New York and Toronto.

Fletcher, R. (1983). "Penalty functions", in *Mathematical Programming: The State of the Art*, (A. Bachem, M. Grötschel and B. Korte, eds.), pp. 87–114, Springer-Verlag, Berlin, Heidelberg, New York and Tokyo.

Gabay, D. (1982). Reduced quasi-Newton methods with feasibility improvement for nonlinearly constrained optimization, *Math. Prog. Study* **16**, pp. 18–44.

Gill, P. E., Golub, G. H., Murray, W. and Saunders, M. A. (1974). Methods for modifying matrix factorizations, *Math. Comp.* **28**, pp. 505–535.

Gill, P. E. and Murray, W. (1972). Quasi-Newton methods for unconstrained optimization, *J. Inst. Maths. Applics.* **9**, pp. 91–108.

Gill, P. E. and Murray, W. (1974). Newton-type methods for unconstrained and linearly constrained optimization, *Math. Prog.* **28**, pp. 311–350.

Gill, P. E. and Murray, W. (1978). Numerically stable methods for quadratic programming, *Math. Prog.* **14**, pp. 349–372.

Gill, P. E., Murray, W., Saunders, M. A. and Wright, M. H. (1984a). User's guide for QPSOL (Version 3.2): a Fortran package for quadratic programming, Report SOL 84-6, Department of Operations Research, Stanford University, California.

Gill, P. E., Murray, W., Saunders, M. A. and Wright, M. H. (1984b). User's guide for NPSOL (Version 2.1): a Fortran package for nonlinear programming, Report SOL 84-7, Department of Operations Research, Stanford University, California.

Gill, P. E., Murray, W., Saunders, M. A. and Wright, M. H. (1984c). Procedures for optimization problems with a mixture of bounds and general linear constraints, *ACM Transactions on Mathematical Software* **10**, pp. 282–298.

Gill, P. E., Murray, W., Saunders, M. A. and Wright, M. H. (1984d). Software and its relationship to methods, Report SOL 84-10, Department of Operations Research, Stanford University, California.

Gill, P. E., Murray, W., Saunders, M. A., Stewart, G. W. and Wright, M. H. (1985a). Properties of a representation of a basis for the null space, Report SOL 85-1, Department of Operations Research, Stanford University, California.

Gill, P. E., Murray, W., Saunders, M. A. and Wright, M. H. (1985b). The design and implementation of a quadratic programming algorithm, to appear, Department of Operations Research, Stanford University, California.

Gill, P. E., Murray, W. and Wright, M. H. (1981). *Practical Optimization*, Academic Press, London and New York.

Goldfarb, D. and Idnani, A. (1983). A numerically stable dual method for solving strictly convex quadratic programs. *Math. Prog.* **27**, pp. 1–33.

Han, S.-P. (1976). Superlinearly convergent variable metric algorithms for general nonlinear programming problems, *Math. Prog.* **11**, pp. 263–282.

Harris, P. M. J. (1973). Pivot selection methods of the Devex LP code, *Math. Prog.* **5**, pp. 1–28. [Reprinted in *Math. Prog. Study* **4** (1975), pp. 30–57.]

Hock, W. and Schittkowski, K. (1981). Test examples for nonlinear programming, *Lecture Notes in Economics and Mathematical Systems*, Volume 187, Springer Verlag, Berlin, Heidelberg and New York.

McCormick, G. P. (1983). *Nonlinear programming*, Wiley, Wiley-Interscience.

Murray, W. (1971). An algorithm for finding a local minimum of an indefinite quadratic program, Report NAC 1, National Physical Laboratory, England.

Murray, W. and Wright, M. H. (1978). Methods for nonlinearly constrained optimization based on the trajectories of penalty and barrier functions, Report SOL 78–23, Department of Operations Research, Stanford University.

Murray, W. and Wright, M. H. (1982). Computation of the search direction in constrained optimization algorithms, *Math. Prog. Study* **16**, pp. 63–83.

Murtagh, B. A. and Saunders, M. A. (1978). Large-scale linearly constrained optimization, *Math. Prog.* **14**, pp. 41–72.

Nocedal, J. and Overton, M. (1983). Projected Hessian updating algorithms for nonlinearly constrained optimization, Report 95, Department of Computer Science, Courant Institute of Mathematical Sciences, New York University, New York.

Powell, M. J. D. (1974). "Introduction to constrained optimization", in *Numerical Methods for Constrained Optimization* (P. E. Gill and W. Murray, eds.), pp. 1–28, Academic Press, London and New York.

Powell, M. J. D. (1977). "A fast algorithm for nonlinearly constrained optimization calculations", in *Numerical Analysis, Dundee, 1977* (G. A. Watson, ed.), pp. 144–157, Springer-Verlag Lecture Notes in Mathematics, Volume 630, Berlin, Heidelberg and New York.

Powell, M. J. D. (1983a). "Variable metric methods for constrained optimization", in *Mathematical Programming: The State of the Art*, (A. Bachem, M. Grötschel and B. Korte, eds.), pp. 288–311, Springer-Verlag, Berlin, Heidelberg, New York and Tokyo.

Powell, M. J. D. (1983b). ZQPCVX a Fortran subroutine for convex quadratic programming, Report DAMTP 83/NA17, Department of Applied Mathematics and Theoretical Physics, University of Cambridge, England.

Robinson, S. M. (1974). Perturbed Kuhn-Tucker points and rates of convergence for a class of nonlinear programming algorithms, *Math. Prog.* **7**, pp. 1–16.

Schittkowski, K. (1981). The nonlinear programming method of Wilson, Han, and Powell with an augmented Lagrangian type line search function, *Numerische Mathematik* **38**, pp. 83–114.

Schittkowski, K. (1983). On the convergence of a sequential quadratic programming method with an augmented Lagrangian line search function, Math. Operationsforsch. u. Statist., Ser. Optimization **14**, pp. 197–216.

Stewart, G. W. (1973). *Introduction to matrix computations,* Academic Press, London and New York.

Stoer, J. (1985). Foundations of recursive quadratic programming methods for solving nonlinear programs. This volume.

Tone, K. (1983). Revisions of constraint approximations in the successive QP method for nonlinear programming, *Math. Prog.* **26**, pp. 144–152.

Wilson, R. B. (1963). *A Simplicial Algorithm for Concave Programming*, Ph. D. Thesis, Harvard University.

Wright, M. H. (1976). Numerical Methods for Nonlinearly Constrained Optimization, Ph. D. Thesis, Stanford University.

COMPARATIVE PERFORMANCE EVALUATION, EXPERIMENTAL DESIGN, AND GENERATION OF TEST PROBLEMS IN NON-LINEAR OPTIMIZATION

F.A. Lootsma

Department of Mathematics and Informatics
Delft University of Technology
P.O. Box 356, 2600 AJ Delft, Netherlands.

1. INTRODUCTION

The key issues in performance evaluation and computational testing were raised during the NATO Advanced Research Institute on Nonlinear Programming, Cambridge, UK, July 1981 (see Powell (1982)). Briefly summarized, they are as follows:

a. Objectives of Testing. What do we want to test: fool-proof commercial optimization codes with all their user-oriented features and safety provisions to serve the general user, or the underlying algorithmic ideas to support future research of algorithm designers? Should we concern ourselves with complete optimization codes for solving (classes of) nonlinear optimization problems, or with the "building bricks" such as linear searches, LP subroutines, and QP subroutines?

b. Selection of Test Problems. The question of how to select artificial or real-life test problems is a point of major concern, and various ideas have been brought forward: the selection on the basis of particular *properties* (dimension, sparsity, ill-conditioning, ...), and the selection of certain *categories* of test problems (convex, linearly constrained, posynomial problems, sums of squares, ...).

c. Performance Criteria. There is a growing consensus on the performance criteria to be considered (particularly efficiency, robustness, ease of use, and program organization), but quantification of the performance remains cumbersome. Efficiency can be expressed in execution times (highly machine-dependent) or equivalent function evaluations (ignoring the overheads of the algorithm). Both efficiency and robustness are very sensitive to parameters which control the course of the computations (e.g. tuning of nested iterative processes). Ease of use (documentation, input/output facilities, numerical differentiation, graceful termination) and program organization (modularity, portability) are not easy to assess.

d. The Computational Environment. The test results are highly dependent on the computers and compilers that have been used. The development of optimization algorithms is seriously challenged by the appearance of parallel computers. Interactive optimi-

NATO ASI Series, Vol. F15
Computational Mathematical Programming
Edited by K. Schittkowski

zation software will be required, both for inexperienced and sophisticated users so that they can follow the course of the computations and the degree of improvement.

e. Analysis of Test Results. A neglected area has hitherto been the formal analysis of the test results (statistical analysis, multi-criteria analysis, significance tests), and the ranking and rating of codes and methods under various performance criteria.

2. COMPARATIVE STUDIES

The leading comparative studies in non-linear optimization (see Table 1) are mainly concerned with the computational implementations of the following general methods:

- Reduced Gradients (RG),
- Quadratic Approximations (QA),
- Penalty Functions (PF),
- Augmented Lagrangians (AL),

and with a method for a particular class of non-linear programming problems:

- Geometric Programming (GP).

What the authors observe, is the behaviour of the codes, but they are mainly interested in the performance of the underlying methods; see, for instance, the titles of Fattler e.a. (1982), Mulvey (1982), Rijckaert and Martens (1978), Sandgren and Ragsdell (1982), Staha and Himmelblau (1973), where the words "methods", "techniques", and "algorithms" appear, not the word "codes" as in Schittkowski (1980). And indeed, we have the impression that the published results reveal some sort of clustering which is due to the methods. So, we may speculate that it is possible to rank and rate the non-linear optimization algorithms on the basis of their performance under certain decision criteria.

Table 1 briefly summarizes the scope of each computational study: the total number of codes, the total number of test problems, the number of runs per code (the number of starting points, usually one per test problem, but two in Schittkowski (1980), and roughly fourteen in Fattler et. al. (1982)), and the largest test problem with regard to the dimension. The studies are clearly restricted to small problems. First-order and second-order derivatives are supplied in various ways: analytically (A) in computational studies with test problems of a special structure, numerically (N) in studies using finite-difference approximations to the derivatives, or both (N,A) accordingly as the codes accept analytical derivatives or finite-difference approximations.

Table 1

Number of codes and number of test problems used in comparative studies of nonlinear optimization software

Study and year	Number of codes						Derivatives	Total codes	Total problems	Runs per code	Max n
	RG	QA	PF	AL	GP	XX					
Colville 1968	4	1	4			25	N,A	34	8	8	16
Staha and Himmelblau 1973	1		1			2	N,A	4	24	31	100
Eason and Fenton 1974			17				N	17	10	10	7
Dembo 1978	2			1	5	2	A	10	8	13	16
Sarma et al. 1978			1		4		A	5	16	21	24
Rijckaert and Martens 1978				1	16		A	17	24	120	13
Schittkowski 1980	4	3	6	12		1	N,A	26	185	370	20
Fattler et al. 1982	1		1	1	6	1	N,A	10	42	588	30
Sandgren and Ragsdell 1982	4		27	1		3	N	35	30	30	48

3. TEST PROBLEMS

Real-life test problems have the advantage of being found in practical circumstances. Usually, however, it is not easy to establish their topological properties. Even the theoretical solution may be unknown.

Generators of artificial test problems are based on particular models of the objective and constraint functions. Rosen and Suzuki (1965) originally proposed a quadratic model. Schittkowski (1980) developed a test battery with geometric-programming problems, using the model of a signomial. The coefficients of artificial test problems are determined to guarantee that the first-order necessary conditions as well as the second-order sufficient conditions for optimality are satisfied at a prespecified solution. Ideally, the purpose of controlled testing is to establish the relative performance of codes or algorithms on a graded set of test problems, that is, on problems with increasing size, increasing degree of degeneracy (the ratio of the largest and the smallest Lagrangian multipliers corresponding to the active constraints), increasing degree of ill-conditioning (the ratio of the largest and the smallest eigenvalue of the Hessian matrix of the Lagrangian function at a Kuhn-Tucker point), etc. With these "performance profiles" the actual selection of codes or algorithms for particular purposes is a multi-criteria decision problem (Dembo and Mulvey (1978)).

A serious objection against the above mode of operation is that many real-life engineering problems differ substantially from the artificial test problems: the function evaluations are expensive and inaccurate, the derivatives are not explicitly available, their evaluations are also expensive and inaccurate. Hence, several test batteries contain a mixture of real-life and artificial test problems selected in a haphazard manner.

Nevertheless, algorithms are compared experimentally because theoretical studies do not produce decisive answers. The information in the comparative studies can now be used to prepare particular decisions. We will illustrate this by addressing the following decision (see also Lootsma (1982)): what is the best method for solving geometric-programming problems in research, development and engineering - is it the geometric-programming (GP) method or some general method for non-linear optimization (RG, QA, PF, or AL)?

4. PERFORMANCE CRITERIA

The formulation of performance criteria in an actual decision problem is a complicated process. In the first step we usually follow a top-down approach: we formulate criteria on the basis of logical arguments, and we also introduce criteria that have previously been employed in similar situations (see Crowder, Dembo and Mulvey (1978),

Crowder and Saunders (1980)). In doing so we obtained:

F_1: Domain of Applications, the type of problems for which the method has been designed (problems with continuous functions, for instance, or with convex functions, problems with functions having continuous second derivatives, problems with special functions like posynomials or signomials, etc.).

F_2: Robustness, the power to solve problems in the domain of applications with the required accuracy (the power is usually measured in the percentage of problems that have "reasonably" been solved, even if the required accuracy is not fully obtained).

F_3: Efficiency, the effort (usually measured in execution time, number of function evaluations and/or number of equivalent function evaluations) necessary to solve problems in the domain of applications with the required accuracy.

F_4: Capacity, the maximum size (measured in the number of variables and the number of constraints) of the problems that can generally be solved.

These criteria have the advantage of being generally accepted; they can be anticipated in future comparative studies, but they may be ineffective in a particular decision problem. It is worth noting (see the end of sec. 3) that we compare methods, not computer codes, so the well-known criteria "ease of use" (referring to user-oriented features such as the number of input parameters to be set, the length of the data-preparation phase, the facilities for numerical differentiation, proper diagnostics in cases of slow or no convergence, and output facilities) and "program organization" (the language, the modularity, and the portability) are not applicable here.

The second step, a bottom-up approach, is usually triggered by striking properties of the alternatives under consideration (the algorithms mentioned in sec. 2). And there are striking properties, at least in our opinion: geometric programming is much more difficult to explain in an engineering environment (codes for non-linear optimization are no black boxes; the user needs a fair amount of understanding of the method to make it operational), and the leading reduced-gradient methods in the comparative studies are much longer than the other ones (a factor of 5). Hence, the following criteria emerge.

F_5: Conceptual simplicity: inversely proportional to the amount of sophistication in a method (the employment of duality theory, for instance). Simple methods sometimes have a widespread popularity in research and development laboratories: they can easily be adapted to the handling of special or large problems, and sometimes they readily show how to scale the variables and the constraints.

F_6: Shortness of code, the number of lines of coding in a high-level programming language. By this criterion we penalize structured codes with many numerical safeguards, but a short code is suitable for incorporation in other codes and usable on small computers.

These criteria are typically ad hoc: relevant in an actual decision problem, but unexpected for code designers and authors of comparative studies. In a comparison of gradient methods for unconstrained minimization, for instance, the storage requirements would emerge as an effective criterion to distinguish the Fletcher-Reeves conjugate-gradient method from the variable-metric and the Newton-type methods.

After the second step the formulation of criteria is not necessarily finished. It may happen, during a decision process, that criteria are reformulated or dropped and that new criteria emerge as a result of the on-going deliberations.

5. THE EVALUATION PROCEDURE

In answering the question formulated at the end of sec. 3 we are in fact running up against a two-level decision problem. At the first level, we try to establish the relative weights of the performance criteria. At the second level, we estimate the relative weights of the alternatives under each of the performance criteria separately. The weighing is preferably carried out via a method of pairwise comparisons because straightforward ranking and rating introduces many inconsistencies (an issue which is nicely illustrated by a wine-tasting party). Lastly, multiplying the weights of the methods by the weights of the respective criteria and adding the products, we obtain a final score for each alternative.

In a method of pairwise comparisons, stimuli (criteria, alternatives, for example) are presented in pairs to one or more referees (decision makers). The basic experiment is the comparison of two stimuli S_i and S_j by a referee. He may either have no preference for one of the two, or a graded preference for one over the other. The gradation in the comparative judgement of the k-th referee is usually expressed by the estimate r_{ijk} of the weight ratio V_i/V_j. We approximate the weights V_i and V_j of the respective stimuli S_i and S_j by the normalized solution $\bar{v}_i$ and $\bar{v}_j$ of the least-squares problem of minimizing

$$\sum_{i<j} \sum_{k} (\ln r_{ijk} - \ln v_i + \ln v_j)^2 . \qquad (1)$$

The solution represents the weights assigned by the group of referees. The individual weights are obtained by the minimization of

$$\sum_{i<j} (\ln r_{ijk} - \ln v_i^{(k)} + \ln v_j^{(k)})^2 , \qquad (2)$$

for each referee separately. Further details about the scaling of qualitative judgements (the assignment of numerical values r_{ijk}) may be found in Lootsma (1982).

Operating at the first level of the decision process, we mailed the questionnaire of Figure 1 (with the pairs of performance criteria in random order) to the organizers and the invited speakers of the NATO ASI on Computational Mathematical Programming. With the responses (11 out of 17) we obtained the group weights and the individual weights displayed in Table 2. Interesting conclusions can be drawn from it. Robustness is almost uniformly the most important criterion. Second (at least for the group, the individual opinions vary) is capacity; it is surprising, then, to observe that computational studies are virtually based on small test problems only ($n \leq 30$). Efficiency is the third criterion, notwithstanding the predominant concern for it in computational testing. Shortness of code does not play a significant role in non-linear optimization, which is probably carried out mainly on larger computers. In summary, it is urgent to include medium-size test problems ($30 < n \leq 100$) in future computational studies. We note in passing that application of cluster analysis to the 11 vectors of individual weights has not been successful (Dr. G. Timmer, Delft University of Technology); the results, highly dependent on the cluster techniques that have been employed, did not yield any additional explication of the opinions.

To illustrate a possible modus operandi at the second level of the decision process, we refer to a previous paper (Lootsma (1982)). The weights of the algorithms under the respective criteria appear in Table 9 of the cited article. They can be used to calculate final scores of the algorithms. The results are displayed in Table 3. We have tacitly assumed that the group weights of the criteria are acceptable in the first attempt to identify the critical issues and to arrive at a consensus.

We do not expect the reader to accept QA methods immediately for the solution of GP problems. Table 3 is a basis for further discussions. Highly subjective, qualitative criteria have been employed so far; it may be beneficial to repeat the calculations with criteria allowing for straightforward quantification of the performance: robustness, efficiency, capacity and shortness of code. It depends entirely on the decision makers which criteria they finally accept in order to arrive at a joint conclusion.

Our evaluation has a particular drawback: in order to obtain weights for the criteria and the optimization methods, we assigned numerical values to certain performance ratios. In general, however, we only have a fuzzy notion of these ratios (this is due to the choice of test problems, the setting of parameters which control the computational process, and the choice of stopping rules); the numerical values of the performance ratios are obviously too precise, and the final scores of the methods under consideration suggest an accuracy which is out of proportion. Hence, we also

Multi Criteria Analysis Nonlinear Programming Methods

F.A.Lootsma

FOR OFFICE USE ONLY

1-4 | 1

Compare the impact of the following pairs of CRITERIA and mark the appropriate box.

56 78

FOR OFFICE USE ONLY	Criterion	has MUCH HIGHER impact	has SOMEWHAT HIGHER impact than	has ABOUT EQUAL impact as		has SOMEWHAT LOWER impact than	has MUCH LOWER impact	Criterion
3- 2	Efficiency							Robustness
2- 5	Robustness							Concept.simplicity
4- 2	Capacity							Robustness
4- 3	Capacity							Efficiency
2- 6	Robustness							Length of code
1- 2	Generality							Robustness
5- 6	Concept.simplicity							Length of code
5- 3	Concept.simplicity							Efficiency
1- 6	Generality							Length of code
4- 6	Capacity							Length of code
1- 5	Generality							Concept.simplicity
3- 1	Efficiency							Generality
1- 4	Generality							Capacity
5- 4	Concept.simplicity							Capacity
3- 6	Efficiency							Length of code

Figure 1. Computer-generated questionnaire for pairwise comparison of performance criteria

Performance criteria	Group weights	Individual weights (obtained from organizers and invited speakers)										
		1	2	3	4	5	6	7	8	9	10	11
1. Generality	12.7	19.3	16.0	8.7	11.7	10.1	7.0	12.6	15.2	10.3	12.3	9.0
2. Robustness	34.5	35.8	27.4	34.3	44.7	30.0	20.8	26.2	34.0	39.3	35.9	31.7
3. Efficiency	18.5	15.3	19.0	31.5	15.3	18.9	39.2	26.2	6.7	21.1	10.6	18.3
4. Capacity	21.9	7.7	27.4	14.1	20.0	32.7	19.1	26.2	31.2	11.3	28.7	29.1
5. Conc. Simpl.	7.3	11.6	4.1	5.6	5.2	4.2	7.0	4.4	8.2	13.4	8.8	8.1
6. Shortness of Code	5.2	10.3	6.0	5.8	3.1	4.2	7.0	4.4	4.7	4.5	3.6	3.7

Table 2. Group weights and individual weights of performance criteria for non-linear optimization algorithms, calculated via pairwise-comparison method.

Performance criteria	1 RG	2 QA	3 PF	4 AL	5 GP	6. Group weights of criteria	7 RG	8 QA	9 PF	10 AL	11 GP
1. Generality	.235	.235	.235	.235	.059	.127	.030	.030	.030	.030	.007
2. Robustness	.224	.321	.143	.173	.139	.345	.077	.111	.049	.060	.048
3. Efficiency	.275	.384	.065	.106	.171	.185	.051	.071	.012	.020	.032
4. Capacity	.200	.200	.200	.200	.200	.219	.044	.044	.044	.044	.044
5. Conc. Simpl.	.218	.122	.379	.201	.081	.073	.016	.009	.028	.015	.006
6. Shortness of Code	.051	.366	.194	.194	.194	.052	.003	.019	.010	.010	.010
Final Score						1.000	.221	.284	.173	.179	.147

Table 3. Final scores of non-linear optimization algorithms. The columns 1-5 contain the weights of the algorithms under each performance criterion separately, the columns 7-11 the same weights multiplied by the group weight (column 6).

concerned ourselves with an evaluation procedure where fuzzy numbers are assigned to the performance ratios (Lootsma (1984)). For reasons of simplicity, we used fuzzy numbers with triangular membership functions, and we employed the algebraic manipulations of van Laarhoven and Pedrycz (1983) who simplified the rules of Dubois and Prade (1978, 1980).

The above evaluation is based on the philosophy that high-quality test batteries, although they produce contradictory results, should be taken equally serious. The authors represent a wide community of sophisticated users. We do not believe that the divergent results and experiences will ever be overruled by one, huge test battery. The older comparative studies will always remain a valuable source of information, whereas new studies (Rijckaert and Walraven (1984)) will gradually change our insight.

The deterministic and the fuzzy evaluation procedure are probably applicable in other areas of optimization and numerical analysis as well. The crucial question is whether the computational studies in those areas provide sufficient material for an evaluation under the criteria which are felt (at least by the scientific community) to be predominant.

ACKNOWLEDGEMENT

It is a pleasure to acknowledge Mr. F. Vos (Delft University of Technology) for his excellent support in multi-criteria decision analysis.

REFERENCES AND LITERATURE

[1] Colville, A.R., A Comparative Study on Nonlinear Programming Codes. Report 320-2949, IBM Scientific Center, New York, 1968.

[2] Crowder, H.P., Dembo, R.S., and Mulvey, J.M., Reporting Computational Experiments in Mathematical Programming. Math. Progr. 15, 291-315, 1978.

[3] Crowder, H.P., and Saunders, P.B., Results of a Survey on MP Software Indicators. Committee on Algorithms Newsletter (Math. Progr. Society), January 1980, 2-7.

[4] Dembo, R.S., Current State of the Art of Algorithms and Computer Software for Geometric Programming. J. Opt. Th. Appl. 26, 149-183, 1978.

[5] Dembo, R.S., and Mulvey, J.M., On the Analysis and Comparison of Mathematical Programming Algorithms and Software. In W.W. White (ed.), Computers and Mathematical Programming. NBS Special Publication 502, US Government Printing Office, Washington DC, 106-116, 1978.

[6] Dubois, D., and Prade, H., Operations on Fuzzy Numbers. Int. J. Systems Sci. 9, 613-626, 1978.

[7] Dubois, D., and Prade, H., Fuzzy Sets and Systems, Theory and Applications. Academic Press, New York, 1980.

[8] Eason, E.D., and Fenton, F.G., A Comparison of Numerical Optimization Methods for Engineering Design. Trans. ASME, Ser. B. Vol. 96, 196-200, 1974.

[9] Fattler, J.E., Sin, Y.T., Root, R.R., Ragsdell, K.M., and Reklaitis, G.V., On the Computational Utility of Posynomial Geometric Programming Solution Methods. Math. Progr. 22, 163-201, 1982.

[10] Hock, W., and Schittkowski, K. Test Examples for Nonlinear Programming Codes. Lecture Notes Ec. Math. Systems 187, Springer, Berlin, 1981.

[11] Laarhoven, P.J.M. van, and Pedrycz, W., A Fuzzy Extension of Saaty's Priority Theory. Fuzzy Sets and Systems 11, 229-242, 1983.

[12] Lootsma, F.A., Performance Evaluation of Nonlinear Optimization Methods via Multi-Criteria Decision Analysis and via Linear Model Analysis. In M.J.D. Powell (ed.), Nonlinear Optimization 1981. Academic Press, London, 419-453, 1982.

[13] Lootsma, F.A., Performance Evaluation of Nonlinear Optimization Methods via Pairwise Comparison and Fuzzy Numbers. Paper presented at the NATO ASI on Comp. Math. Progr., Bad Windsheim, FR Germany, 1984.

[14] Mulvey, J. (ed.), Evaluating Mathematical Programming Techniques. Lecture Notes Ec. Math. Systems 199, Springer, Berlin, 1982.

[15] Powell, M.J.D., Nonlinear Optimization 1981. Academic Press, London, 1982.

[16] Rijckaert, M.J., and Martens, X.M., Comparison of Generalized Geometric Programming Algorithms. J. Opt. Th. Appl. 26, 205-242, 1978.

[17] Rijckaert, M.J., and Walraven, E.J.C., Reflections on Geometric Programming. Paper presented at the NATO ASI on Comp. Math. Progr., Bad Windsheim, F.R. Germany, 1984. Cath. Univ. Louvain, Dept. of Chem. Eng., Louvain, Belgium.

[18] Rosen, J.B., and Suzuki, S., Construction of Nonlinear Programming Test Problems, Comm. ACM 8, 113, 1965.

[19] Saaty, Th.L., The Analytic Hierarchy Process, Planning, Priority Setting, Resource Allocation. McGraw-Hill, New York, 1980.

[20] Sandgren, E., and Ragsdell, K.M., On Some Experiments which delimit the Utility of Nonlinear Programming Methods for Engineering Design. Math. Progr. Study 16, 118-136, 1982.

[21] Sarma, P.V., Martens, X.M., Reklaitis, G.V., and Rijckaert, M.J., A Comparison of Computational Strategies for Geometric Programs. J. Opt. Th. Appl. 26,

185-203, 1978.

[22] Schittkowski, K., Nonlinear Optimization Codes. Lecture Notes Ec. Math. Systems 183, Springer, Berlin, 1980.

[23] Staha, R.L., and Himmelblau, D.M., Evaluation of Constrained Nonlinear Programming Techniques. Report, Dept. of Chem. Eng., University of Texas at Austin, Texas 78712, 1973.

[24] White, W.W. (ed.), Computers and Mathematical Programming. NBS Special Publication 502, US Government Printing Office, Washington, DC, 1978.

ON CONVERTING OPTIMAL CONTROL PROBLEMS INTO NONLINEAR PROGRAMMING PROBLEMS

Dieter Kraft
Deutsche Forschungs- und Versuchsanstalt
für Luft- und Raumfahrt
Institut für Dynamik der Flugsysteme

D-8031 Oberpfaffenhofen, W-Germany

Abstract Direct solutions of the optimal control problem are considered. Two discretization schemes are proposed which are based on the parameterization of the control functions and on the parameterization of the control and the state functions, leading to direct shooting and direct collocation algorithms, respectively. The former is advantageous for problems with unspecified final state, the latter for prescribed final state and especially for stiff problems. The sparsity of the Jacobian matrix of the constraints and the Hessian matrix of the Lagrangian must be exploited in the direct collocation method in order to be efficient. The great advantage of the collocation approach lies in the availability of analytical gradients.

Keywords Nonlinear Systems, Optimal Control, Piecewise Polynomial Approximation, Direct Shooting, Direct Collocation, Nonlinear Optimization

1. Introduction

Recently much progress has been made in the development of efficient and robust numerical methods for the solution of nonlinear programming problems

$$\min_{x \in R^n} \quad F(x)$$

$$\text{subject to} \quad c_i(x) = 0, \qquad i=1,\ldots,m_e,$$

$$c_i(x) \leq 0, \qquad i=m_e+1,\ldots,m,$$

by finding the Kuhn-Tucker points of the set of equalities defined by the first-order necessary conditions. [1] Generally Newton's method or a quasi-Newton method is applied to their iterative solution. The conceptual analogy in optimal control has long been known, [2,3] and leads to two-point boundary-value problems which are solved by shooting-, discretization-, or collocation algorithms. [4] These methods are of the indirect type because they revert to the first-order necessary conditions. [5] In indirect methods, a second vector- differential equation is intro-

NATO ASI Series, Vol. F15
Computational Mathematical Programming
Edited by K. Schittkowski

duced, defining the multipliers which are adjoined to the differential constraints in the first-order necessary conditions.
To avoid this disadvantage a number of authors have proposed the direct application of nonlinear programming techniques to the solution of optimal control problems. [6-18] The most common approach for converting the optimal control problem into a nonlinear programming problem is to approximate the control variables by choosing an appropriate structure with finitely many free parameters and to consider the state variables as functions of the control variables through the differential equations. Hereby, the problem is solved in the reduced space of control parameters, and the state variables are obtained by the solution of an initial value problem. This relates the method strongly to shooting methods, and it is therefore called direct shooting.
An alternative approach is to discretize both control and state variables. This leads to a problem with higher dimension, but avoids the initial value problem as the differential equations are approximated by collocation. A further advantage is a structural sparsity pattern, both in the Hessian matrix of the Lagrangian function and in the Jacobian matrix of the constraint functions if the discretization model is appropriatly (locally) chosen. From similar reasons as with direct shooting, this method will be termed direct collocation.
Any nonlinear programming code can be taken to solve the resulting problems, but it is strongly recommended to use algorithms based on the Lagrangian function in connection with sequential quadratic programming. [19-21]
In the present paper, the optimal control problem formulation is given in a fairly general setting in section 2. Here also first order necessary conditions are given, which lead to a better understanding of the problem and a clarification of some algorithmic details. No proofs will be given as they would be considerably too lengthy and technical. Neustadt (1976), [22] Ioffe and Tichomirov (1979), [23] or Cesari (1983) [3] are thorough mathematical sources. Section 3 contains the necessary elements from approximation theory that will be used for the problem reformulation. The transformations of the optimal control problem into nonlinear programming formulations, as indicated above, are presented in section 4, together with algorithmic details, such as gradient generation and step size control. Numerical examples with the approaches are considered elsewhere. [24, 25]
Finally we introduce some notation. We will be working with column vectors and matrices, the transposed of which will be denoted by ^{T}. $(\cdot|\cdot)$ is the scalar product. f_x stands for the (column vector of) partial derivatives of a real-valued scalar function $f(x)$ with respect to (w.r.t.) the elements of a vector x. $\dot{x}(t)$ indicates differentiation w.r.t. the independent variable t. Equations are numbered serially in each section. If we refer to an equation in another section that equation is supplied with the respective section number.

2. Formulation of the Optimal Control Problem

2.1. Functionals, Constraints, Boundary Conditions and Optimal Control Processes.

We shall consider functions depending on one independent variable $t \in [a,b] =: I$. These functions can be arranged in state variables $x \in R^n$ and control variables $u \in R^m$. x is assumed to be absolutely continuous, u to be bounded and measurable. If a and/or b are allowed to be free (for optimization) we have a free-time problem, otherwise we have a fixed time problem. Generally, we are concerned with finding minima of three types of functionals of the form

(i) Mayer problem:

$$J_M(x(\cdot),u(\cdot)) = g(a,x(a),b,x(b)), \tag{1}$$

(ii) Lagrange problem:

$$J_L(x(\cdot),u(\cdot)) = \int_a^b f_o(t,x(t),u(t))dt, \tag{2}$$

(iii) Bolza problem:

$$J_B(x(\cdot),u(\cdot)) = J_M(x(\cdot),u(\cdot)) + J_L(x(\cdot),u(\cdot)). \tag{3}$$

All three functionals are equivalent to each other. [22]

The pair $(x(\cdot),u(\cdot))$ is linked by a differential constraint

$$\dot{x}(t) = f(t,x(t),u(t)), \tag{4}$$

called the state equation or dynamic equation. The pair is called a control process. Every solution $x(\cdot)$ of (4) corresponding to a control $u(\cdot)$ is called a phase trajectory or trajectory (for short).
Boundary conditions for the process (4) are given in the following general functional relation

$$r(a,x(a),b,x(b)) = 0, \tag{5}$$

which are often separated in practical applications

$$\begin{aligned} r_0(a,x(a)) &= 0, \\ r_1(b,x(b)) &= 0. \end{aligned} \tag{6}$$

The trajectory and the control are subject to further constraints of the form

$$c(t,x(t),\dot{x}(t),u(t)) \leq 0, \quad \forall\, t \in I, \tag{7}$$

$$u(t) \in U \subset R^m, \quad \forall\, t \in I, \text{ a.e. } . \tag{8}$$

In (7), equality is considered as a special case of inequality. If the inequalities in (7) are independent of $\dot{x}$ and u

$$c(t,x(t)) \leq 0, \tag{9}$$

these constraints are called state or phase constraints; if they are independent of x and $\dot{x}$

$$c(t,u(t)) \leq 0, \tag{10}$$

they are called control constraints; if they appear in the form

$$c(t,x(t),u(t)) \leq 0, \tag{11}$$

they are named mixed control and state constraints.

A triple (I,x(t),u(t)) is admissible if the control process (x(t),u(t)) satisfies (4) a.e. in I, and furthermore if admissible controls (8) generate via (4) admissible trajectories (5) and (7). This triple is called an admissible control process.

An admissible control process $(I^0,x^0(t),u^0(t))$ is an optimal control process, if there exists a number $\varepsilon > 0$ such that for every other admissible control process (I,x(t),u(t)) we have

$$|a-a^0| \leq \varepsilon, \quad |b-b^0| \leq \varepsilon, \quad \|x(t)-x(t)^0\| \leq \varepsilon, \tag{12}$$

and

$$J(x^0(\cdot),u^0(\cdot)) \leq J(x(\cdot),u(\cdot)), \tag{13}$$

for all $t \in I \cap I^o$. (13) is called a strong local minimum of the Mayer-, Lagrange-, or Bolza-problem, depending on the used functional. If, in addition to (12), $\|u(t) - u^o(t)\| \leq \varepsilon$ then (13) is a weak local minimum. Our problem is to find optimal control processes.

2.2. First-Order Necessary Conditions

2.2.1. Lagrange-Problem without State Constraints

To introduce some additional notions we begin with a classical problem from the calculus of variations:
minimize

$$J(x(\cdot),u(\cdot)) = \int_a^b f_o(t,x(t),u(t))dt, \tag{14}$$

subject to the constraints

$$\begin{aligned} \dot{x}(t) &= f(t,x(t),u(t)), \\ r_o(x(a)) &= 0, \\ r_1(x(b)) &= 0. \end{aligned} \tag{15}$$

We assume the following differentiability requirements: $f_o, f : R \times R^n \times R^m \to R$, and $r_o, r_1 : R^n \to R^{\rho_0,\rho_1}$ are continuously differentiable with respect to all variables throughout their domain. The control process $(x(t),u(t))$ belongs to $C_1^n(I) \times C^m(I)$. With $H: = \lambda_o f_o(t,x(t),u(t)) + (\lambda(t) | \dot{x}(t) - f(t,x(t),u(t)))$ the Lagrange function of problem (14)-(15) is defined.
With this we are able to state

Theorem 1. If the pair $(x^o(\cdot),u^o(\cdot))$ is a weak local minimum of problem (14)-(15), then there exist Lagrange multipliers $\lambda_o \in R_+$, $\mu_o \in R^{\rho_0}$, $\mu_1 \in R^{\rho_1}$, $\lambda \in C_1^n(I)$, which do not vanish simultaneously, such that

(i) the Euler equation of H with respect to x

$$\left(- \frac{d}{dt} H_{\dot{x}} + H_x\right)\Big|_{(x^o(t),u^o(t))} = 0, \tag{16}$$

with boundary conditions

$$H_{\dot{x}}\big|_{(x^o(a),u^o(a))} = r^T_{ox(x^o(a))}\mu_o,$$

$$H_{\dot{x}}\big|_{(x^o(b),u^o(b))} = -\, r^T_{1x(x^o(b))}\mu_1, \tag{17}$$

is satisfied, and

(ii) the Euler equation of H with respect to u

$$H_u\big|_{(x^o(t),u^o(t))} = 0 \tag{18}$$

is satisfied.

Remarks. (i) Equations (16) and (18) are known as the Euler-Lagrange equations. (ii) Equations (16) can be written explicitly in the form

$$\dot{\lambda}(t) = -f_x^T(t,x^o(t),u^o(t))\lambda(t) + \lambda_o f_{ox}(t,x^o(t),u^o(t)), \tag{19}$$

and are called adjoint equations. Their boundary conditions (17) are the transversality conditions. (iii) In explicit form equation (18) is $f_u^T(t,x^o(t),u^o(t))\lambda(t) = \lambda_o f_{ou}(t,x^o(t),u^o(t))$ from which we can obtain $u^o(t)$ by solving a (generally nonlinear) system of equations. Substituting for $u^o(t)$ in (14) and (19) leads to the two-point boundary value problem for $x(t)$ and $\lambda(t)$, mentioned in the introduction.

2.2.2. The Maximum Principle for Problems with State Constraints

In this section we are concerned with problems of the type:
minimize

$$J(x(\cdot),u(\cdot)) = \int_a^b f_o(t,x(t),u(t))dt \tag{20}$$

subject to the constraints

$$\dot{x}(t) = f(t,x(t),u(t)) \tag{21}$$

$$r_o(x(a)) = 0, \quad r_1(x(b)) = 0, \tag{22}$$

$$u(t) \in U, \ \forall\, t \in I, \tag{23}$$

$$c(t,x(t)) \leq 0, \ \forall\, t \in I. \tag{24}$$

The time interval is assumed to be fixed. A free time interval necessitates further technicalities that go beyond the scope of this paper. [23]

There is a fundamental difference in the differentiability assumptions of the functions involved in problem (20)-(24), in that we relax continuous differentiability w.r.t. t and u. We define the Hamiltonian function

$$H(t,x(t),u(t),\lambda_o,\lambda(t)) := (\lambda(t) | f(t,x(t),u(t))) - \lambda_o f_o(t,x(t),u(t))$$

and

$$\mathcal{H}(t,x(t),\lambda_o,\lambda(t)) := \sup_{u \in U} H(t,x(t),u(t),\lambda_o,\lambda(t)),$$

and state the following

Theorem 2 (Maximum Principle). Let $(x^o(\cdot),u^o(\cdot))$ be an optimal process of problem (20)-(24). Then there exist multipliers $\lambda_o \in R_+$, $\mu_o \in R^{\rho 0}$, $\mu_1 \in R^{\rho 1}$, $\lambda(\cdot) : I \to R^n$, and nonnegative regular measures σ_i, concentrated on $I_i = \{t \in I : c_i(t,x^o(t)) = 0\}$, $i = 1,\ldots,k$, not all of them vanishing simultaneously, such that

(i) $\lambda(\cdot)$ is solution of the following integral equation

$$\lambda(t) = - r_{1x}^T(x^o(b))\mu_1 + \int_t^b H_x(\tau,x^o(\tau),u^o(\tau),\lambda_o,\lambda(\tau))d\tau - \sum_{i=1}^{k} \int_t^b c_{ix}(\tau,x^o(\tau))d\sigma_i, \tag{25}$$

with

$$\lambda(a) = r_{ox}^T(x^o(a))\mu_o, \tag{26}$$

and

(ii) for almost every $t \in I$

$$H(t,x^o(t),u^o(t),\lambda_o,\lambda(t)) = \mathcal{H}(t,x^o(t),\lambda_o,\lambda(t)) \tag{27}$$

is satisfied.

Remarks: (i) As in theorem 1, equations (25) are called adjoint equations.

(ii) While in theorem 1 $\lambda(\cdot)$ is an absolutely continuous function, in the case of state constraints, $\lambda(\cdot)$ may be discontinuous because of the integrals w.r.t. the measures σ_i in equations (25). As the measures are regular the $\lambda(\cdot)$ are functions of bounded variation and continuous from the left.

3. Piecewise Polynomial Approximation

3.1. The Approximation Problem

For the numerical algorithms which will be given in the next sections we need some elements of approximation theory. We give some definitions and state some results for piecewise-polynomial approximation.

Let u(t) be a real-valued function definied on the interval I. We want to construct a piecewise-cubic polynomial function v(t) which interpolates u(t) at the points

$$a = t_1 < t_2 < \dots < t_{N+1} = b. \qquad (1)$$

We choose the interior interpolation points $t_2, \dots t_N$ as breakpoints for v(t), and on each interval $I_i := [t_i, t_{i+1}]$, v(t) is constructed as a cubic polynomial $p_i(t)$, $i=1,\dots,N$:

$$p_i(t) = a_i + b_i(t-t_i) + c_i(t-t_i)^2 + d_i(t-t_i)^3. \qquad (2)$$

From the interpolation properties

$$v(t_i) = u(t_i), \quad i=1,\dots,N+1, \qquad (3)$$

we have

$$p_i(t_i) = u(t_i), \quad p_i(t_{i+1}) = u(t_{i+1}), \quad i=1,\dots,N, \qquad (4)$$

and thus v(t) is guaranteed to be continuous on I

$$p_{i-1}(t_i) = p_i(t_i), \quad i=2,\dots,N. \qquad (5)$$

Two more degrees of freedom are left to determine the coefficients of p(•) uniquely. We choose conditions that guarantee continuity of certain derivatives of $p(t_i)$, namely first derivatives (piecewise-cubic Hermite approximation), or second derivatives (cubic spline approximation).

3.2. Piecewise-Cubic Hermite Approximation

Here, besides conditions (4) the following is required

$$\dot{p}_i(t_i) = \dot{u}(t_i), \qquad \dot{p}_i(t_{i+1}) = \dot{u}(t_{i+1}), \qquad i=1,\dots,N, \tag{6}$$

which leads to continuity of the derivatives of v(t) on I:

$$\dot{p}_{i-1}(t_i) = \dot{p}_i(t_i), \qquad i=2,\dots,N. \tag{7}$$

With this requirement, the coefficients of p(•) can be determined with $h_i = t_{i+1} - t_i$, and $u_i = u(t_i)$, etc.

$$\begin{aligned} a_i &= u_i, \\ b_i &= \dot{u}_i, \\ c_i &= \frac{3}{h_i^2}(u_{i+1} - u_i) - \frac{1}{h_i}(\dot{u}_{i+1} + 2\dot{u}_i), \\ d_i &= -\frac{2}{h_i^3}(u_{i+1} - u_i) + \frac{1}{h_i^2}(\dot{u}_{i+1} + \dot{u}_i). \end{aligned} \tag{8}$$

The error in piecewise-cubic Hermite approximation v(t) - u(t) is estimated by

<u>Theorem 3</u>. Provided that $u(\cdot) \in C^4(I)$ and $v(\cdot) = p_i(\cdot)$, $i=1,\dots,N$, where $p_i(\cdot)$ is the polynomial of degree 3 which coincides with $u(\cdot)$ at t_i and t_{i+1} in value and first derivative, then

$$\|v(t) - u(t)\| \leq \frac{1}{384} \max_{\xi \in I} |u^{(4)}(\xi)| \, (\max_i h_i)^4. \tag{9}$$

For a proof of this theorem see de Boor (1978). [26]

Piecewise-cubic Hermite approximation is a local approximation, which means that, for any fixed t, the function values v(t) depend only on a few of the data points, whose abscissae are close to t. This property will lead to structural sparsity of the matrices involved in the control approximation problem. Piecewise-cubic Hermite approximation requires the knowledge of derivatives $\dot{u}(t_i)$, which are readily available if u(•) is for example the solution of some differential equation. Otherwise one is forced to resort to some reasonable approximation to $\dot{u}(t_i)$, as for instance in piecewise-cubic Bessel approximation, [26] another local approximation scheme, the approximation order of which is one lower than in theorem 3.

3.3. Cubic Spline Approximation

Cubic spline approximation will result in continuity of second derivatives of v(t) on I:

$$\ddot{p}_{i-1}(t_i) = \ddot{p}_i(t_i), \quad i=2,\dots,N, \tag{10}$$

which implies, in view of equation (2),

$$2c_{i-1} + 6d_{i-1}h_{i-1} = 2c_i, \quad i=2,\dots,N. \tag{11}$$

Substituting for c_i and d_i from (8), and using abbreviations $s_i = \dot{p}_i(t_i)$ gives for $i=2,\dots,N$

$$h_i s_{i-1} + 2(h_{i-1} + h_i)s_i + h_{i-1}s_{i+1} = 3(h_i[t_{i-1},t_i]u + h_{i-1}[t_i,t_{i+1}]u), \tag{12}$$

where

$$[t_i,t_{i+1}]\,u = \frac{u(t_i) - u(t_{i+1})}{t_i - t_{i+1}} \tag{13}$$

is the first divided difference of $u(\cdot)$ at t_i, t_{i+1}. This is a tridiagonal system of N-1 linear equations in the N+1 unknowns s_i, $i=1,\dots,N+1$, with strictly row diagonally dominant coefficient matrix, which will be solved by Gaussian elimination without pivoting. s_1 and s_{N+1} introduce two degrees of freedom into system (12); for a choice of these boundary conditions consult Powell (1981). [27]

A bound for the approximation error is given in

Theorem 4. Let $u(\cdot)$ be four times continuously differentiable and let $v(\cdot) = p_i(\cdot)$, $i=1,\dots,N$, be constructed as in (12), then

$$\|v(t) - u(t)\| \le \frac{5}{384} \max_{\xi \in I}|u^{(4)}(\xi)|\,(\max_i h_i)^4. \tag{14}$$

This theorem is proved in Hall & Meyer (1976). [28]

The error bound in (14) is only 5 times as big as that in (9), even though piecewise-cubic Hermite approximation uses twice as much information about the function $u(\cdot)$ to be approximized. On the other hand, note that this approximation scheme is not a local one.

4. Approximation of the Control Process

4.1. Approximation of the Control Vector

4.1.1. Parameters, Performance Index and Constraints

We begin our study of approximating the control process for the solution of the constrained optimal control problem with a discretization of the control vector $u(\cdot)$. The interval I is partitioned into a (not necessarily equidistant) grid

$$\{a = t_1 < t_2 < \ldots. < t_{N+1} = b\}, \qquad (1)$$

and on this grid the finite set of control parameters is defined

$$y = \{u_1(t_1),\ldots,u_1(t_{N+1}),u_2(t_1),\ldots,u_m(t_{N+1})\}. \qquad (2)$$

The data (2) together with the grid (1) approximate $u(t)$ by cubic spline functions $v(t)$, $v \in R^m$, $\forall\, t \in I$. The approximate trajectory $\tilde{x}(t)$ is generated by solving an initial value problem [29, 30] with dynamic equation (2.4)

$$\dot{\tilde{x}}(t) = f(t,\tilde{x}(t),v(t)) \qquad (3)$$

and appropriate initial conditions $\tilde{x}(a)$ from the relations (2.5). If the boundary conditions are not in the separated form (2.6), $\tilde{x}_o = \tilde{x}(a)$ is treated as a parameter, augmenting the set y, [24] and leading to approximate boundary conditions

$$r_i(a,\tilde{x}_o,b,\tilde{x}(b)) = r_i(y) = 0, \qquad i=1,\ldots,p. \qquad (4)$$

The performance functional, e.g. (2.3), is evaluated simultaneously with the approximate trajectory, leading to an approximate cost function

$$F(y) \qquad (5)$$

which is to be minimized.

State and/or control constraints, e.g. (2.11), are transformed into end conditions by introducing auxiliary state variables $\tilde{x}_{n+i}$, $i=1,\ldots,q$, satisfying

$$\dot{\tilde{x}}_{n+i} = \max(o,c_i(t,\tilde{x}(t),v(t))) \qquad \tilde{x}_{n+i}(a) = 0, \qquad i=1,\ldots,q, \qquad (6)$$

and

$$\tilde{x}_{n+i}(b) = r_{p+i}(y) \leq 0, \quad i=1,\dots,q. \tag{7}$$

By this we have approximated a rather general optimal control problem by a nonlinear programming problem (4)-(7) in standard form:
find a local minimizer of

$$\begin{array}{lll} & F(y) & \\ \text{subject to} & r_i(y) = 0, & i=1,\dots,p, \\ \text{and} & r_i(y) \leq 0, & i=p+1,\dots,p+q. \end{array} \tag{8}$$

4.1.2. Gradient of Performance Index and Jacobian Matrix of Contraints

4.1.2.1. Difference Approximations

For the most efficient numerical methods that solve problem (8) the gradient of the performance index F and the constraints r_i are necessary. [19] The most straight-forward and conceptually simplest way to generate the gradient g(y) of a multivariate function f(y), $y \in R^n$, $f \in \{F, r_i\}$, which cannot be differentiated analytically, is to replace the exact gradient by a finite-difference approximation, e.g. first-order forward differences

$$g(y) = (f(y + h_i e_i) - f(y))/h_i, \quad i=1,\dots,n, \tag{9}$$

with e_i the ith axis vector. Often the choice of the step-lengths h_i is crucial. [31] A careful balance between truncation and cancellation error is necessary. For the evaluation of (9) the integration of n additional trajectories (3) is mandatory.

4.1.2.2. Error-Integrals

A second method which leads to semi-analytical gradients uses information from the error in the optimality conditions of section 2.2. While this method clearly consumes more man-time, in some cases it saves computer-time. [24] It is also prone to mistakes in evaluating the partials of the dynamic matrix of the adjoint variables.
The formal procedure is the following

$$g_i(y) = \int_a^b H_u \cdot \dot{v}_i(t)dt, \quad i=1,\dots,n, \tag{10}$$

with H defined as for theorem 1 or the maximum principle. For each element $f \in \{F, r_i\}$ a separate set of adjoint equations (2.16) has to be integrated with boundary values

$$\lambda^j(b) = f_x^j\big|_b, \quad j=1,\dots,p+q+1. \tag{11}$$

$\dot{v}_i(t)$ is the derivative of the approximating control w.r.t. y_i

$$\dot{v}_i(t;y) = (v_i(t;y + h_i e_i) - v_i(t;y))/h_i, \quad i=1,\dots,n, \tag{12}$$

where by $v(\cdot;y)$ in (12) the dependence of v on the data y is indicated. A proof of this procedure (10)-(12) is given in Kraft (1980). [32]

4.2. Approximation of the Control Vector and the State Vector

4.2.1. Parameters, Performance Index and Constraints

We continue our study of approximating the control process for the solution of the constrained optimal control problem with a discretization of the control vector $u(\cdot)$ and the state vector $x(\cdot)$. At this stage we have the following finite set of parameters on the grid (1)

$$y = \{x_1(t_1), x_2(t_1), \dots, x_n(t_1),\ u_1(t_1)\dots,\ u_m(t_1), \dot{u}_1(t_1), \dots, \dot{u}_m(t_1),$$
$$x_1(t_2), \dots, \dot{u}_m(t_{N+1})\},$$

abbreviated as

$$y = \{x_1, u_1, \dot{u}_1, x_2, u_2, \dot{u}_2, \dots, x_{N+1}, u_{N+1}, \dot{u}_{N+1}\}. \tag{13}$$

On each subinterval $[t_i, t_{i+1}]$, $i=1,\dots,N$, $x(\cdot)$ and $u(\cdot)$ are approximated by piece-wise-cubic Hermite (p.c.H.) polynomials $p_i(\cdot)$ componentwise with coefficients given by (3.8), where at the breakpoints we determine $\dot{x}_i$ by evaluating the r.h.s. of the differential equation

$$\dot{x}_i = f(t_i, x_i, u_i). \tag{14}$$

Thus we get an approximation which is continuously differentiable on I and satisfies the differential equation at the breakpoints. To test the approximation within the subintervals the defect at $\bar{t}_i := (t_i + t_{i+1})/2$ is calculated [33]

$$r_i = r_i(x_i, u_i, \dot{u}_i, x_{i+1}, u_{i+1}, \dot{u}_{i+1}) := \dot{p}_i(\bar{t}_i) - f(\bar{t}_i, \bar{x}_i, \bar{u}_i), \tag{15}$$

with $\bar{x}_i$ and $\bar{u}_i$ the values of their approximating p.c.H. polynomials at $\bar{t}_i$

$$\begin{aligned}
\bar{x}_i &= \frac{1}{2}(x_i + x_{i+1}) + \frac{h_i}{8}(f_i - f_{i+1}), \\
\dot{p}_i(\bar{t}_i) &= \frac{3}{2h_i}(x_{i+1} - x_i) - \frac{1}{4}(f_i + f_{i+1}), \\
\bar{u}_i &= \frac{1}{2}(u_i + u_{i+1}) + \frac{h_i}{8}(\dot{u}_i - \dot{u}_{i+1}), \\
\dot{\bar{u}}_i &= \frac{3}{2h_i}(u_{i+1} - u_i) - \frac{1}{4}(\dot{u}_i + \dot{u}_{i+1}),
\end{aligned} \tag{16}$$

with $h_i = t_{i+1} - t_i$ and $f_i = f(t_i, x_i, u_i)$, $i=1,\dots,N$. The aim of the algorithm is to drive the defect to zero

$$r_i = 0, \qquad i=1,\dots,N, \tag{17}$$

together with boundary conditions $r(a, x_o, b, x_{N+1})$. The performance criterion is either evaluated directly in the Mayer problem or it is changed to a Mayer problem. The state and/or control constraints can either be treated as hard constraints by the method of (6) and (7) or as soft constraints by satisfying (2.11), say, only at the breakpoints. In practical application these constraints mostly reduce to upper and lower bounds on the variables y, that each efficient numerical code for nonlinear programming treat as seperate constraints.

In the p.c.H. approximation it is necessary to enlarge the set of parameters by the $\dot{u}_i$. This is often welcome in optimal control when there are constraints on the control derivatives. If this is not the case it might be more efficient to work, with the p.c. Bessel approximation. If so the sparsity structure of the gradients will be changed, and the grid must be refined to get the same accuracy as in p.c.H. approximation. Computational experience is completely lacking in this area.

4.2.2. Gradient of Performance Index and Jacobian Matrix of Constraints

A great advantage of the approach in the preceding section is the availability of analytical gradients of the involved functions.
The gradient of the performance index is obvious. The gradient of the constraints can be obtained as follows

$$\frac{\partial r_i}{\partial x_i} = -\frac{3}{2h_i} I - \frac{1}{4}\frac{\partial f_i}{\partial x_i} - \frac{\partial \bar{f}_i}{\partial \bar{x}_i}\left(\frac{1}{2} I + \frac{h_i}{8}\frac{\partial f_i}{\partial x_i}\right),$$

$$\frac{\partial r_i}{\partial x_{i+1}} = \frac{3}{2h_i} I - \frac{1}{4}\frac{\partial f_{i+1}}{\partial x_{i+1}} - \frac{\partial \bar{f}_i}{\partial \bar{x}_i}\left(\frac{1}{2} I - \frac{h_i}{8}\frac{\partial f_{i+1}}{\partial x_{i+1}}\right),$$

$$\frac{\partial r_i}{\partial u_i} = -\frac{1}{4}\frac{\partial f_i}{\partial u_i} - \frac{1}{2}\frac{\partial \bar{f}_i}{\partial \bar{u}_i} - \frac{h_i}{8}\frac{\partial \bar{f}_i}{\partial \bar{x}_i}\frac{\partial f_i}{\partial u_i},$$

$$\frac{\partial r_i}{\partial u_{i+1}} = -\frac{1}{4}\frac{\partial f_{i+1}}{\partial u_{i+1}} - \frac{1}{2}\frac{\partial \bar{f}_i}{\partial \bar{u}_i} + \frac{h_i}{8}\frac{\partial \bar{f}_i}{\partial \bar{x}_i}\frac{\partial f_{i+1}}{\partial u_{i+1}}, \tag{18}$$

$$\frac{\partial r_i}{\partial \dot{u}_i} = -\frac{h_i}{8}\frac{\partial \bar{f}_i}{\partial \bar{u}_i},$$

$$\frac{\partial r_i}{\partial \dot{u}_{i+1}} = \frac{h_i}{8}\frac{\partial \bar{f}_i}{\partial \bar{u}_i},$$

and thus the Jacobian matrix of the differential constraints together with the boundary constraints has the following staircase form [46]

$$\frac{\partial r}{\partial y} =$$

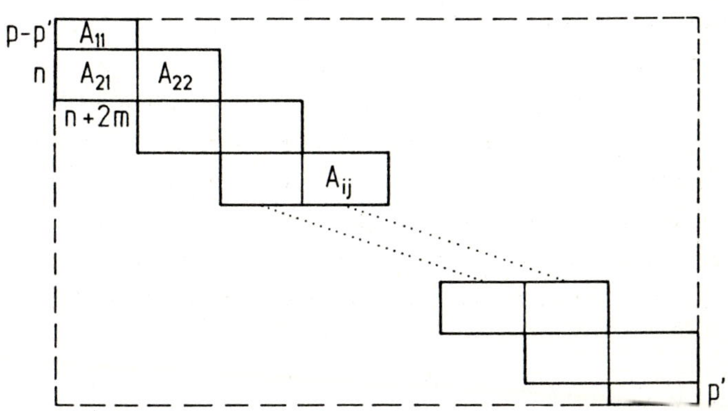

where the A_{ij} have dimension n x (n + 2m), and $\bar{f}_i$ is $f_i(\bar{t}_i,\bar{x}_i,\bar{u}_i)$ in (18).
A similar sparsity pattern is inherent in the Hessian matrix of the Lagrangian function of problem (8), and it is mandatory for the nonlinear programming solver to exploit this structure in large application problems.

Note that by (18), this is the only method which allows the (analytical) evaluation of gradients by evaluating $\partial f/\partial x$ and $\partial f/\partial u$. In indirect methods these matrices have to be available already for the computation of the involved trajectories.

4.3 Stepsize Control

For an efficient calculation of the solution of optimal control problems it is necessary to use a stepsize control, i.e. a control to locate the discretization appropriately. Many solution proposals have been made for this problem; [35] a very popular one is the equidistribution of the approximation error. [26] We follow an approach originated by Fehlberg (1969) [36] for the numerical treatment of initial value problems. [37] The special application to indirect collocation is due to Meyer (1979). [33]
With $\bar{t}_i := (t_i + t_{i+1})/2$ the interval I_i is halved into $I_i^1 := [t_i,\bar{t}_i]$ and $I_i^2 = [\bar{t}_i,t_{i+1}]$, i=1,...,N, and on both intervals an additional cubic polynomial according to (3.2) is defined:

$$
\begin{aligned}
p_{i1}(t_i) &= x_i, & p_{i1}(\bar{t}_i) &= p_i(\bar{t}_i),\\
\dot{p}_{i1}(t_i) &= \dot{x}_i, & \dot{p}_{i1}(\bar{t}_i) &= f(\bar{t}_i,x(\bar{t}_i),u(\bar{t}_i)),\\
p_{i2}(\bar{t}_i) &= p_i(\bar{t}_i), & p_{i2}(t_{i+1}) &= x_{i+1},\\
\dot{p}_{i2}(\bar{t}_i) &= f(\bar{t}_i,x(\bar{t}_i),u(\bar{t}_i)), & \dot{p}_{i2}(t_{i+1}) &= \dot{x}_{i+1},\\
\bar{p}_i(t) &= p_{i1}(t),\ t\in I_i^1, & \bar{p}_i(t) &= p_{i2}(t),\ t\in I_i^2.
\end{aligned}
\tag{19}
$$

Note that no additional evaluation of the right hand sides of the differential equation is necessary as $f(\bar{t}_i,\cdot,\cdot)$ is already available from the collocation conditions.
From theorem 3 we have the following

$$\|x(t) - p_i(t)\| = O(h_i^4), \quad t \in I_i, \tag{20}$$

or

$$p_i(t) = x(t) + \tau_1(t)h_i^{\,4} + \tau_2(t)h_i^{\,5} + \dots, \tag{21}$$

and

$$\bar{p}_i(t) = x(t) + \tau_1(t)h_i^{\,4}/2^4 + \tau_2(t)h_i^{\,5}/2^5 + \dots \, . \tag{22}$$

With (21) and (22) we do an extrapolation step [38] and get

$$\hat{p}_i(t) := (2^4\bar{p}_i(t) - p_i(t))/(2^4 - 1) = x(t) - \tau_2(t)h_i^{\,5}/(2^4 - 1)/2, \tag{23}$$

which indicates that $\hat{p}$ is a fifth order approximation:

$$\|x(t) - \hat{p}_i(t)\| = O(h^5), \quad t \in I_i. \tag{24}$$

With the assumption

$$x(t_{NEW}) = \hat{p}_i(t_{NEW}), \quad x(t_{OLD}) = \hat{p}_i(t_{OLD}), \tag{25}$$

we obtain

$$h_{NEW} = h_{OLD} \left(\max_{j=1}^{n} \left| \frac{p_i^j(t_{NEW}) - \hat{p}_i^j(t_{NEW})}{p_i^j(t_{OLD}) - \hat{p}_i^j(t_{OLD})} \right| \right)^{1/4} .$$

We estimate an upper bound EST for the quality of the approximation obtained with the old discretization. Since

$$\max_{t \in I_i} | p_i(t) - \hat{p}_i(t) | \leq | p_i(\hat{t}) - \hat{p}_i(\hat{t}) | ,$$

is valid componentwise for all $j=1,\dots,n$, for $\hat{t}_i = t_i + (t_{i+1} - t_i)/3$ the bound is

$$EST = \max_{j=1}^{n} \left| \frac{16}{135} (x_{i+1}^{\,j} - x_i^j) - \frac{8}{405} h_i \, (\dot{x}_{i+1}^{\,j} + \dot{x}_i^j + 4f^j(\bar{t}_i, x_i(\bar{t}_i), u_i(\bar{t}_i))) \right| . \tag{26}$$

For a prescribed approximation tolerance TOL

$$\max_{j=1}^{n} | p_i^j(t_{NEW}) - \hat{p}_i^j(t_{NEW}) | \leq \text{TOL} \tag{27}$$

the stepsize ratio

$$r = \frac{h_{NEW}}{h_{OLD}} = \left(\frac{\text{TOL}}{\text{EST}} \right)^{1/4} \tag{28}$$

allows the insertion of k equidistant breakpoints in the interval I_i with k calculated from

$$2^{-k} \leq r \leq 2^{-k+1}.$$

p(•) is evaluated at the new break points to give the complete trajectory.

5. Software Considerations

The implementation of the algorithms of section 4.1 is described in Kraft (1980).[39] The nonlinear programming subproblem solver follows the lines given in Powell (1978)[40] as implemented by Schittkowski (1981)[41] with least squares routines given by Lawson & Hanson (1974).[42] Computational experiments with alternative quadratic programming solvers[43-45] are now being performed.
For the implementation of the algorithms of section 4.2 only an experimental version exists. The computational results obtained thus far are very promising. The main task for the future is to implement a staircase quadratic programming subroutine.

References

1. Bertsekas, D.P., Constrained Optimization and Lagrange Multiplier Methods, Academic Press, New York, 1982.
2. Hestenes, M.R., Calculus of Variations and Optimal Control Theory, Wiley, New York, 1966.
3. Cesari, L., Optimization - Theory and Applications, Springer, New York, 1983.
4. Childs, B., Scott, M., Daniel, J.W., Denman, E., Nelson, P. (eds.), Codes for Boundary-Value Problems in Ordinary Differential Equations, Springer, Berlin, 1979.
5. Bryson, A.E., Ho, Y.C., Applied Optimal Control, Ginn & Company, Waltham, 1969.
6. Canon, M.D., Cullum, C.D., Polak, E., Theory of Optimal Control and Mathematical Programming, McGraw-Hill, New York, 1970.
7. Tabak, D., Kuo, B.C., Optimal Control by Mathematical Programming, Prentice-Hall, Englewood Cliffs, 1971.
8. Brusch, R.G., Schappelle, R.H., "Solution of Highly Constrained Optimal Control Problems Using Nonlinear Programming", AIAA Journal 11 (1973) 135-136.
9. Kamm, J.L., Johnson, I.L., "Nonlinear Programming Approach for Optimizing Two-Stage Lifting Vehicle Ascent to Orbit", Automatica 9 (1973) 713-720.
10. Neuman, C.P., Sen, A., "A Suboptimal Control Algorithm for Constrained Problems Using Cubic Splines", Automatica 9 (1973) 601-613.
11. Gill, P.E., Murray, W., "The Numerical Solution of a Problem in the Calculus of Variations", in: Bell, D.J. (ed.), Recent Mathematical Developments in Control, Academic Press, London, 1973.
12. Rader, J.E., Hull, D.G., "Computation of Optimal Aircraft Trajectories Using Parameter Optimization Methods", J. Aircraft 12 (1975) 864-866.
13. Hager, W.W., "Rates of Convergence for Discrete Approximations to Unconstrained Control Problems", SIAM J. Numer. Anal. 13 (1976) 449-472.
14. Kraft, D., "Optimierung von Flugbahnen mit Zustandsbeschränkungen durch Mathematische Programmierung", DGLR Jahrbuch, München, 1976.
15. Mantell, J.B., Lasdon, L.S., "A GRG Algorithm for Econometric Control Problems", Annals of Economic and Social Measurement 6 (1978) 581-597.
16. Sargent, R.W.H., Sullivan, G.R., "The Development of an Efficient Optimal Control Package", in: Stoer, J. (ed.), Optimization Techniques, Springer, Berlin, 1978.
17. Jonson, H., "A Newton Method for Solving Non-Linear Optimal Control Problems with General Constraints", Linköping Studies in Science and Technologie 104, Linköping, 1983.
18. Di Pillo, G., Grippo, L., Lampariello, F., "A Newton-Type Computing Technique for Optimal Control Problems", Optimal Control Applications & Methods 5 (1984) 149-166.
19. Powell, M.J.D., "Variable Metric Methods for Constrained Optimization", in: Bachem, A., Grötschel, M., Korte, B. (eds.), Mathematical Programming - The State of the Art, Springer, Berlin, 1983.
20. Schittkowski, K., Nonlinear Programming Codes, Springer, Berlin, 1980.
21. Stoer, J., "Foundation of Recursive Quadratic Programming Methods for Solving Nonlinear Programs", this volume.
22. Neustadt, L.W., Optimization, Princeton University Press, Princeton, 1976.
23. Ioffe, A.D., Tichomirov, V.M., Theorie der Extremalenaufgaben, Deutscher Verlag der Wissenschaften, Berlin, 1979. Also: Theory of Extremal Problems, North Holland, Amsterdam, 1979.

24. Kraft, D., "Finite-Difference Gradients versus Error-Quadrature Gradients in the Solution of Parameterized Optimal Control Problems", Optimal Control Applications & Methods 2 (1981) 191-199.
25. Kraft, D., "Optimal Control of a Cryogenic Windtunnel", to appear as DFVLR Research Report, 1985.
26. de Boor, C., A Practical Guide to Splines, Springer, New York, 1978.
27. Powell, M.J.D., Approximation Theory and Methods, Cambridge University Press, Cambridge, 1981.
28. Hall, C.A., Meyer, W.W., "Optimal Error Bounds for Cubic Spline Interpolation", J. Approx. Theory 16 (1976) 105-122.
29. Shampine, L.F., Gordon, M.K., Computer Solution of Ordinary Differential Equations, W. H. Freeman & Company, San Francisco, 1975.
30. Shampine, L.F., Watts, H.A., "Practical Solution of Ordinary Differential Equations by Runge-Kutta Methods", Sandia Laboratories, Report SAND 76-0585, Albuquerque, 1976.
31. Gill, P.E., Murray, W., Saunders, M.A., Wright, M.H., "Computing Forward-Difference Intervals for Numerical Optimization", SIAM J. Sci. Stat. Comput. 4 (1983) 310-321.
32. Kraft, D., "Comparing Mathematical Programming Algorithms Based on Lagrangian Functions for Solving Optimal Control Problems", in: Rauch, H.E. (ed.), Control Applications of Nonlinear Programming, Pergamon Press, New York, 1980.
33. Maier, M., "Die numerische Lösung von Halbleitermodellen mit Hilfe des Kollokationsverfahrens PITOHP unter Verwendung einer automatischen Schrittweitenkontrolle", Thesis, TU München, 1979.
34. Dickmanns, E.D., Well, K.H., "Approximate Solution of Optimal Control Problems Using Third Order Hermite Polynomial Functions", in: Marchuk, G.I. (ed.), Optimization Techniques, Springer, Berlin, 1975.
35. Russel, R.D., Christiansen, J., "Adaptive Mesh Selection Strategies for Solving Boundary Value Problems", SIAM J. Numer. Anal. 15 (1978) 59-80.
36. Fehlberg, E., "Klassische Runge-Kutta-Formeln fünfter und siebenter Ordnung mit Schrittweiten-Kontrolle", Computing 4 (1969) 93-106.
37. Stoer, J., Bulirsch, R., Introduction to Numerical Analysis, Springer, New York, 1980.
38. Reinsch, C., Einführung in die numerische Mathematik, Mimeographed Lecture Notes, TU München, Abt. Mathematik, 1977. [3]
39. Kraft, D., "FORTRAN-Programme zur numerischen Lösung optimaler Steuerungsprobleme", DFVLR-Mitt. 80-03, Köln, 1980.
40. Powell, M.J.D., "A Fast Algorithm for Nonlinearly Constrained Optimization Calculations", in: Watson, G. (ed.), Numerical Analysis, Springer, Berlin, 1978.
41. Schittkowski, K., "The Nonlinear Programming Method of Wilson, Han, and Powell with an Augmented Lagrangian Type Line Search Function", Numer. Math. 38 (1981) 83-127.
42. Lawson, C.L., Hanson, R.J., Solving Least Squares Problems, Prentice Hall, Englewood Cliffs, 1974.
43. Gill, P.E., Murray, W., Saunders, M.A., Wright, M.H., "User's Guide for SOL/QPSOL: A Fortran Package for Quadratic Programming", TR SOL 83-7, Stanford University, Stanford, 1983.
44. Powell, M.J.D., "ZQPCVX A Fortran Subroutine for Convex Quadratic Programming", DAMTP /1983/NA17, University of Cambridge, Cambridge, 1983.
45. Goldfarb, D., Idnani, A., "A Numerically Stable Dual Method for Solving Strictly Convex Quadratic Programs", Math. Progr. 27 (1983) 1-33.
46. Fourer, R., "Staircase Matrices and Systems", SIAM Review 26 (1984) 1-70.

A STOCHASTIC APPROACH TO GLOBAL OPTIMIZATION

A.H.G. Rinnooy Kan *

C.G.E. Boender * **

G.Th. Timmer * **

* Econometric Institute, Erasmus University Rotterdam, The Netherlands

** Department of Mathematics and Informatics, Delft University of Technology, The Netherlands

NATO ASI Series, Vol. F15
Computational Mathematical Programming
Edited by K. Schittkowski

1. INTRODUCTION

Let $f : \mathbb{R}^n \to \mathbb{R}$ be a real valued smooth objective function. The area of nonlinear programming is traditionally concerned with methods that find a local optimum (say local minimum) of f, i.e. a point $x^* \in \mathbb{R}^n$ such that there exists a neighbourhood B of x^* with

$$f(x^*) \leq f(x) \qquad \forall\, x \in B. \tag{1}$$

In general, however, several local optima may exist and the corresponding function values may differ substantially. The global optimization problem is to find the global optimum (say global minimum) x_* of f, i.e. to find a point $x_* \in \mathbb{R}^n$ such that

$$f(x_*) \leq f(x) \qquad \forall\, x \in \mathbb{R}^n. \tag{2}$$

For computational reasons one usually assumes that a convex and compact set $S \subset \mathbb{R}^n$ is specified in advance, which contains the global minimum as an interior point. None the less, the problem to find

$$y_* = f(x_*) = \min_{x \in S} f(x) \tag{3}$$

remains essentially one of unconstrained optimization, and as such forms the subject of this paper.

So far only few solution methods for the global optimization problem have been developed, certainly in comparison with the multitude of methods that aim for a local optimum. The relative difficulty of global optimization as compared to local optimization is easy to understand. It is well known that under the assumption that f is twice continuously differentiable, all that is required to test if a point is a local minimum is knowledge of the first and second order derivatives at this point. If the test does not yield a positive result, the smoothness properties of f ensure that a neighbouring point can be found with a lower function value. Thus, a sequence of points can be constructed that converges to a local minimum.

Such local tests are obviously not sufficient to verify global optimality. Indeed, the global optimization problem as stated in (3) is inherently unsolvable [Dixon 1978]: for any continuously differentiable function f, any point $\bar{x} \in S$ and any neighbourhood B of $\bar{x}$, there exists a function f' such that (i) f+f' is continuously differentiable, (ii) f+f' is equal to f in all points outside B, and (iii) the global minimum of f+f' is attained in $\bar{x}$. As B can be chosen arbitrarily small, it immediately follows that it requires an unbounded number of function evaluations to guarantee that the global minimum x_* will be found.

Of course, this argument does not apply when one is satisfied with an approximation of the global minimum. In particular, for the case that a point within distance ε from x_* is sought, enumerative strategies exist that only require a finite number of function evaluations. These strategies, however, are of limited practical use. Thus, either a further restriction of the class of objective functions or a further relaxation of what is required of an algorithm will be inevitable in what follows.

Subject to this first conclusion the methods developed to solve the global optimization problem can be divided in <u>deterministic</u> and <u>stochastic</u> methods.

Some deterministic methods will be reviewed in Section 2. If a rigid guarantee is desired for these methods, the previous argument indicates that additional assumptions about f are unavoidable. The most popular such assumption is that a <u>Lipschitz constant</u> L is given, i.e. for all $x_1, x_2 \in S$

$$|f(x_1) - f(x_2)| \leq L\|x_1-x_2\|, \tag{4}$$

where $\|.\|$ denotes the Euclidean distance. The upper bound on the rate of change of f implied by this Lipschitz constant can be used in various ways to perform an exhaustive search over S. In practice, however, it is impossible to verify whether a function satisfies such a Lipschitz condition or not. In addition, the computational effort required by these methods tends to be formidable and forbidding.

Better computational results are obtained by methods that exploit the continuous differentiability of f. As mentioned before, this property allows for the construction of a sequence of points converging to a local optimum. As there exists no local test to verify global optimality, these deterministic methods try to find the global minimum by locating all local minima. No such method, however, can truly guarantee that all local minima of f are really found. Thus, as we shall see, their superior computational results are obtained at the expense of more (possibly implicit) assumptions about f or of no certainty of success.

Generally, far better results - both theoretically and computationally - have been obtained by stochastic methods [Rinnooy Kan & Timmer 1984, Timmer 1984]. In most stochastic methods, two phases can be usefully distinguished. In the global phase, the function is evaluated in a number of randomly sampled points. In the local phase , the sample points are manipulated, e.g. by means of local searches, to yield a candidate global minimum.

Generally in turning to stochastic methods, we do sacrifice the possibility of an absolute guarantee of success. However, under mild conditions on the sampling distribution and on f, the probability that a feasible solution within distance ε of x_* is sampled will be seen to approach 1 as the sample size increases [Solis & Wets 1981]. If the sample points are drawn from a uniform distribution over S and if f is continuous, then an even stronger result holds: the sample point with lowest function value converges to the global minimum value with probability 1 (or almost surely). Thus, the global phase can yield an asymptotic guarantee with probability 1, and is therefore essential for the reliability of the method. However, a method that only contains a global phase will be found lacking in efficiency. To increase the latter while maintaining the former is one of the challenges in global optimization.

Stochastic methods will be discussed in Section 3. The most promising methods appear to be variants of the so-called Multistart technique where points are sampled iteratively from a uniform distribution over S (global phase), after which local minima are found by applying a local search procedure to these points (local phase).

In practice, the number of local minima of an objective function is usually unknown. A fortiori, it is uncertain if a sample of observed local minima includes the global one. Thus, in this approach there is typically a need for a proper stopping rule. A theoretical framework which provides a solution to this problem is developed in [Boender 1984]. It turns out to be possible, for example, to compute a Bayesian estimate of the number of local minima not yet identified, so that the sequence of sampling and searching can be stopped if the estimated number of local minima is equal to the number of minima identified.

Multistart is still lacking in efficiency because the same local minimum may be located several times. If we define the region of attraction R_{x^*} of a local minimum x^* to be the set of points in S starting from which a given local search procedure converges to x^*, then ideally, the local search procedure should be started exactly once in every region of attraction. Several new algorithms designed to satisfy this criterion are presented in [Timmer 1984].

The method discussed in Section 3.3 temporarily eliminates a prespecified fraction of the sample points whose function values are relatively high. The resulting reduced sample consists of groups of mutually relatively close points that correspond to the regions with relatively small function values. Within each group the points are still distributed according to the original uniform distribution. Thus, these groups can be identified by clustering techniques based upon tests on the uniform distribution. Only one local search procedure will be started in each group [Boender et al. 1982].

Unfortunately, the resulting groups do not necessarily correspond to the regions of attraction of f. It is possible that a certain group of points corresponds to a region with relatively small function values which contains several minima. Therefore, the method which is based on the reduced sample may fail to find a local minimum although a point is sampled in its region of attraction. A better method is described in Section 3.4. Here, the function value is used explicitly in the clustering process. A very simple method results, for which both the probability that the local search procedure is started unnecessarily, and the probability that the local search is not started although a new local minimum would have been found, approach 0 with increasing sample size. In some sense the results proven for this method can be seen to be the strongest possible ones.

The results of some computational experiments are reported in Section 4.

2. DETERMINISTIC METHODS

2.1. Finite exact methods

We first consider exact methods that provide an absolute guarantee that the global minimum will be found in a finite number of steps.

Space covering methods exploit the availability of a Lipschitz constant L (cf. (4)) to perform an exhaustive search over S. A conceptually simple method of this type has been proposed by Evtushenko [Evtushenko 1971]. Suppose that f has been evaluated in $x_1,\ldots,x_k$ and define $M_k = \min\{f(x_1),\ldots,f(x_k)\}$. If the spheres V_i (i=1....,k) are chosen with centre x_i and radius $r_i = (f(x_i) - M_k + \varepsilon)/L$, then for any $x \in V_i$

$$f(x) \geq f(x_i) - Lr_i = M_k - \varepsilon. \tag{5}$$

Hence, if the spheres V_i (i=1,...,k) cover the whole set S, M_k differs less than ε from y_*. Thus, this result converts the global minimization problem to the problem of covering S with spheres. In the simple case of 1-dimensional optimization where S is an interval $\{x \in \mathbb{R} |\ a \leq x \leq b\}$, this covering problem is solved by choosing $x_1 = a + \varepsilon/L$ and

$$x_k = x_{k-1} + \frac{2\varepsilon + f(x_k) - M_k}{L} \qquad k = 2,3,\ldots \tag{6}$$

The method obviously stops if $x_k \geq b$.

A generalization for higher dimensional problems (n>1) consists of covering S with hypercubes whose edgelength is $2r_i/\sqrt{n}$, i.e. cubes inscribed in the spheres V_i.

Note that the efficiency of the method depends on the value of M_k. Since the distances between the iteration points increase with decreasing M_k, it may be worthwhile to improve M_k using a local minimization procedure.

A different method, for which it is not necessary to specify any a priori accuracy ε, is proposed in [Shubert 1972]. Here a bound on the accuracy is calculated at each iteration. The method consists of iteratively updating a piecewise linear function, which has directional derivaties equal to L or -L everywhere and which forms a lower bound on f that improves with each iteration. The method was orignally designed for 1-dimensional problems, but can be generalized to higher dimensional problems.

Initially, f is evaluated at some arbitrary point x_1. A piecewise linear function $\psi_1(x)$ is defined by

$$\psi_1(x) = f(x_1) - L\|x - x_1\|. \tag{7}$$

Now an iterative procedure starts, where in iteration k ($k \geq 2$) a global minimum of $\psi_{k-1}(x)$ on S is chosen as the point where f is next evaluated. A new piecewise linear function $\psi_k(x)$ is constructed by a modification of $\psi_{k-1}(x)$.

$$\psi_k(x) = \max\{f(x) - L\|x - x_k\|, \psi_{k-1}(x)\} \qquad (k = 2,3,\ldots) \tag{8}$$

Hence,

$$\psi_{k-1}(x) \leq \psi_k(x) \leq f(x), \tag{9}$$

$$\psi_k(x_i) = f(x_i) \qquad (i = 1,\ldots,k). \tag{10}$$

In each iteration, the piecewise linear approximation for f will improve. The method is stopped when the difference between the global minimum of $\psi_k(x)$, which is a lower bound on the global minimum of f, and the best function value found is small enough.

To conclude the description of this method, note that $\psi_k(x)$ is completely determined by the location and the value of its minima. If $\psi_k(x)$ is decribed in terms of these parameters it is no problem to find one of its global minima.

Although the space covering techniques are intuitively appealing they have two major drawbacks. Firstly, the number of function evaluations required by these methods tends to be formidable. To analyse this number, let S be a hypersphere with radius r, so that

$$m(S) = \frac{r^n \pi^{n/2}}{\Gamma(1 + \frac{n}{2})}, \tag{11}$$

where Γ denotes the gamma function.

Furthermore, let c be the maximum of f over S and suppose that f has been evaluated in k points $x_1,\ldots,x_k$. The function value in a point x can only be known to be greater than the global minimum value y_* if the function has been evaluated in a point x_i within distance $(f(x_i) - y_*)/L$ of x. Hence, the hyperspheres with radii $(f(x_i) - y_*)/L$ centered at the points x_i, $i = 1,\ldots,k$, must cover S to be sure that the global minimum has been found. The joint volume of these k hypersphere is smaller than

$$\frac{k\left(\frac{c-y_*}{L}\right)^n \pi^{n/2}}{\Gamma(1+\frac{n}{2})} \, . \tag{12}$$

Thus, for the k hyperspheres to cover S we require

$$k > \left(\frac{r}{c-y_*}\right)^n L^n . \tag{13}$$

Unless the derivative of f in the direciton of the global minimum equals -L everywhere, L is greater than $\frac{r}{c-y^*}$, and the computational effort required increases exponentially with n.

A second drawback of the space covering techniques is that the Lipschitz constant has to be known or estimated before starting the minimization. Overestimating L raises the cost considerably (cf. (13)), while underestimating L might lead to failure of the method. In most practical cases, however, obtaining a close estimate of L poses a problem comparable in difficulty with the original global optimization problem. Both drawbacks seem inherent to the approach chosen.

Surprisingly good computational results have been obtained by a similar enumerative technique in which upper and lower bounds on f over a subset of S (say, a hypercube) are computed by <u>interval arithmetic</u> [Hansen 1980]. This approach presupposes that f is given as a (not too complicated) mathematical expression. This is the case for all the standard testproblems - though not always in practice - and on those problems the straightforward branch-and-bound procedure based on the above idea has performed very well indeed.

In addition to the enumerative methods mentioned above, an absolute guarantee of success can also be achieved for certain very special classes of functions, most notably polynomials.

If f is a one dimensional <u>polynomial</u>, then a <u>deflation technique</u> has been proposed by [Goldstein & Price 1971].

Consider the <u>Taylor series</u> around a local minimum x* of a one dimensional function f.

$$f(x) = f(x^*) + \frac{f^{(2)}(x^*)}{2!}(x-x^*)^2 + \frac{f^{(3)}(x^*)}{3!}(x-x^*)^3 + \ldots + \frac{f^{(k)}(x^*+\theta(x-x^*))}{k!}(x-x^*)^k, \tag{14}$$

where $0 \leq \theta \leq 1$ and $f^{(i)}(.)$ is the i-th order derivative of f. Now let

$$f_1(x) = \frac{f(x) - f(x^*)}{(x - x^*)^2} . \tag{15}$$

If f is a polynomial of degree m, then $f_1(x)$ is a polynomial of degree m-2. If, in addition, it can be shown that the global minimum of $f_1(x)$ is positive, then x* is the global minimum of f. In case there is a point $\bar{x}$ for which $f_1(x)$ is negative, then $f(\bar{x}) < f(x^*)$ and x* is not the global minimum. In the latter case one can proceed using the Taylor series around a new local minimum which can be found by applying P to $\bar{x}$. To determine whether the global minimum is positive, we proceed iteratively considering $f_1(x)$ as the new basic function. If f(x) is a one dimensional polynomial, then this is a finite and rapidly converging process. For a more general function, however, there is no reason to assume that the problem of showing that the global minimum of $f_1(x)$ is positive is easier than the original problem.

Recently <u>piecewise linear homotopy</u> methods [Todd 1976, Allgower & Georg 1980] have proven to be useful in identifying all roots of polynomials, which is related to identifying all minima. Using a <u>labeling rule</u> it is possible to determine N points, such that all roots of a one dimensional polynomial of degree N will be found as the result of a <u>simplicial path following algorithm</u> applied to each of these points [Kuhn et al. 1984]. This can be implemented efficiently: it only takes $O(N^3 \log(N/\varepsilon))$ evaluations of f to find a point which is within ε distance of a root of f. For details we refer to [Kuhn et al. 1984].

Polynomials are not the only class of functions for which methods have been proposed that exploit the specific features of that class. For instance, successively closer approximations of f, for which the global minimum can be easily calculated, can be determined if f is separable into convex and concave terms [Falk & Solund 1971, Solund 1971], if a convex envelope of f can be found [McCormick 1976], and if f can be written as a finite sum of products of a finite number of uniform continuous functions of a single argument [Beale & Forrest 1978].

2.2. Heuristic methods

We now turn to heuristic methods that only offer an empirical guarantee (i.e., they may fail to find the global optimum). These methods apply a local search procedure to different starting points to find the local minima of f.

The tunneling method attempts to solve the global optimization problem by performing local searches such that each time a different local minimum is reached [Levy & Gomez 1980].

The method consists of two phases. In the first phase (minimization phase) the local search procedure is applied to a given point x_0 in order to find a local minimum x*. The purpose of the second phase (tunneling phase) is to find a point x different from x*, but with the same function value as x*, which is used as a starting point for the next minimization phase. This point is obtained by finding a zero of the tunneling function

$$T(x) = \frac{f(x) - f(x^*)}{\|x-x_m\|^{\lambda_0} \prod_{i=1}^{\ell} \|x-x_i^*\|^{\lambda_i}}, \tag{16}$$

where $x_1^*,\ldots,x_\ell^*$ are all local minima with a function value equal to f(x*) found in previous iterations. Subtracting f(x*) from f(x) eliminates all points satisfying f(x) > f(x*) as a possible solution. The term $\Pi_{i=1}^{\ell}\|x-x_i^*\|$ is introduced to prevent the algorithm from choosing the previously found minima as a solution. To prevent the zero finding algorithm to converge to a stationary point of

$$\frac{f(x) - f(x^*)}{\prod_{i=1}^{\ell} \|x-x_i^*\|^{\lambda_i}} \tag{17}$$

which is not a zero of (16), the term $\|x-x_m\|^{\lambda_0}$ is added, with x_m chosen appropriately.

If the global minimum has been found, then (16) will become positive for all x. Therefore the method stops if no zero of (16) can be found.

The tunneling method has the advantage that, provided that the local search procedure is of the descent type, a local minimum with smaller function value is located in each iteration. Hence, it is likely that a point with small function value will be found relatively quickly. However, a major drawback of the method is that it is impossible to be certain that the search for the global minimum has been sufficiently thorough. In essence, the tunneling method only reformulates the problem: rather than solving the original minimization problem, one now must prove that the tunneling function does not have a zero. This, however, is once again a global problem which is strongly related to the original one. The information gained during the foregoing iterations is of no obvious use in solving this new global problem; which therefore appears to be as hard to solve as the original one. Thus, lacking any sort of guarantee, the method is at best of some heuristic value.

The same is true for the trajectory method due to Branin [Branin 1972, Branin & Hoo 1972], based on the construction (by numerical integration) of the path along which the gradient of f points in constant direction. This method is known to fail on certain functions [Treccani 1975], and it is not clear under which conditions convergence to a global minimum can be assured.

3. STOCHASTIC METHODS

Stochastic methods are asymptotically exact, i.e. they offer an a symptotic guarantee in some probabilistic sense. The methods can usefully be separated into two different phases.

In the global phase, the function is evaluated in a number of randomly sampled points. In the local phase, the sample points are manipulated, for example by means of local searches, to yield a candidate solution.

The global phase is necessary because there is no local improvement strategy which, starting from an arbitrary point, can be guaranteed to converge to the global minimum. As we have seen in Section 1, a global search over S, which in the long run locates a point in every subset of S of positive measure, is required to ensure the reliability of the method. But, although the local improvement techniques cannot guarantee that the global minimum will be found, they are efficient tools to find a point with relatively small function value. Therefore, the local phase is incorporated to improve the efficiency of the method. Because the local phase generally complicates the formal analysis considerably, we will start our survey with a method consisting only of a global phase.

3.1. Pure Random Search

The simplest stochastic method for global optimization consists only of a global phase. Known confusingly as Pure Random Search [Brooks 1958, Anderssen 1972], the method involves no more than a single step.

Pure Random Search

Step 1. Evaluate f in N points, drawn from a uniform distribution over S. The smallest function value found is the candidate solution for y_*.

The proof that Pure Random Search offers an asymptotic guarantee in a probabilistic sense is based on the observation that the probability that a uniform sample of size N contains at least one point in a subset $A \subset S$ is equal to [Brooks 1958]

$$1 - \left(1 - \frac{m(A)}{m(S)}\right)^N, \tag{18}$$

where m(.) denotes the Lebesgue measure. Thus Pure Random Search locates an element close to the global minimum with a probability approaching to 1 as N increases. In fact, if we let $\underline{y}_N^{(1)}$ be the smallest function value found in a sample of size N, then it can be proved that $\underline{y}_N^{(1)}$ converges to the global minimum value y_* with probability 1 [cf. Devroye 1978, Rubinstein 1981].

We also observe that (18) implies that

$$\frac{\log(1-\alpha)}{\log(1-\delta)} \tag{19}$$

sample points are required to find an element of a set A with probability α, provided that $m(A)/m(S) = \delta$. This result can be used to provide a stopping rule for this method in the obvious manner.

3.2. Multistart

In view of the extreme simplicity and the resulting poor computational quality of Pure Random Search, several extensions have been proposed that also start from a uniform sample over S (hence, the results of the foregoing section can be applied including the asymptotic guarantee), but that at the same time involve local searches from some or all points in the sample. In this section we will discuss the prototype of these methods which is known as Multistart. In this approach a local search procedure P is applied to each point in the random sample; the best local minimum found in this way is our candidate for the global minimum x_*.

Multistart

Step 1. Draw a point from the uniform distribution over S.

Step 2. Apply P to the new sample point.

Step 3. The local minimum x^* identified with the lowest function value is the candidate value for x_*. Return to Step 1, unless a stopping criterion is satisfied.

Let us consider the issue of a proper stopping criterion for this method. In the sequel we will show that the stopping rules developed for Multistart remain valid for more efficient variants of this folklore approach.

Recall that the region of attraction R_{x^*} of a local minimum x^*, given a particular local search routine P, is defined as the subset of points in S starting from which P will arrive at x^* [Dixon & Szegö 1975, 1978]. Furthermore, let k be the number of local minima of f, and denote the relative size of the i-th region of attraction by θ_i $(i=1,\ldots,k)$. If these values are given, we have several stopping criteria at our disposal. We may terminate the Multistart method, for example, if the number of different local minima observed is equal to k or if the total size of the observed regions of attraction is greater than some prespecified value.

In practice, $k,\theta_1,\ldots,\theta_k$ are frequently unknown. The sampled minima, however, clearly provide information about their values. The crucial observation that enables us to learn about the values of $k,\theta_1,\ldots,\theta_k$ is that since the starting points of the Multistart method are uniformly distributed over S, a local minimum has a fixed probability of being found in each trial that is equal to the relative size of its region of attraction. This implies that, given a number of local searches N, the observed local minima are a sample from a multinomial distribution whose cells correspond to the local minima: the number of cells is equal to the unknown number k of local minima of f and the cell probabilities are equal to the unknown relative sizes θ_i (i=1,...,k) of the regions of attraction. However, since it is unknown in what way S is subdivided in regions of attraction, it is impossible to distinguish between samples of local minima that are identical up to a relabeling of the minima. We therefore have to rely on the generalized multinomial distribution that has been studied in great detail in [Boender 1984]. It is now standard statistical practice to use an observed sample of local minima to make inferences about the unknown parameters $k,\theta_1,\ldots,\theta_k$. In a Bayesian approach, in which the unknowns are themselves assumed to be random variables with a uniform prior distribution, it can be proved that, given that W different local minima have been found in N searches, the optimal Bayesian estimate of the unknown number of local minima k is given by the integer E nearest to

$$W.\frac{N-1}{N-W-2} \quad (N \geq W-3). \tag{20}$$

(cf [Boender 1984]). Hence, the Multistart method can (for instance) be stopped when E = W.

This theoretical framework which was initiated in [Zielinski 1981] is an attractive one, the more so since it can easily be extended to yield optimal Bayesian stopping rules that incorporate assumptions about the costs and potential benefits of further local searches and weigh these against each other probabilistically. Several loss structures and corresponding stopping rules are described in [Boender 1984].

3.3. Single Linkage

In spite of the reliability of Multistart, the method is lacking in efficiency, which stems from the fact that each local minimum, particularly the ones with a large region of attraction, will generally be found several times. From efficiency considerations only, the local search procedure P should ideally be invoked no more than once in each region of attraction. Computationally successful adaptations of Multistart in that direction are provided by clustering methods [Becker & Lago 1970; Törn 1978; Boender et al. 1982; Timmer 1984]. Clustering methods also generate points iteratively in S according to the uniform distribution. Now, however, only a prespecified fraction q containing the points with the lowest function values are retained in the sample. Let f_q be the largest function value in the reduced sample and define $R_q \subset S$ as the set of all points in S whose function value does not exceed f_q. R_q will consist of a number of disjoint components that together contain all the points from the reduced sample: a nonempty set of all reduced sample points that are contained in one component of R_q is called a cluster. Ideally, the clusters should be in 1-1 correspondence with the regions of attraction whose intersection with R_q is nonempty. Then, one local search from the best point in each cluster will suffice to find the set of local minima with function value smaller than f_q, which obviously includes the global minimum.

In the Single Linkage global optimization algorithm [Timmer 1984], clusters are efficiently identified by exploiting the fact that the points in the reduced sample are uniformly distributed over R_q. Clusters are created one by one, and each cluster is initiated by a seedpoint. Selected points of the reduced sample are added to the cluster until a termination criterion is met. Under conditions to be specified, the local search procedure is started from one point in the cluster.

Before we state the algorithms we need some additional notation. Fix $\tau > 0$ and let S_τ denote the points in S whose distance to the boundary of S is at least τ. Furthermore, let X^* be the set of detected local minima, and given $\upsilon > 0$, let $X^*_\upsilon = \{x \in S \mid \|x-x^*\| < \upsilon, \text{ for any } x^* \in X^*\}$. Henceforth it is assumed that (i) all local minima of f occur in the interior of S_τ, (ii) a positive constant ε can be specified such that the distance between any two stationary points of f exceeds ε, (iii) the local search procedure P always finds a local minimum x^*, and (iv) P is strictly descent, i.e. starting from any $x \in S$ P converges to a local minimum $x^* \in S$ such that there exists a path in S from x to x^* along which the function values are nonincreasing. We now describe the Single Linkage algorithm, given N uniform points in S.

Single Linkage

Step 1. (Determine reduced sample). Determine the reduced sample by taking qN sample points with the smallest function values. Let W be the number of elements of the set of local mninima X*. Set j := 1.

Step 2. (Determine seed points). If all reduced sample points have been assigned to a cluster, stop.
If $j \leq W$, then choose the j-th local minimum in X* as the next seedpoint; go to Step 3.
Determine the point $\bar{x}$ which has the smallest function value among the unclustered reduced sample points; $\bar{x}$ is the next seedpoint.
If $\bar{x} \in S_\tau$ and if $\bar{x} \in X^*_\upsilon$, then apply P to $\bar{x}$ to find a local minimum x*.

Step 3. (Form cluster). Initiate a cluster from a seedpoint which is determined in Step 2. Add reduced sample points which are within the critical distance r_N from a point already in the cluster until no more such points remain. Let j := j+1 and go to Step 2.

The sample is expanded and the above procedure repeated until the stopping rule applies.

Several observations are in order. First of all, in [Timmer 1984] it is proved that if the critical distance r_N is chosen equal to

$$\pi^{-\frac{1}{2}}\left(\Gamma(1+\tfrac{n}{2})m(S)\frac{\sigma \log N}{N}\right)^{1/n} \tag{21}$$

with $\sigma > 2$ then the probability that a local search is started tends to 0 with increasing N; of $\sigma > 4$, then, even of the sampling continues forever, the total number of local searches ever started is finite with probability 1. In addition, whenever the critical distance tends to 0 for increasing N, then in every component in which a point has been sampled a local minimum will be found with probability 1.

Secondly, the stopping rules developed for Multistart can be applied to the clustering method provided that the number of trials is taken equal to the number of points qN in the reduced sample rather than the number of local searches, the number of local minima is taken equal to the number of local minima whose function value is not greater than f_q and the cell probabilities are taken to be equal to the relative Lebesgue measure of the intersections of the regions of attraction with R_q. In applying these rules, we do have to assume that the way R_q changes (slightly) with the sample size N does not affect the analysis. More importantly, we also have to assume that each local minimum with function value smaller than f_q whose region of attraction does contain at least one point from the reduced sample is actually found, i.e. that the methods identify the same local minima that would be found by performing a local search from each of the qN points in the reduced sample. This assumption is unfortunately not justified for the Single Linkage: a component way contain several local minima, of which we are only guaranteed to find one asymptotically.

3.4. Multi Level Single Linkage

The method described in Section 3.3 only makes minimal use of the function values of the sample points. These function values are used to determine the reduced sample, but the clustering process applied to this reduced sample hardly depends on the function values. Instead, the clustering process concentrates on the location of the reduced sample points. As a result, the method cannot distinguish between different regions of attraction which are located in the same component of R_q. The function value of a sample point x evidently can be of great importance if one wishes to predict to which region of attraction x belongs, because the local search procedure which defines these regions is known to be strictly descent. Hence, x cannot belong to the region of attraction of a local minimum x*, if there is no descent path from x to x*, i.e. a path along which the function values are monotonically decreasing. Furthermore, x does certainly belong to the region of attraction R_{x^*}, if there does not exist a descent path from x to any other minimum than x*.

Obviously, it is impossible to consider all descent paths starting from x. Instead, we will (implicitly) consider all r_N-descent sequences, where a r_N-descent sequence is a sequence of sample points, such that each two successive points are within distance r_N of each other and such that the function values of the points in the sequence are monotonically decreasing. It will turn out that if the sample size increases and if r_N tends to 0, then every descent path can be conveniently approximated by such a sequence of sample points.

For a better understanding of the remainder of this section it is advantageous to consider the following algorithm first. Let W be the number of local minima known when the procedure is started.

<u>Step 1</u>. Initiate W different clusters, each consisting of one of the local minima present.

<u>Step 2</u>. Order the sample points, such that $f(x_i) < f(x_{i+1})$, $1 \leq i \leq N-1$. Set $i := 1$.

<u>Step 3</u>. Assign the sample point x_i to every cluster which contains a point within distance r_N.
If x_i is not assigned to any cluster yet, then start a local search at x_i to yield a local minimum x^*. If $x^* \notin X^*$, then add x^* to X^*, set $W := W+1$ and initiate the W-th cluster by x^*. Assign x_i to the cluster that is initiated by x^*.

<u>Step 4</u>. If $i = N$, then stop. Else, set $i := i+1$ and go to Step 3.

Note that a sample point x can only be linked to a point with smaller function value that is within distance r_N (provided that a local search has not been applied unnecessarily, and the starting point is added to the resulting minimum for that reason only). Moreover (under the same provision), if x is assigned to a cluster which is initiated by a local minimum x^*, then there exists an r_N-descent sequence connecting x and x^*. The sample point x can be assigned to several clusters, if there exist r_N-descent sequences from x to each of the corresponding local minima.

Unfortunately, even if r_N tends to 0, then the fact that there exists an r_N-descent sequence from x to a local minimum x^*, does not necessarily imply that $x \in R_{x^*}$. If P is applied to x, then it is still possible that it will follow another descent path, and find another (possibly undetected) local minimum. However, as we will see later, this cannot happen if x is located in the interior of a component which includes some local minimum as its only stationary point.

To understand the advantage of this approach over Single Linkage, let us consider the one dimensional example in Figure 1.

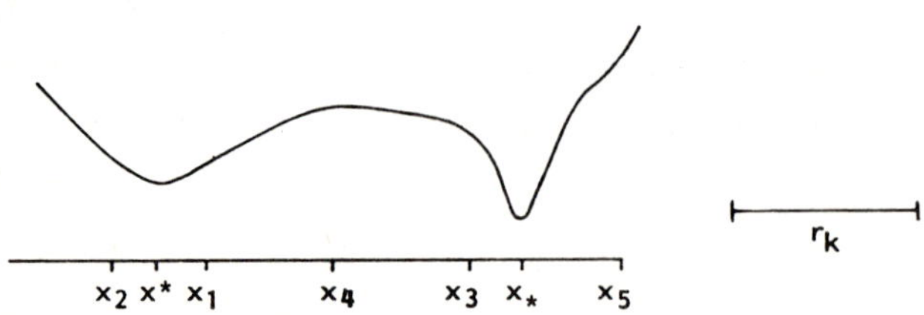

<u>FIGURE 1</u>

Suppose that $x_1,\ldots,x_5$ are reduced sample points which are ordered according to their function value. Both Single Linkage and the above procedure will start by applying P to x_1. Single Linkage will then assign all points $x_1,\ldots,x_5$ to the cluster which is initiated by the local minimum x^*, thus missing the global minimum x_*. The above procedure will assign x_2 to the cluster which is initiated by x^*. But at the moment that x_3 is considered, it is not possible to link x_3 to x^*, since $\|x_3-x^*\| > r_N$. Thus, P will be applied to x_3 and the global minimum x_* is located.

Intuitively speaking, any two local minima will always be separated by a region with higher function values, so that the above procedure will locate every local minimum in the neighbourhood of which a point has been sampled if r_N is small enough.

Since the function values are used in an explicit way in the clustering process, it is no longer necessary to reduce the sample. Note that it is not even essential to actually assign the sample points to clusters. For every sample point x, the decision whether P should be applied to x does not depend on the cluster structures; the decision only depends on the fact whether or not there exists a sample point z with $f(z) < f(x)$ within distance r_N of x. We now turn to an algorithm in which the superfluous clustering is omitted altogether.

Multi Level Single Linkage [Timmer 1984]

Step 1. For every $i = 1,\ldots,N$ apply P to the sample point x_i except if $x_i \in (S-S_\tau) \cup X^*_\upsilon$ or if there is a sample point x_j with $f(x_j) < f(x_i)$ and $\|x_j-x_i\| \leq r_N$.
Add new local minima encountered during the local search to X^*.

For this method it can be proved [Timmer 1984] that if r_N is chosen according to (21) with $\sigma > 0$, and if x is an arbitrary sample point, then the probability that P is applied to x tends to 0 with increasing N. If $\sigma > 2$, the probability that a local search is applied tends to 0 with increasing N. If $\sigma > 4$, then, even if the sampling continues forever, the total number of local searches ever started is finite with probability 1. Furthermore, if r_N tends to 0, than any local minimum x^* will be found within a finite number of iterations with probability 1.

Obviously, this final asymptotic correctness result justifies application of the stopping rules developed for Multistart to Multi Level Single Linkage. We refer the reader to [Timmer 1984] for a more extensive discussion of the Multi Level Single Linkage method. (Technical reports describing further details will also shortly be available from the authors.)

4. COMPUTATIONAL EXPERIMENTS

In this section we shall discuss the computational performance of the methods described in Sections 3.3 and 3.4 on a number of test problems. For this purpose the algorithms were coded in Fortran IV and run on the DEC 2060 computer of the Computer Institute Woudestein.

To be able to compare our methods with other existing ones, the unconstrained methods have been tested on the standard set of test functions [Dixon & Szegö 1978], which is commonly used in global optimization. Since all test functions are twice continuously differentiable, we used the VA10AD variable metric subroutine from the Harwell Subroutine Library as the local search procedure in all (unconstrained) experiments.

To obtain an impression of the numerical performance of the Single Linkage methods we applied them to four independent samples of size 1000. For all three methods we reduced the sample to 100 points (q=0.1) and set σ equal to 4. Furthermore, we chose both υ and τ to be equal to zero in all experiments, thus neglecting the set $S-S_\tau$ and X^*_υ. If, however, a local search was performed resulting in a so far undetected minimum, then we replaced the starting point of the search by the newly detected minimum, to prevent a local search from being started close to this minimum in every succeeding iteration.

The average results of the four runs are listed in Table 1.

Table 1.

Samples of size 1000

Function		Single Linkage	Multi Level Single Linkage
GP:	l.m.	3	3
	l.s.	3	3
	f.e.	163	91
BR:	l.m.	3	3
	l.s.	3	3
	f.e.	157	65
H3:	l.m.	2	2
	l.s.	2	4
	f.e.	161	112
H6:	l.m.	2	2
	l.s.	5	10
	f.e.	585	986
S5:	l.m.	5	5
	l.s.	5	5
	f.e.	324	211
S7:	l.m.	6(*)	6(*)
	l.s.	6	6
	f.e.	429	281
S10:	l.m.	7(*)	8(*)
	l.s.	7	8
	f.e.	439	346

l.m.: number of local minima found

l.s.: number of local searches performed

f.e.: number of function evaluations required (not including the 1000 function evaluations required to determine the function values of the sample points)

(*): Global minimum was not found in one of the four runs

In one of the four runs the methods did not find the global minimum of both the S7 and the S10 test function. The reasons for this are twofold. Firstly, the global minimum of these functions is relatively close to other local minima. Secondly, one of the four samples happened to be a very unfortunate one: the regions of attraction surrounding the region of attraction of the global minimum contained sample points whose function values were smaller than the smallest function value attained in a sample point in the region of attraction of the global minimum. (Note that in the case of the S7 test function, the global minimum was the only minimum that was not found).

It is possible to implement the methods such that the global minimum of every test function is found in each of the four runs. For instance, this will be achieved if a steepst descent step is performed from every reduced sample point and the methods are applied to the resulting transformed sample. A small value of σ (e.g. $\sigma=2$) will also cause the methods to find the global minimum of every test function in each of the four runs. However, both changes will increase the number of function values required.

The number of local searches started unnecessarily is the largest for the test functions H3 and H7. This is due to the fact that these functions are badly scaled.

The computational experiments are continued with Multi Level Single Linkage. This method has been compared with a few leading contenders whose computational behaviour is described in [Dixon & Szegö 1978]. In this reference methods are compared on the basis of two criteria: the number of function evaluations and the running time required to solve each of the seven test problems. To eliminate the influence of the different computer systems used, the running time required is measured in <u>units of standard time</u>, where one unit corresponds to the running time needed for 1000 evaluations of the S5 test function in the point (4,4,4,4).

Since both the number of function evalutions and the units of standard time required are sensitive to the peculiarities of the sample at hand, the results reported for Multi Level Single Linkage are the average outcome of four independent runs again. As before we chose $\tau = \upsilon = 0$ and $\sigma = 4$ in our implementation of Multi Level Single Linkage. However, we now applied Multi Level Single Linkage to 20% of the sample points ($q=0.2$) (the reason that we set q equal to 0.1 before was that a major reduction of the sample is necessary for successful application of Single Linkage). Furthermore, it did not seem reasonable to apply Multi Level Single Linkage to samples of fixed size. After an initial sample of size 100, we increased the sample and applied Multi Level Single Linkage iteratively until the expected number of minima was equal to the number of different local minima observed (cf.20).

In Table 3 and Table 4 we summarize the computational results of the methods listed in Table 2 (except for Multi Level Single Linkage, the results are taken from [Dixon & Szegö 1978]).

Table 2.

Methods

A	Trajectory method [Branin & Hoo 1972]
B	Random direction method [Bremmerman 1970]
C	Controlled Random Search [Price 1978]
D	Method proposed in [Törn 1976, 1978] based on concentration of the sample and density clustering
E	Method based on reduction, density clustering and a spline approximation of the distribution function ϕ of f [De Biase & Frontini 1978]
F	Multi Level Single Linkage

Table 3.

Number of function evaluations

Method	GP	BR	H3	H6	S5	S7	S10
A	-	-	-	-	5500	5020	4860
B	300	160	420L	515	375L	405L	336L
C	2500	1800	2400	7600	3800	4900	4400
D	2499	1558	2584	3447	3649	3606	3874
E	378	597	732	807	620	788	1160
F	148	206	197	487	404	432(*)	564

L : the method did not find the global minimum

(*): the global minimum was not found in one of the four runs

Table 4.

Number of units standard time

	Function						
	GP	BR	H3	H6	S5	S7	S10
Method							
A	-	-	-	-	9	8.5	9.5
B	0.7	0.5	2L	3	1.5L	1.5L	2L
C	3	4	8	46	14	20	20
D	4	4	8	16	10	13	15
E	15	14	16	21	23	30	30
F	0.15	0.25	0.5	2	1	$1^{(*)}$	2

L : the method did not find the global minimum

(*): the global minimum was not found in one of the four runs

As before Multi Level Single Linkage did not find the global minimum of the S7 test function in one of the four runs. Again, this failure could have been prevented by chosing σ to be equal to 2. In that case, the computational results of the method obtained on the test functions GP, BR, H3, H7 and S5 turn out to be comparable to the numbers given in Table 3 and Table 4. However, the number of function evaluations required for the functions S7 and S10 increase by a factor of 2 and 3 respectively. This is due to the fact that all minima of both functions are found in an early stage if σ equals 2. However, the sample must then be increased considerably before our stopping criterion is satisfied.

Since the stopping rules involved in the methods listed in Table 2 are totally different, the comparison between the methods can never be entirely fair: there is always a trade-off between reliability and computational effort that is hard to measure consistently. However, we feel confident that Multi Level Single Linkage is one of the most reliable and efficient methods presently available.

Acknowledgement. This research was partially supported by the Netherlands Foundation for Mathematics SMC with financial aid from the Netherlands Organisation for Advancement of Pure Research (ZWO). The research of the first author was also partially supported by NSF Grant no. ECS - 831-6224, and by NATO Senior Scientist Fellowship.

REFERENCES

Allgower, E. and K. Georg (1980), Simplicial and continuation methods for approximating fixed points and solutions of systems of equations. Siam Review 22, 28-84.

Anderssen, R.S. (1972), Global optimization, In R.S. Anderssen, L.S. Jennings and D.M. Ryan (eds.). Optimization (University of Queensland Press).

Beale, E.M.L. and J.J.H. Forrest (1978), Global optimization as an extension of integer programming. In [Dixon & Szegö 1978].

Becker, R.W. and G.V. Lago (1970), A global optimization algorithm. In Proceedings of the 8th Allerton Conference on Circuits and Systems Theory.

Boender, C.G.E., A.H.G. Rinnooy Kan, L. Stougie and G.T. Timmer (1982), A stochastic method for global optimization. Mathematical Programming 22, 125-140.

Boender, C.G.E. (1984), The Generalized Multinomial Distribution: A Bayesian Analysis and Applications. Ph.D. Dissertation, Erasmus Universiteit Rotterdam (Centrum voor Wiskunde en Informatica, Amsterdam).

Branin, F.H. (1972), Widely convergent methods for finding multiple solutions of simultaneous nonlinear equations. IBM Journal of Research Developments, 504-522.

Branin, F.H. and S.K. Hoo (1972), A method for finding multiple extrema of a function of n variables. In F.A. Lootsma (ed.), Numerical Methods of Nonlinear Optimization (Academic Press, London).

Bremmerman, H. (1970), A method of unconstrained global optimization. Mathematical Biosciences 9, 1-15.

Brooks, S.H. (1958), A discussion of random methods for seeking maxima. Operations Research 6, 244-251.

De Biase, L. and F. Frontini (1978), A stochastic method for global optimization: its structure and numerical performance. In [Dixon & Szegö 1978].

Devroye, L. (1978), Progressive global random search of continuous functions. Mathematical Programming 15, 330-342.

Dixon, L.C.W., J. Gomulka and G.P. Szegö (1975), Towards global optimization. In [Dixon & Szegö 1975].

Dixon, L.C.W. (1978), Global optima without convexity. Technical Report, Numerical Optimisation Centre Hatfield Polytechnic, Hatfield, England.

Dixon, L.C.W. and G.P. Szegö (eds.) (1978), Towards Global Optimization 2 (North-Holland, Amsterdam).

Evtushenko, Y.P. (1971), Numerical methods for finding global extrema of a nonuniform mesh. U.S.S.R. Computing Machines and Mathematical Physics 11, 1390-1404.

Falk, J.E. and R.M. Solund (1969), An algorithm for separable nonconvex programming. Management Science 15, 550-569.

Goldstein, A.A. and J.F. Price (1971), On descent from local minima. Mathematics of Computation 25, 569-574.

Hansen, E. (1980), Global optimization using interval analysis - the multi-dimensional case. Numerische Mathematik 34, 247-270.

Kuhn, H.W., Z. Wang and S. Xu (1984), On the cost of computing roots of polynomials. Mathematical Programming 28, 156-164.

Levy, A. and S. Gomez (1980), The tunneling algorithm for the global optimization problem of constrained functions. Technical Report, Universidad National Autonoma de Mexico.

McCormick, G.P. (1976), Compatibility of global solutions to factorable non-convex programming, Part I - convex underestimating problems. Mathematical Programming 10, 147-175.

Price, W.L. (1978), A controlled random search procedure for global optimization. In [Dixon & Szegö 1978a].

Rinnooy Kan, A.H.G. and G.T. Timmer (1984), Stochastic methods for global optimization. To appear in the American Journal of Mathematical and Management Sciences.

Rubinstein, R.Y. (1981), Simulation and the Monte Carlo Method (John Wiley & Sons, New York).

Shubert, B.O. (1972), A sequential method seeking the global maximum of a function. Siam Journal on Numerical Analysis 9, 379-388.

Solis, F.J. and R.J.E. Wets (1981), Minimization by random search techniques. Mathematics of Operations Research 6, 19-30.

Solund, R.M. (1971), An algorithm for separable nonconvex programming problems 2. Management Science 17, 759-773.

Timmer, G.T. (1984), Global Optimization: A Stochastic Approach. Ph.D. Dissertation, Erasmus Universiteit Rotterdam (Centrum voor Wiskunde en Informatica, Amsterdam).

Todd, M.J. (1976), The Computation of Fixed Points and Applications Springer Verlag, Berlin).

Törn, A.A. (1976), Cluster analysis using seed points and density determined hyperspheres with an application to global optimization. In Proceeding of the third International Conference on Pattern Recognition, Coronado, California.

Törn, A.A. (1978), A search clustering appraoch to global optimization. In [Dixon & Szegö, 1978].

Treccani, G. (1975), On the convergence of Branin's method: a counter example. In [Dixon & Szegö 1975].

Zielinski, R. (1981), A stochastic estimate of the structure of multi-extremal problems. Mathematical Programming 21, 348-356.

ALGORITHMIC PROCEDURES FOR STOCHASTIC OPTIMIZATION

Roger J.B. Wets
IIASA, A-2361 Laxenburg and
Chr. Michelsen Institute, N-5036 FANTOFT

For purposes of preliminary discussion, it is convenient to identify stochastic optimization problems with:

$$\text{find } x \in R^n \text{ that minimizes } z = E\{f(x,\tilde{\xi})\}$$

where $\tilde{\xi}$ is a random N-vector with distribution function, P, $f:R^n \times R^N \to R \cup \{+\infty\}$ is a lower semicontinuous function, possibly convex, where dom $f(.\xi)$ $= \{x \mid f(x,\xi) \text{ is finite}\}$, corresponds to the set of acceptable choices for x when ξ is the observed value of the random vector $\tilde{\xi}$, and

$$E\{f(x,\tilde{\xi})\} = \int f(x,\xi)\, dP(\xi).$$

To simplify matters, we may even take f finite, it will not affect much the discussion of the numerical obstacles, that must be overcome to solve stochastic optimization problems. With

$$F(x) := \int f(x,\xi)\, dP(\xi),$$

we see that our original problem is equivalent to the deterministic optimization problem:

$$\text{find } x \in R^n \text{ that minimizes } z = F(x).$$

Thus, in principle, any algorithmic procedure developed for nonlinear optimization problems could be used. In fact, we have:

$$F \text{ is convex if } f(.,\tilde{\xi}) \text{ is convex a.s.,}$$

and, assuming that dom $f(.,\xi)$ is independent of ξ, then

$$\partial F(x) = \int \partial f(x,\xi)\, dP(\xi).$$

Moreover, in general

NATO ASI Series, Vol. F15
Computational Mathematical Programming
Edited by K. Schittkowski

$$\nabla F(x) = \int \nabla f(x,\xi)\, P(d\xi) \text{ if } f(.,\, \underset{\sim}{\xi}) \text{ is a.s. differentiable}$$

although usually F is nondifferentiable and then with an appropriate definition of the subgradient-set, we have

$$\partial F(x) \subset \int \partial f(x,\xi)\, dP(\xi)$$

Related formulas can also be found for second order derivatives, so that, in theory at least, all what we need to do is to include an integration subroutine in a standard nonlinear (nondifferentiable) optimization package to have state-of-the-art software for stochastic optimization problems. And indeed this would work very well if $\underset{\sim}{\xi}$ is a random varialbe i.e. N=1 -- excellent integration subroutines are available in this case -- or even possibly when N=2 or 3 and the analytic description of f is not too complicated. However, most applications that are of interest have N much larger that 2 or 3, in some cases all the coefficients of a given problem have stochastic components that cannot be ignored in which case N could be quite large. Except for certain specific distribution functions, such as for gamma or normal distribution functions and then with $N \leq 4$, the only known multidimensional integration subroutines available rely on Monte-Carlo techniques, involving generating pseudo- or quasi-random numbers. And for these methods to be effective we need ready access to functions values or to (sub)gradients, etc, and, as we shall see later, in stochastic optimization this is the exception rather than the rule.

Thus, in one way or another, we must design solution procedures that do not rely on multidimensional integration subroutines. Excluding certain specific classes of problems, such as stochastic programs with simple recourse and some stochastic programs with probabilistic constraints, where the properties of the problem at hand make it possible to carry out at low cost the required integration, the suggested solution strategies can be divided in two major categories: (i) "descent" methods that rely on directions determined by statistical estimates of the subgradients of F, and (ii) approximation methods that replace to original distribution function P by a discrete distribution P^ν involving a sufficiently small number of probability mass points, so that $\int f(x,\xi)\, dP^\nu(\xi)$ -- now corresponding to a finite sum -- is numerically feasible.
The remainder of this presentation is devoted to a brief description of the major features these solutions procedures and of their actual or potential implementations.

A. Stochastic quasigradient methods.

Let us consider the case:

find $x \in C$ that minimizes $E\{g(x,\underset{\sim}{\xi})\}$

where

C is a closed convex set of R^n,

$g(.,\xi)$ is convex for all ξ

and the random vector $\underset{\sim}{\xi}$ is as defined above. In terms of our original formulation we would set

$$f(x,\xi) = \begin{cases} g(x,\xi) & \text{if } x \in C \\ +\infty & \text{otherwise.} \end{cases}$$

The algorithm generates a sequence $\{x^1, x^2, \ldots\}$ of points of C through the recursive formula

$$x^{\nu+1} = \text{prj}_C(x^\nu - \rho_\nu h^\nu)$$

where prj_C denotes projection on the set C, $\{\rho_\nu, \nu=1, \ldots\}$ is a sequence of scalars and

h^ν is a stochastic quasigradient of G at x^ν

with

$$G(x) := \int g(x,\xi)\, dP(\xi).$$

By stochastic quasigradient, one means a realization of a random n-vector $\underset{\sim}{h}^\nu$ satisfying

$$E\{\underset{\sim}{h}^\nu \mid x^1, \ldots, x^\nu\} \in \partial G(x^\nu).$$

Typically

$$h^\nu \in \partial g\,(x^\nu, \xi^\nu)$$

with the $\{\xi^\nu,\ \nu=1, \ldots\}$ independent random samples of $\underset{\sim}{\xi}^\nu$, or more generally

$$h^\nu = (1/L) \sum_{l=1}^{L} h^{l\nu}$$

where for all l, $h^{l\nu} \in \partial g(x^\nu,\ \xi^{\nu l})$ and the $\{\xi^{\nu l};\ l=1, \ldots, L,\ \nu=1, \ldots\}$ are independent random samples.

The sequence of feasible solutions $\{x^\nu,\ \nu=1, ..\}$ converges with probability 1 to an optimal solution -- assuming naturally that it exists -- provided that the scalars ρ_ν are chosen so as to satisfy

$$\rho_\nu \geq 0,\ \sum_\nu \rho_\nu > \infty \text{ and } \sum_\nu \rho_\nu^2 < \infty\ ;$$

$\rho_\nu = 1/\nu$ is such a sequence. The proof can be derived from a modified super-martingale convergence argument.

In the implementation of this method we must contend with three possible stumbling blocks:

the projection on C,
the choice of the step size,
the stopping criterion.

The <u>projection</u> of a point on a closed set C is easy only if C is "simple" by which we mean that C is a bounded interval, a sphere, ... The most general case that we know how to handle quite efficiently, has the set C as the intersection of a bounded interval and 1 nonlinear (or linear) constraint of the type

$$\sum_{j=1}^{n} a_j(x_j) + a_0 \leq 0$$

where a_j is convex and differentiable with $a'_j > 0$ on the bounded interval. If C is a polyhedron it may be possible to develop a technique based on the observation that for ν sufficiently large, the x^ν are liable to be quite close to each other and thus project in all likelihood on the same face of C. If C is a general convex set then each projection involves minimizing a quadratic function on a convex set. For all practical purposes this complicated projection operation would make every step of the algorithm very expensive as soon as we approach the boundary of C. The objections we might have about using this method on these grounds, could be overcome by relying on penalization approximates -- see the literature on nonlinear programming -- such as for example

$$f^{\nu}(x,\xi) = g(x,\xi) + \nu \,\mathrm{dist}(x,C)$$

and h^{ν} is then a stochastic quasigradient of

$$F^{\nu}(x) = \int f^{\nu}(x,\xi)\, dP(\xi).$$

However, experimental results have shown that due to the steepness of the subgradients, penalization has a tendency to destabilize the method whenever the optimal solution lies on the boundary of C.

The choice of the <u>step-size</u> ρ_{ν} is in principle prescribed by the convergence requirements. However, since in practice only the short run properties of the sequence $\{x^{\nu}, \nu=1, \ldots\}$ are of interest, there is at present a gap between theory and practice where the choice of the step-size is usually guided by some adaptive rule that tries to estimate the progress made during the last M iterations, $M \geq 1$. Some preliminary results that begin to fill this gap having recently been obtained.

Finding a <u>good stopping criterion</u> is still very much an open question. As already mentioned earlier, in stochastic optimization problem evaluating $F = E\{f(.,\tilde{\xi})\}$ may be quite expensive -- and this is why we rely on the method of stochastic quasigradients in the first place -- so it is out of question to use value comparisons between F at x^{ν} and $x^{\nu+1}$. The quantity

$$F_a(x^{\nu}) = (1/M+1) \sum_{l=\nu-M}^{M} f(x^l,\xi^l)$$

has been suggested as an estimate for $F(x^{\nu})$. The algorithm is to terminate when no improvement is observed in the value of F_a. The fact that we never really know if we have or have not reached an optimal, or nearly optimal, or sufficiently optimal solution is Achille's heel of this class of methods. Finding stopping criteria based on probabilistic error bounds, and the related question of step-size, is an area ripe for research and experimentaion.

B. Approximate solutions by discretization.

If P^{ν} is a distribution function that approximates the given distribution P we may hope that an optimal solution x^{ν} of the approximating stochastic optimization problem:

$$\text{find } x \in R^n \text{ that minimizes } F^{\nu}(x) = \int f(x,\xi)\, P^{\nu}(d\xi),$$

will be an approximate solution of the original problem. And, in fact this is the case provided f is not too exotic and P^{ν} is not chosen so selectively that it generates the unusual. Every distribution function P can be approximated as closely as desired by a piecewise constant distribution function P^{ν} which corresponds to assigning probabilities

$$p_1, p_2, \ldots, p_L$$

to a finite collection of vectors

$$\xi^1, \xi^2, \ldots, \xi^L.$$

Moreover, it can usually be shown that the approximation error, measured by the quantity

$$|F(x^{\nu}) - \inf F|,$$

is a function of the goodness of fit of P^{ν} to P, even proportional to it in the polyhedral case, i.e. when $f(.,\xi)$ is a convex piecewise linear function. If we are satisfied with an approximate solution -- and often we shall not have any alternative -- and we choose a discrete distribution P^{ν} close enough to P, we could solve the problem

$$\text{find } x \in R^n \text{ that minimizes } F^{\nu}(x) := \textstyle\sum_{k=1}^{L} p_k\, f(x,\xi^k).$$

No longer is there any need for a multidimentional integration scheme, gradients and values for F^{ν} only involve computing a finite sum. However, we should not be lulled into believing that in this way we have licked the multidimensional integration problem. Unless $N \leq 3$, the number L of points that we need to approximate P sufficiently closely so as to guarantee an acceptable error bound for the solution may be truly astronomical. For example if N=10 and we have 10 independent random variables taking on each 10 possible values or we have approximated each marginal distribution function by a discrete distribution with 10 density points, then L=10 billion ! Thus even if the orginal problem itself involves a discrete distribtuion we may shy away from solving such types of problems.

The alternative is to choose a very rough approximate of P involving only a small number of density points, and this even if P itself is a discrete distribution, and hope that the resulting solution x^{ν} is nonetheless a good approximate. This actually works (!), at least with our limited computational experience. There is actually some basic justification for this: The optimal solutions of stochastic optimization problems exhibit surprising stability properties with respect to perturbations of the distribution function of the random variables.

However, we can no longer rely on the proximity of P and P^{ν} to obtain error bounds, this must be obtained through other means. What could be done is to choose a pair of discrete distributions P^l and P^u in such a way that, if we solve

$$\text{find } x \in R^n \text{ that minimizes } F^l(x) := \sum_{k=1}^{L} p_k^l\, f(x,\xi^{lk})$$

we obtain a lower approximate for the original problem, and if we solve

$$\text{find } x \in R^n \text{ that minimizes } F^u(x) := \sum_{k=1}^{L'} p_k^u\, f(x,\xi^{uk})$$

we obtain an upper approximate, i.e. we have

$$\inf F^l \leq \inf F \leq \inf F^u$$

This of course gives an error estimate that can be used as a termination criterion. If we feel that the error bound provided by these approximates is not tight enough we can refine either P^l or P^u or both, to obtain a better bracketing of the optimal value. In fact we could design a solution procedure that systematically refines the approximating distribution while carrying out the steps of the algorithm.

The design of discrete distributions P^l and P^u with the desired properties, either relies on convexity or concavity properties of f(x,.), i.e. with respect to the random parameter, or we try to identify a class of distributions Σ that contains P and such that for all x, or at least for some region in the neighborhood of an optimal solution,

$$P^l \in \operatorname{argmin}_{Q\in\Sigma} \int f(x,\xi)\, dQ(\xi),$$

and

$$P^u \in \operatorname{argmax}_{Q\in\Sigma} \int f(x,\xi)\, dQ(\xi).$$

If the distributions in Σ are restricted to a fixed compact support and we choose to define Σ as the class of distribution functions that have the same moments up to order r as P, then P^l and P^u as defined above are discrete distributions having about as many points of support as the number of moments that we want to match. The bounds obtained through convexity or concavity of $f(x,.)$ rely on Jensen's inequality and the fact that $\sup_{\xi} f(x,\xi)$ is attained at an extreme point of the (convex hull of) the support of P. Assuming convexity, it yields

$$f(x,E\tilde{\xi}) \leq E\{f(x,\tilde{\xi})\}$$

Thus,

P^l which assigns probability 1 to $E\tilde{\xi}$

yields a lower bound since then $F^l(x) = f(x,E\tilde{\xi})$. On the other hand if

$$\xi^u \in \operatorname{argmax}\{f(x,\xi) \mid \xi \in \text{support of } P\}$$

then with

P^u which assigns probability 1 to ξ^u,

we have an upper bound since $F^u(x) = f(x,\xi^u)$.

All these bounds can be substantially refined by partitioning the support of the distribution function P and taking conditional moments or conditional extreme points instead of moments or extreme points as here above.

All we have done so far is lay the ground work to justify limiting our attention to

$$\text{find } x \in R^n \text{ that minimizes } w = \sum_{k=1}^{L} p_k\, f(x,\xi^k)$$

in the development of solution techniques for stochastic optimization problems, with L relatively small, maybe a few hundreds or thousands. We could now rely on standard linear or nonlinear optimization techniques for solving this class of problems, and this would work well enough (and in some cases we actually can proceed in this manner), except that in most applications the function f is quite difficult to evaluate, the same being true about subgradients as well as other related quantities. To see this we need to examine a little bit more closely the type of functions f that we have to deal with in stochastic optimization.

As a first example, let us consider a simple version of stochastic programs with probabilistic constraints:

$$\text{find } x \in R^n_+ \text{ such that } Ax = b,\ P(Tx) \geq \alpha$$

$$\text{and } z = cx \text{ is minimized.}$$

The constraint $P(Tx) \geq \alpha$ -- recall P is here the distribution function -- means that with probability α we want the values of $\tilde{\xi}$ to be less than Tx. The function

$$f(x,\xi) = \begin{cases} cx \text{ if } x \geq 0,\ Ax = b,\ \alpha - P(Tx) \leq 0, \\ + \text{ otherwise} \end{cases}$$

does not really depend on ξ but to check if $x \in \text{dom} f(.,\xi)$, or equivalently if x is a feasible solution of the stochastic program, we must evaluate the integral

$$\int_0^{Tx} dP(\xi),$$

which can be replaced by a finite sum to obtain upper and lower bounds. We could then refine the approximation in the neighborhood of the suspected optimal value of Tx to obtain tighter bounds. The most efficient and reliable algorithm for solving such problems appears to be a primal-dual procedure that works as follows: with

$$\rho(v) := \inf[vx \mid P(x) \geq \alpha]$$

we can show that

find $u \in R^{m1}$, $v \in R_{+}^{m2}$ such that $c = uA + vT$

and $w = ub + \rho(v)$ is maximized,

is dual to our stochastic program, at least when

$\{\chi | P(\chi) \geq \alpha\}$ is convex,

the function ρ is then concave. Suppose (u^k, v^k) is a feasible solution of this dual program, let

$\chi^k \in \text{argmin}\ \rho(v^k)$

and

$x^k \in \text{argmin}\ [cx |\ Ax = b,\ Tx \geq \chi^k,\ x \geq 0]$

This last problem is a linear program. Let $(\bar{u}^k, \bar{v}^k)$ be the simplex multipliers associated with the constraints at the optimum. If (u^k, v^k) matches $(\bar{u}^k, \bar{v}^k)$, we are done since then we satisfy the optimality conditions. Otherwise note that $(\bar{u}^k, \bar{v}^k)$ is an extreme point of the dual feasible region and the direction

$(\bar{u}^k - u^k,\ \bar{v}^k - v^k)$

is a direction of ascent for the dual problem. A new point (u^{k+1}, v^{k+1}) is selected between

(u^k, v^k) and $(\bar{u}^k, \bar{v}^k)$

that improves the dual objective, and the procedure is repeated until an optimal solution of the dual problem is reached: the corresponding x^k solves the original program. In fact only convergence can be claimed. The touchy part in this algorithm from a numerical viewpoint is the minimization of ρ which requires evaluating $P(\chi)$.

As a second example, we take f to be the essential objective function of a (linear) stochastic program with recourse (with random right hand sides), namely

$$f(x,\xi) = \begin{cases} cx + \inf_y [qy \mid Wy = \xi - Tx,\ y \geq 0] & \text{if } Ax = b,\ x-0, \\ +\infty & \text{otherwise} \end{cases}$$

The stochastic program is a model for the following decision process: we choose an activity level x subject to certain deterministic constraints $Ax = b$, $x \geq 0$, and generate an output Tx before we can observe the value ξ of the random vector $\tilde{\xi}$. If there is any discrepancy between ξ and Tx, we make it up by selecting a recouse decision y at cost qy such that $Wy = \xi - Tx$, $y \geq 0$. The penalty for not matching exactly the random outcome ξ with the output Tx can be calculated by solving a linear programming problem. There are of course a myriad of variants of this model.

Unless the problem has specific structural properties, the standard solution procedure is a partial decomposition method to which one usually refers as the L-shaped algorithm. Let

$$\psi(x) = E\{Q(x,\xi)\} = \int Q(x,\xi)\, dP(\xi),$$

where

$$Q(x,\xi) = \inf_y\ [qy \mid Wy = \xi - Tx,\ y \geq 0].$$

The method consists of 3 steps that can be interpreted as follows. In step 1 we solve an approximation to

$$\text{find } x \in R^n_+ \text{ with } Ax = b \text{ that minimizes } z = cx + \psi(x)$$

obtained by outer-linearization. The two types of additional linear constraints that appear in this linear program come from

(i) feasibility cuts -- generated in Step 2 -- that restrict x to the region where $\psi(x) < +\infty$, i.e. which render the recourse problem feasible for all possible values of $\tilde{\xi}$, and

(ii) optimality cuts -- generated in Step 3 -- that refine the linear approximation to ψ at least in the neighborhood of the optimal solution.

We give here a coarse version of this algorithm. At the outset set all counting parameters $\nu = s = t = 0$.

<u>Step 1</u>. Set $\nu = \nu+1$ and solve the linear program:

$$\begin{aligned} &\text{find } x \geq 0,\ \theta \in R \text{ such that} \\ &Ax = b \\ &D_k x \geq d_k,\ k=1, \ldots, s \\ &E_k x + \theta \geq e_k,\ k=1, \ldots, t \\ &cx + \theta = z \quad \text{is minimized} \end{aligned}$$

Let (x^ν, θ^ν) be an optimal solution. If there are no constraints involving θ, we set $\theta^\nu = -\infty$ and the variable θ is ignored in the linear program.

<u>Step 2</u>. For all possible realizations ξ of $\tilde{\xi}$ solve the linear programs

$$\begin{aligned} &\text{find } y \geq 0,\ v^+ \geq 0,\ v^- \geq 0 \text{ such that} \\ &Wy + Iv^+ - Iv^- = \xi - Tx^\nu \text{ and} \\ &\quad ev^+ + ev^- = w^1 \text{ is minimized.} \end{aligned}$$

If for some ξ the corresponding value $w^1 > 0$, let σ^ν be the simplex multipliers associated to an optimal solution, and define

$$D_{s+1} = \sigma^\nu T, \text{ and } d_{s+1} = \sigma^\nu \xi.$$

Return to Step 1 adding this feasibility cut and set $s = s+1$, $\nu = \nu+1$. If for all possible vectors ξ, $w^1 = 0$ then go to Step 3.

<u>Step 3</u>. For all possible realizations ξ of $\tilde{\xi}$ solve the linear program:

$$\begin{aligned} &\text{find } y \geq 0 \text{ such that} \\ &Wy = \xi - Tx, \text{ and} \\ &qy = w^2 \text{ is minimized} \end{aligned}$$

Let $\pi(\xi)$ be the multipliers associated with the optimal solution, of course they depend on ξ. Define

$$E_{t+1} = E\{\pi(\tilde{\xi})\}T$$

$$e_{t+1} = E\{\pi(\tilde{\xi})\tilde{\xi}\}$$

and set

$$w^{2\nu} = e_{t+1} - E_{t+1}x^{\nu}.$$

If $\theta^{\nu} \geq w^{2\nu}$ we stop, x^{ν} is an optimal solution. Otherwise return to step 1 improving the outer-linearization to ψ by adding the optimality cut

$$E_{t+1}x + \theta \geq e_{t+1}$$

and set $t = t+1$, $\nu = \nu+1$.

We see that in carrying out the steps of the algorithm we are up against two major difficulties. The first one is: given $\pi(.)$ compute E_{t+1} and e_{t+1}. This has been dealt with in the earlier part of this section, and we shall assume that an acceptable discrete approximation of the distribution function P has been found that renders this calculation numerically feasible. The second difficulty is that even if the probability mass is only carried by a finite number of vectors

$$\xi^1, \ldots, \xi^L$$

and L is not too large -- of the order of 50, or 100, maybe even 1000 -- we need to solve in Step 3 L linear programs. (Step 2 allows usually for much further simplification, it often suffices to solve the linear program that appears there for a very limited number of vectors of type ξ to check if x^{ν} is feasible.)

To solve a large number of linear programs (with constant technology matrix) we rely on a discrete parametric analysis technique which goes under the name of <u>bunching</u>. Given x^{ν}, let B be a submatrix of W that is optimal for some $(\xi^1 - Tx^{\nu})$. Then from the optimality conditions for linear programming it follows that this basis will also be optimal for all vectors ξ^k such that

$$B^{-1}(\xi^k - Tx^{\nu}) \geq 0.$$

Since B^{-1} is already available, verifying the preceeding inequality involves substantially less work than solving a whole collection of linear programs. Moreover, because of the nature of the problem at hand it is reasonable to expect that only a small number of bases in the neighborhood of B, i.e. which can be obtained from B by a small number of pivot steps, should be sufficient to bunch all vectors $\xi^1, \ldots, \xi^L$; a bunch is the collection of vector associated to a basis by verifying if the linear inqualities above are satisfied.

Efficient bunching, the favorite one these days is a trickling down procedure that creates a tree of neighboring bases rooted at the optimal basis that corresponds to the vector $E\{\xi\}-Tx$, brings the carrying out of Step 3 of the L-shaped algorithm in the realm of possibilites. Special subroutines have been built for the case when the linear programs: for k=1, ..., L

$$\text{find } y \geq 0 \text{ with } Wy = \xi^k - Tx \text{ that minimizes } w^2 = qy$$

are transportation problems or network flow problems that are remarkably efficient, but even for general linear programs much progress has been made at the experimental level.

C. Conclusion.

I have tried to delineate the difficulties that are inherent to stochastic optimization problems and that hamper the development of efficient solution procedures. I have also suggested some strategies for the development of algorithmic procedures. Of course, I have only been able to survey a very limited corner of the on-going work. For recent results please refer to the Mathematical Programming Studies of Stochastic Programming, eds. A. Prekopa and R. Wets, that should appear in early 1985 (North-Holland). For a state-of-the-art description of computational issues in stochastic programming, please refer to the IIASA-Collaboritive Volume on "Numerical Methods in Stochastic Optimization", eds. Y. Ermoliev and R. Wets, that is now being prepared for publication. The algorithm given for stochastic programs with probabilistic constraints is due to E. Komaroni.

Acknowledgment. My views on algorithmic development for stochastic optimization problems have been shaped by too many other workers in the field to name them here, but I would like to mention Larry Nazareth (at IIASA) whose contributions and approach to software development have had a significant influence.

NONDIFFERENTIABLE OPTIMIZATION

J. Zowe
University of Bayreuth
D-8580 Bayreuth, W.-Germany

CONTENTS. *Nondifferentiable optimization* NDO (also called *nonsmooth optimization* NSO) concerns problems in which the functions involved have discontinuous first derivatives. This causes classical methods to fail; hence nonsmooth problems require a new, a nonstandard approach. The paper tries to develop the basic features of the two main direct approaches in NDO, namely the *Subgradient* concept and the *Bundle* concept. Rather than collecting as many results in this area as possible, we will try to motivate and to help understanding the main underlying ideas.

The paper is organized as follows: after a short introduction, recalling some basic definitions from convex analysis, some examples and applications will be considered. Then we will discuss that and why classical methods like steepest descent must fail in the non-smooth situation. Following this, a few words about some indirect approaches to the problem will be said. Chapters 5 and 6 contain the main topic of this paper, the Subgradient and the Bundle idea. Then a numerical comparison of these two proposals with some indirect methods and with a smooth method, the BFGS-method, will follow. Finally a second order model for nonsmooth problems will be discussed; this model is based on the same ε-ideas, which lead to the Bundle concept.

NATO ASI Series, Vol. F15
Computational Mathematical Programming
Edited by K. Schittkowski

1. INTRODUCTION: BASIC DEFINITIONS

This paper studies the minimization of a real functional f on the n-dimensional euclidean space

$$\textit{minimize } f(x) \textit{ on } \mathbb{R}^n . \tag{1.1}$$

There are no constraints in (1.1) but, different from the standard situation, we do not require f to be smooth. More precisely: we are content if the gradient of f exists almost everywhere and, if at every point x where the gradient is not defined, at least the *directional derivative*

$$f'(x;d) := \lim_{t \downarrow 0} \frac{1}{t}[f(x+td)-f(x)] \tag{1.2}$$

exists in every direction d. It is well-known that convex functions are of this kind (see e. g. [29] for this and the following definitions); and for simplifying the presentation we restrict our development to the *convex case*. But let us mention that most of the results remain true under suitable technical assumptions weaker than convexity; see the respective literature.

Typically the function f in (1.1) will be composed of C^1-pieces, i. e., the graph of f is smooth up to some corners where the gradient $\nabla f(x)$ jumps. Consider e. g. the function in one dimension

$$h(x) := \begin{cases} -x & \text{for } x \leqslant 0 \\ x^2 & \text{for } x > 0 \end{cases} . \tag{1.3}$$

Then $\nabla h(x) = -1$ for negative x and $\nabla h(x) = 2x$ for positive x. At x = 0 the gradient is not defined but, obviously, the two limits $\lim_{x \uparrow 0} \nabla h(x) = -1$ and $\lim_{x \downarrow 0} \nabla h(x) = 0$ taken together characterize the (first order) behaviour of h close to the kink x = 0. This leads to the following substitution of the gradient; we define as *subdifferential* of f at x (conv denotes the convex hull)

$$\partial f(x) := \text{conv} \left\{ g \in \mathbb{R}^n \;\middle|\; \begin{array}{l} g = \lim \nabla f(x_i),\ x_i \to x \\ \nabla f(x_i) \textit{ exists, } \nabla f(x_i) \textit{ converges} \end{array} \right\}. \tag{1.4}$$

This definition makes sense since for convex f the gradient exists almost everywhere. The subdifferential is a non-empty convex compact set which shrinks to the gradient in case f is differentiable at x; the elements of $\partial f(x)$ are called *subgradients*. For the above h one gets

$$\partial h(0) = \{g \in \mathbb{R}^1 \mid -1 \leqslant g \leqslant 0\}. \tag{1.5}$$

The directional derivative (1.2) is another tool to characterize the behaviour of a function close to a kink. One easily computes

$$h'(0;d) = \begin{cases} -d & \text{for } d \leqslant 0 \\ 0 & \text{for } d \geqslant 0 \end{cases}. \tag{1.6}$$

As expected, there is a close relation between the subdifferential and the directional derivative. A fundamental theorem from convex analysis states that the directional derivative $f'(x;\cdot)$ is the *support function* of $\partial f(x)$:

$$f'(x;d) = \max_{g \in \partial f(x)} g^T d. \tag{1.7}$$

It is easily checked that (1.7) together with (1.5) gives back (1.6).

In the convex case g is a subgradient at x if and only if $\binom{g}{-1}$ is the normal vector of a hyperplane in $\mathbb{R}^{n+1}$ which supports the graph of f at $\binom{x}{f(x)}$ from below. This leads to another equivalent characterization of the subdifferential:

$$\partial f(x) = \{g \in \mathbb{R}^n \mid g^T(z-x) \leqslant f(z)-f(x) \textit{ for all } z \in \mathbb{R}^n\}. \tag{1.8}$$

(1.8) immediately gives the *necessary and sufficient optimality condition* for the convex problem (1.1):

$$\begin{array}{l} x^* \textit{ is optimal for } (1.1) \\ (\text{i.e. } f(x^*) \leqslant f(x) \textit{ for all } x) \end{array} \iff 0 \in \partial f(x^*). \tag{1.9}$$

Hence the set X* of optimal points for (1.1) is characterized by

$$X^* = \{x^* \in \mathbb{R}^n \mid 0 \in \partial f(x^*)\}.$$

To exclude pathological situations we will often assume that X* is nonempty and bounded.

Suppose we know the gradient of f whenever it exists. Then we can compute at least one subgradient at each point of nondifferentiability; see definition (1.4). Consequently the general assumption for the following seems to be rather natural:

at every x *we know* f(x) *and one (randomly chosen)* $g \in \partial f(x)$. (1.10)

At the end of the next chapter we will come back to assumption (1.10).

Let us add that definition (1.4) makes sense for a general locally Lipschitz continuous function, not necessarily convex. Many of the results to come remain true if this function is directionally differentiable in the sense of definition (1.2) and if again a relation similar to (1.7) holds between the directional derivative and the subdifferential.

2. EXAMPLES AND APPLICATIONS

2.1. In many applications the function to be minimized is given as the maximum of finitely many functions $f_1,\ldots,f_m$:

$$\min_{x\in\mathbb{R}^n} \max_{1\leq i\leq m} f_i(x). \tag{2.1}$$

Even if all f_i's are smooth, the graph of $f := \max_{1\leq i\leq m} f_i$ will have corners where differentiability fails. The above *minimax*-formulation contains a lot of interesting subproblems, e. g. the minimization of a vector-valued function in the ℓ_1- or the ℓ_∞-norm, respectively,

$$\min_{x\in\mathbb{R}^n} \|[f_1(x),\ldots,f_m(x)]^T\|_1, \tag{2.2}$$

$$\min_{x\in\mathbb{R}^n} \|[f_1(x),\ldots,f_m(x)]^T\|_\infty. \tag{2.3}$$

Written in form (2.1) problems (2.2) and (2.3) become

$$\min_{x} \max_{+-} \{\pm f_1(x) \pm f_2(x) \pm \ldots \pm f_m(x)\} \qquad (2.2)'$$

$$\min_{x} \max \{f_1(x), -f_1(x), \ldots, f_m(x), -f_m(x)\}. \qquad (2.3)'$$

In (2.2)' one has to maximize over 2^m possible combinations of + and -. The discrete linear Chebychev approximation is a special case. Here $t_1,\ldots,t_m$ are given points on the real line and $k_1,\ldots,k_n$ are given simple functions; typically m is greater than n. The aim is to replace a nasty function h(t) by a linear combination $\sum_{j=1}^{n} x_j k_j(t)$ such that the maximal distance at the t_i's becomes minimal

$$\min_{x} \max_{1\leq i\leq m} |h(t_i) - \sum_{j=1}^{n} x_j k_j(t_i)|. \tag{2.4}$$

(2.4) reduces to (2.3)' if we put

$$f_i(x) := h(t_i) - \sum_{j=1}^{n} x_j k_j(t_i).$$

Another class of problems which can be subsumed under (2.1) are inequality systems:

$$\textit{find } x \textit{ such that } f_i(x) \leq 0 \textit{ for } i = 1,\dots,m. \tag{2.5}$$

Provided (2.5) is consistent, then we can find a solution by solving the minimax-problem $\min_x \max \{f_1(x),\dots,f_m(x),0\}$.

2.2. The *exact penalty function*

$$\mu f_o(x) + \sum_{i=1}^{k} \max\{0,g_i(x)\} + \sum_{j=1}^{l} |h_j(x)| \tag{2.6}$$

is another interesting example for a nonsmooth problem. It is well-known that, under suitable assumptions and for sufficiently small μ, the solutions of the constrained differentiable problem

$$\begin{aligned}&\min f_o(x)\\ &\text{s. t. } g_i(x) \leq 0 \textit{ and } h_j(x) = 0 \quad (1 \leq i \leq k,\ 1 \leq j \leq l)\end{aligned} \tag{2.7}$$

are also minimizers of (2.6) and vice versa; see Pietrzykowsky [25]. Note that $h_j(x^*) = 0$ for an optimal x^*; hence (2.6) will have a kink at x^*. The loss of the differentiability in (2.6) is the price we have to pay to get rid of the constraints.

2.3. *Duality* is another major source for nonsmoothness. One can associate with (2.7) the dual problem

$$\begin{aligned}&\max f(u,v)\\ &\textit{where } f(u,v) := \inf_x \{f_o(x)+u^Tg(x)+v^Th(x)\} \textit{ for } (u,v) \in \mathbb{R}^k_+ \times \mathbb{R}^l\\ &(g(x) := (g_1(x),\dots,g_k(x)))^T, h(x) := (h_1(x),\dots,h_l(x))^T).\end{aligned} \tag{2.8}$$

Now f is a concave function (hence maximization of f is equivalent to minimization of -f), quite possibly nonsmooth.

2.4. *Decomposition* leads to nonsmooth problems. Suppose e. g. you have a large linear system

$$\begin{aligned}&\min c^Tx + d^Ty\\ &\text{s. t. } Ax + By \leq b\end{aligned}\ ; \tag{2.9}$$

here A,B,c,d,b are suitable matrices and vectors, respectively. Provided the problem has some structure, then one can often easily solve the subproblems in y for every fixed x

$$[g(x) :=] \min \{d^Ty \mid By \leq b - Ax\}.$$

(2.9) reduces to the so-called master problem

$$\min_x \{c^T x + g(x)\}.$$

The function g will have kinks, thus we end up with a smaller but nonsmooth problem.

2.5. *Stiff problems* provide another interesting f is C^1 but the gradient varies rapidly, although continuously, when x varies. Such stiff f is *numerically* nonsmooth. Saying it in another way: NDO can also be viewed as a study for improving classical methods for smooth but stiff problems.

As already said: the role played by the gradient for smooth f will now go over to the subdifferential in the nonsmooth case. We consider ∂f for two of the above examples. Definition (1.4) almost immediately yields for the max-function in (2.1) (provided $f_i \in C$):

$$\partial f(x) = \operatorname{conv} \{\nabla f_j(x) \mid f_j(x) = \max_{1 \leq i \leq m} f_i(x)\}. \tag{2.10}$$

A duality argument gives for the concave function (2.8) (see e.g. [29])

$$\partial f(u,v) = \operatorname{conv} \left\{ \begin{pmatrix} g(x) \\ h(x) \end{pmatrix} \middle| \begin{array}{l} x \text{ minimizes} \\ v_o(x) + u^T g(x) + v^T h(x) \end{array} \right\}. \tag{2.11}$$

It is obvious that a complete knowledge of (2.11) will be unattainable, and that the full knowledge of (2.10) will be too expensive to attain (in applications m may be very large). Hence an idea will be implementable only if it gets along with the knowledge of say one $\nabla f_j(x)$ in (2.10) for some active j and, respectively, $\begin{pmatrix} g(x) \\ h(x) \end{pmatrix}$ in (2.11) for any minimizing x. And this precisely is what we require in (1.10).

3. FAILURE OF CLASSICAL METHODS

3.1. In a classical smooth method one replaces f at x by a *linear* or a *quadratic model* (f'(x) and f"(x) denote the Fréchet-derivatives)

$$f'(x)d \quad [\approx f(x+d) - f(x)] \tag{3.1}$$

$$f'(x)d + \tfrac{1}{2}f''(x)(d,d)\,[\approx f(x+d) - f(x)] \tag{3.2}$$

and minimizes these models. Minimization of model (3.1) on the unit ball leads to steepest descent and minimization of model (3.2) leads to a Newton-type method; compare formula (5.1) and (5.2), respectively. Obviously the above models are no longer defined at a kink and, close to a kink, they do not provide any more an efficient approximation of f. If the minimal x* is a kink (and this is almost the rule for non-smooth f), then for x close to x* the search directions provided by minimizing (3.1) and (3.2), respectively, become almost useless. This is confirmed by all numerical experiments one makes. BFGS, for example, breaks down because of line search difficulties (due to the bad search direction) with not more than 2 or 3 exact figures of f(x*).

It may even be worse. The inappropriate smooth model (3.1) or (3.2) may keep you trapped at a nonoptimal kink. Consider the following example which is a modification of an example due to Wolfe [36]. We study the function

$$f(z_1, z_2) := \begin{cases} 5(9z_1^2+16z_2^2)^{1/2} & \text{for } z_1 \geq |z_2| \\ 9z_1+16|z_2| & \text{for } 0 < z_1 < |z_2|. \\ 9z_1+16|z_2|-z_1^9 & \text{for } z_1 \leq 0 \end{cases} \tag{3.3}$$

The contours of f are plotted in the following figure. One easily verifies that $x^* := (-1,0)^T$ is the global minimum and that f fails to be differentiable only on the ray $z_1 \leq 0$, $z_2 = 0$. Further one can show that f is convex. Now choose a starting point anywhere in the region $z_1 > |z_2| > (9/16)^2|z_1|$. Some arithmetic shows that steepest descent with exact line search follows a polygonal path of successively orthogonal segments; this path is plotted in Figure 1. Hence we have *convergence* to the *nonoptimal* origin $\bar{x} := (0,0)^T$ - obviously a disaster! We will come back to this example later on.

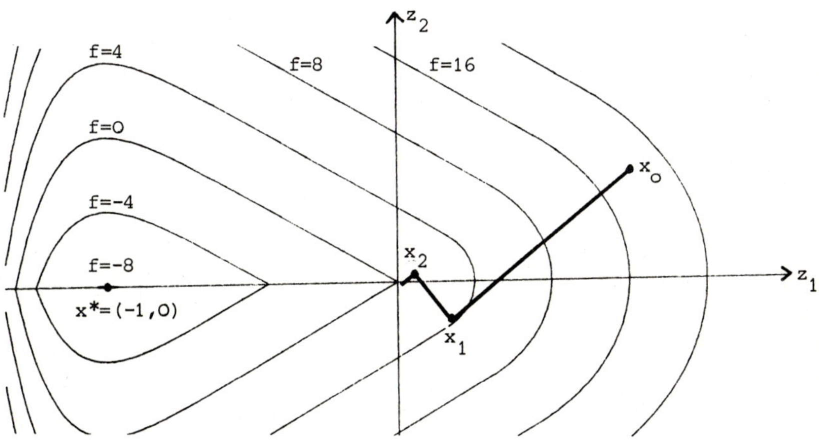

Figure 1: Contours of f and steepest descent path

3.2. Even if we do not suffer from a pitfall as in the above example, there remains a crucial handicap for a smooth method in a nonsmooth context. This is the *lack* of an implementable *stopping rule*. For a C^1-function the gradient will become small in norm, say

$$\|\nabla f(x_k)\| \leqslant \varepsilon \qquad (\varepsilon > 0 \text{ small}), \tag{3.4}$$

when we approach some optimal x^*; this helps to stop the iteration automatically. However, for nonsmooth f a criterion like (3.4) does not make sense, even if the gradient would exist at all iterates. E.g., for the function $f(x) := |x|$ we have $\|\nabla f(x_k)\| = 1$ in (3.4) at each $x_k \neq 0$, x_k may be arbitrarily close to the optimal kink $x^* = 0$. And note that it does not help to iterate til one has found some x_k at which the line search provides a steplength of almost zero. In example (3.3) steepest descent breaks down close to the origin $\bar{x}$ just because of "zero-steplength". But this is only due to the use of a "wrong" model close to $\bar{x}$; at $\bar{x}$ we are still far away from the optimal x^*!

The above shortcomings of classical smooth methods - *convergence to a nonoptimal kink* and the *lack of a stopping criterion* - clearly indicate that a nonsmooth f requires a new approach. Roughly spoken there are two ways of thinking. The first approach is of an *indirect* nature. One tries to eliminate the nondifferentiability and then applies classical methods to the substitute smooth problem. There are two rather popular devices how to proceed. Both ideas, however, are applicable only to the minimax-problem and the ℓ_1-problem (which is a special case of the minimax-problem; compare (2.2), (2.2)'). This indirect approach will be discussed only shortly in the next chapter, since these methods cannot be considered as genuine nonsmooth minimization techniques. In Chapters 5 and 6 we will deal with the much more interesting *direct* approach, the *Subgradient* concept and the *Bundle* concept. Both these ideas attack the problem in a direct way; the methods do not care at all if f is smooth or not and they can deal with arbitrary convex f. Both concepts get along with the minimal information (1.10) on f, the function value and one subgradient at every x. The advantage of the Subgradient methods is their extremely simple structure. However, these methods miss something that is considered as very desirable for minimization methods: they are no descent methods, hence there is no line search etc.; all this results in a very poor convergence rate. The Bundle concept tries to overcome these drawbacks of the Subgradient methods: one gets a descent at every step, there is a line search rule etc.. One has to pay for this with a more complicated structure which demands a more sophisticated implementation.

There are still some other proposals how to deal with nonsmooth f e. g. the so-called *two stage method* (see e. g. Conn [3] or Hornung [14]) or a proposal by Fletcher [6] and Womersley [37] However, just like the indirect approach, these ideas are limited to the minimax-problem, the ℓ_1-problem and elementary compositions of such problems. We will not deal with these proposals.

4. INDIRECT APPROACHES

4.1. Consider the ℓ_1-problem

$$\min_x \sum_{i=1}^{m} |f_i(x)| \tag{4.1}$$

with smooth f_i's. The function $f := \sum_i |f_i|$ will have a kink at x if some f_i vanishes at x. A proposal by El Attar et al. [4] perturbs the problem by adding a small $\varepsilon > 0$ to every f_i, thus obtaining the substitute problem

$$\min_x \sum_{i=1}^{m} {}_+(f_i(x)^2+\varepsilon)^{1/2} .$$

The perturbed $f_\varepsilon := \sum_i {}_+(f_i^2+\varepsilon)^{1/2}$ is obviously a smooth function and any classical smooth method will provide a solution, which in theory converges to a solution of the original problem (4.1) as ε goes to 0. In practice however one runs into serious troubles, since $f_i(x)^2+\varepsilon$ will be almost 0 whenever $f_i(x) = 0$. One has to compute the square-root of an almost zero-term etc.; this is an ill-conditioned problem. The delicate business is that ε has to be chosen a priori and some ε, which is excellent for one problem, may lead to a disaster for the next f. A similar perturbation technique is known for the minimax-problem

$$\min_x \max_{1\leq i\leq m} f_i(x) . \tag{4.2}$$

4.2. A second popular idea transforms (4.1) and (4.2) into constrained problems by introducing additional real variables α and $\alpha_1,\ldots,\alpha_m$, respectively,

$$\min_{x,\alpha_i} \sum_i \alpha_i \quad \text{s.t.} \quad -\alpha_i \leq f_i(x) \leq \alpha_i \quad (1 \leq i \leq m) \tag{4.1)'$$

$$\min_{x,\alpha} \alpha \quad \text{s.t.} \quad f_i(x) \leq \alpha \quad (1 \leq i \leq m). \tag{4.2)'$$

The equivalence of (4.1)' and (4.2)' with (4.1) and (4.2) is easily verified. Obviously (4.1)' and (4.2)' are *differentiable* problems in x, $\alpha_1,\ldots,\alpha_m$ and in x, α; but we have to pay with additional variables, the introduction of constraints etc.. From a more aesthetic point of view this is not very satisfactory, even if this idea might work very well in practice.

5. SUBGRADIENT METHODS

5.1. Suppose for the moment that f is smooth, say C^1 or C^2, and x_k is the current iterate. In a standard first order method one makes a positive step t_k along the negative gradient (compare 3.1):

$$\begin{aligned} x_{k+1} &:= x_k - t_k \nabla f(x_k) \\ t_k &> 0 \ (\textit{line search}). \end{aligned} \tag{5.1}$$

The direction $-\nabla f(x_k)$ is a direction of descent; hence a line search along $x_k - t\nabla f(x_k)$, $t \geq 0$, will provide some $t_k > 0$ such that $f(x_{k+1}) < f(x_k)$. Or, if we want to do better, then we may add some second order information and premultiply $\nabla f(x_k)$ by a matrix H_k which, in some sense, is close to the inverse of the Hessian of f at x_k. The matrix H_{k+1}, needed in the next step, is obtained from H_k by some simple update-procedure:

$$\begin{aligned} x_{k+1} &:= x_k - t_k H_k \nabla f(x_k) \\ t_k &> 0 \ (\textit{line search}) \\ H_{k+1} &:= ..H_k.. \ (\textit{update formula}) \end{aligned} \tag{5.2}$$

For nonsmooth f the gradient at x_k may not exist, but it is obvious what to do. By assumption (1.10) we know at least one subgradient at x_k; hence we will replace the gradient in (5.1) and (5.2), respectively, by a subgradient g_k. Normalizing the search direction ($\|\cdot\|$ denotes the euclidean norm) we obtain as generalization of (5.1)

$$\begin{aligned} x_{k+1} &:= x_k - t_k g_k/\|g_k\| \ \textit{where} \ g_k \in \partial f(x_k) \\ t_k &> 0 \ (\textit{suitable}). \end{aligned} \tag{5.3}$$

The corresponding extension of (5.2) will be studied in Section 5.3.

Two critical items have to be answered for the above Subgradient iteration:

(1) *How can we choose* t_k *in* (5.3) *?*

(2) *Is there any implementable stopping rule for* (5.3) *?*

Let us begin with the second question. In the smooth case the gradient will become small as soon as we are close to an optimal point. This provides us with a convenient stopping rule like (3.4). Unfortunately this is no longer true for nonsmooth f. Consider once more the function

$$f(x) := |x| \quad (= \max\{+x, -x\}) .$$

Then we have the subgradient + 1 or - 1 at $x_k \neq 0$, and this contains no information whatsoever whether we are close to the minimum or not. Suppose, by chance, x_k would be optimal already, i. e. $x_k = 0$ and $\partial f(x_k) = \text{conv}\{-1, +1\}$. Under our general assumption (1.10) we only know but one element of $\partial f(x_k)$, for example the gradient of one of the functions $+x$ or $-x$ (both functions are active at 0), i. e. $+1$ or -1. With this restricted knowledge at x_k we do not even realize that we have already reached the optimum, instead, we will move away again using $g_k = +1$ or -1 in (5.3). This illustrates the difficulty. Different from the smooth situation the subgradient does not provide us with a stopping rule. And, indeed, the *lack of an implementable stopping criterion* is one of the main handicaps of the subgradient idea.

Let us turn to item (1) and consider once more the function (3.3) at a point of nondifferentiability, say $x_k := (0,0)^T$. By definition (1.4) the vector $g_k := (9,16)^T$ is a subgradient at x_k. An inspection of the level lines in Figure 1 tells us that this special $-g_k$ is *not a direction of descent*. There is no $t_k > 0$ such that $f(x_{k+1}) < f(x_k)$ with this g_k in (5.3). Different from (5.1) the steplength t_k cannot be determined via a line search. And note: even if the negative subgradient in (5.3) would be a direction of descent we are, nevertheless, not allowed to make a line search; a line search could lead to the disastrous steepest descent path shown in Figure 1. The nonsmoothness requires that the t_k in (5.3) are chosen *off-line*. The following simple but basic Lemma suggests how to do this in a meaningful way; indeed, it is this result that justifies iteration (5.3).

Lemma 5.1. *Suppose* x_k *is not yet optimal and* x^* *is any optimal point. Then there exists* $T_k > 0$ *such that for iteration* (5.3)

$$\|x_{k+1} - x^*\| < \|x_k - x^*\| \quad \textit{for all } t_k \in (0, T_k) . \tag{5.4}$$

Proof. By definition of the euclidean norm we have for arbitrary positive t_k ($\langle\cdot,\cdot\rangle$ denotes the inner product in $\mathbb{R}^n$)

$$\|x^* - x_{k+1}\|^2 = \|x^* - (x_k - t_k g_k/\|g_k\|)\|^2$$

$$= \langle (x^* - x_k) + t_k g_k/\|g_k\|, (x^* - x_k) + t_k g_k/\|g_k\| \rangle$$
$$= \|x^* - x_k\|^2 + 2t_k \langle g_k/\|g_k\|, x^* - x_k \rangle + t_k^2 \langle g_k/\|g_k\|, g_k/\|g_k\| \rangle .$$

We obtain

$$\|x^* - x_{k+1}\|^2 = \|x^* - x_k\|^2 + 2t_k b_k + t_k^2$$
$$\textit{with } b_k := \langle g_k/\|g_k\|, x^* - x_k \rangle . \quad (5.5)$$

Since g_k is a subgradient at x_k and since x_k is not yet optimal

$$b_k \leqslant (f(x^*) - f(x_k))/\|g_k\| < 0, \quad (5.6)$$

and thus

$$2t_k b_k + t_k^2 < 0 \quad \textit{for} \quad 0 < t_k < T_k := -2b_k . \quad (5.7)$$

This, together with (5.5), proves the assertion. □

The above Lemma has a nice geometric interpretation. (5.6) says that the angle $|\alpha|$ between $-g_k$ and $x^* - x_k$ is less than $\pi/2$ i. e. the ray $x_k - tg_k/\|g_k\|$, $t \geqslant 0$, points into the open halfspace with boundary point x_k and "normal vector" $x^* - x_k$. The existence of $T_k > 0$, satisfying (5.4), is obvious.

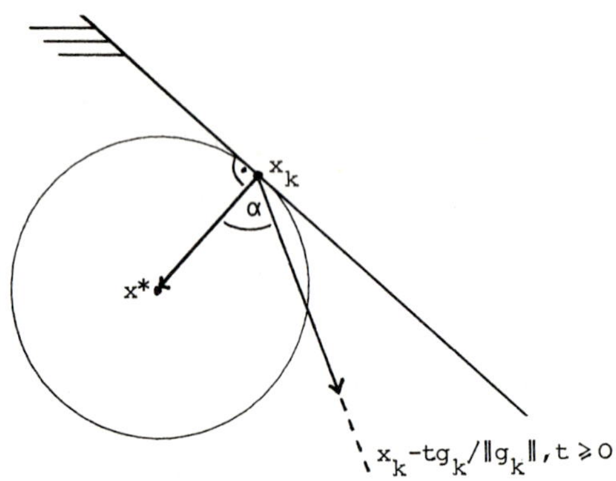

Figure 2

Unfortunately we do not know the value of b_k in (5.5) and thus we cannot compute the T_k in (5.7). We only know that the steplength t_k in (5.3) should be small to ensure the inequality of (5.4). This leads to the idea to choose the t_k as elements of a zero-sequence. However, there is still a second item which has to be considered for the t_k-choice. Let x_o be the starting point for iteration (5.3) and put

$\Lambda := \sum_{j=o}^{\infty} t_j$. Then for each iteration index k

$$\begin{aligned} \|x_o - x_k\| &\leq \|x_o - x_1\| + \|x_1 - x_2\| + \ldots + \|x_{k-1} - x_k\| \\ &= t_o + t_1 + \ldots + t_{k-1} \leq \Lambda. \end{aligned} \tag{5.8}$$

In other terms: we stay all the time in a ball with radius Λ around the starting point x_o. This suggests to choose the small t_k such that, nevertheless, the sum of the t_k is large, say $+\infty$. Then also some optimal x^*, far away from the starting point x_o, will not be out of reach. This leads to a somewhat more detailed variant of (5.3):

$$\begin{aligned} &x_{k+1} := x_k - t_k g_k / \|g_k\| \quad \textit{with} \quad g_k \in \partial f(x_k) \\ &t_k \quad \textit{such that} \quad t_k \downarrow 0 \quad \textit{and} \quad \sum_{k=o}^{\infty} t_k = \infty. \end{aligned} \tag{5.9}$$

The following beautiful result was proved for iteration (5.9) by Ermoliev [5] and Poljak [26].

<u>Theorem 5.2</u>. *Suppose the set* X^* *of optimal points is nonempty and bounded. Then for arbitrary starting point* x_o *a subsequence of the* x_k, *provided by* (5.9), *converges to some* $x^* \in X^*$

$$x_{k_j} \longrightarrow x^* \quad \textit{for} \quad j \to \infty. \tag{5.10}$$

The proof of 5.2 is rather technical. Put $f^* := \inf f(x)$ and use the t_k-choice in (5.9) to prove that for every $j \in \mathbb{N}$ there exists some x_{k_j} such that $f(x_{k_j}) \leq f^* + \frac{1}{j}$, hence $\lim_j f(x_{k_j}) = f^*$. A closer look shows that, by boundedness, of X^* the x_{k_j} lie in a compact set and thus the x_{k_j} (or a suitable subsequence) converge to some $\bar{x}$, i. e., $f(\bar{x}) = \lim_j f(x_{k_j}) = f^*$. It follows $\bar{x} \in X^*$.

Iteration (5.9) is of utmost simplicity. In particular, no line search is needed. The dark side of the picture: one cannot expect but a poor convergence speed if the t_k are chosen off-line, i. e., if the t_k are not adapted to the special f and the iteration points x_k. Let us suppose for a moment that the x_k, given by (5.9), would tend to x^* with *geometric convergence rate* (also called R-*linear convergence*), i. e.,

$$\begin{aligned} &\textit{there exists } M > 0 \textit{ and } 0 < q < 1 \textit{ such that} \\ &\|x_k - x^*\| \leq Mq^k \textit{ for all } k. \end{aligned} \tag{5.11}$$

Then for all k

$$t_k = \|x_{k+1} - x_k\| \leqslant \|x_{k+1} - x^*\| + \|x^* - x_k\| \leqslant M(q+1)q^k.$$

Summing up over k we obtain

$$\sum_{k=o}^{\infty} t_k \leqslant M(q+1)\sum_{k=o}^{\infty} q^k = M(q+1)/(1-q).$$

We get a contradiction since by choice of the t_k the left-hand side is $+\infty$. Consequently we have the disappointing supplement to the convergence result 5.2.:

Theorem 5.3. *The convergence rate in* (5.10) *is of less than geometric type.*

The following Figure and Table show the x_k we obtain for function (3.3), when choosing the step size $\frac{2.175}{k+1}$, $k = 0,1,\ldots$, and the starting point from Figure 1. The conditions $t_k \downarrow 0$ and $\Sigma t_k = \infty$ are satisfied and thus, by Theorem 5.2., the x_k (or a suitable subsequence) must converge to the unique minimizer $x^* = (-1,0)^T$. However, the convergence speed is exceedingly poor. After two steps we are already close to the optimal point x^*, but then the iterates start to oscillate around x^*. This is the typical behaviour which is caused by the "blindness" of the Subgradient method: the t_k are chosen (and have to be chosen!) independently of the special situation.

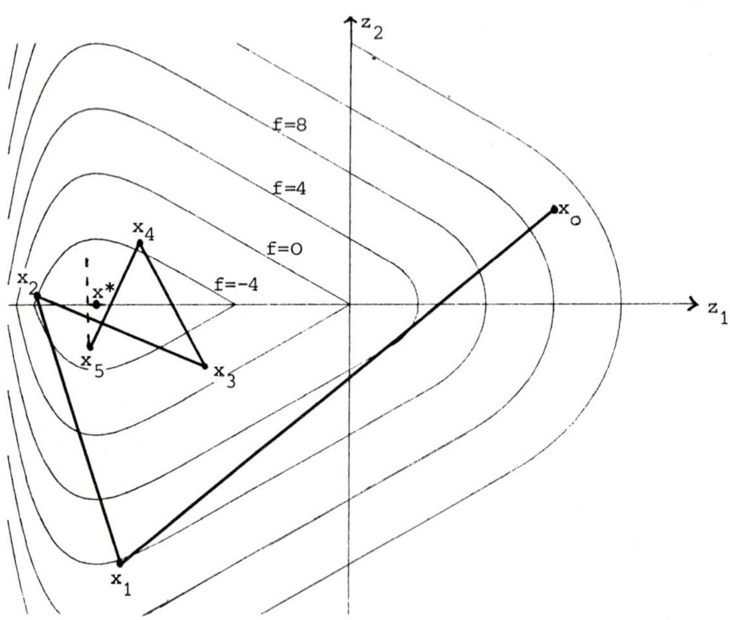

Figure 3: Contours of f and Subgradient path

k	$\|x_k - x^*\|$	$f(x_k) - f(x^*)$	k	$\|x_k - x^*\|$	$f(x_k) - f(x^*)$
0	2.1276	22.0000	19	$2.0325_{10}-2$	$3.2520_{10}-1$
1	1.0047	16.2920	20	$8.8425_{10}-2$	1.4148
2	$2.3693_{10}-1$	4.0973	⋮	⋮	⋮
3	$4.9793_{10}-1$	6.7450	39	$4.9540_{10}-4$	$7.9263_{10}-3$
4	$2.9386_{10}-1$	4.5383	40	$5.3880_{10}-2$	$8.6207_{10}-1$
5	$1.6075_{10}-1$	2.5719	⋮	⋮	⋮
6	$2.0175_{10}-1$	3.2280	59	$1.0656_{10}-4$	$1.7049_{10}-3$
7	$1.0896_{10}-1$	1.7434	60	$3.6143_{10}-2$	$5.7830_{10}-1$
⋮	⋮	⋮	⋮	⋮	⋮
			79	$2.7086_{10}-4$	$4.3337_{10}-3$
			80	$2.3896_{10}-2$	$3.8233_{10}-1$

Table 1: Subgradient path

The above Table shows that the off-line-choice of the t_k which, on one hand enforces the convergence (5.10), on the other hand makes this method almost useless for practical purposes. (Definitely some applied mathematicians would not consider the above behaviour as convergence at all!). In the next section we will shortly discuss a special case, which allows to make the *qualitative* statement of Lemma 5.1 more precise and provides us with a *quantitative* selection rule for t_k. This choice will guarantee a monotone decrease for $\|x_k - x^*\|$, which was not the case in the above example.

5.2. Suppose for the moment that

we know $f(x^*)$ *for some optimal* x^*.

Then we can compute $2(f(x_k) - f(x^*))/\|g_k\|$. We obtain a positive number (in case x_k is not yet optimal) equal or less than the T_k in (5.7). Hence, inequality (5.4) holds in particular for the special

$$t_k := \lambda(f(x_k) - f(x^*))/\|g_k\| \text{ with } 0 < \lambda < 2. \tag{5.12}$$

The computable steplength (5.12) incorporates the knowledge we have of f, x_k and x^*. This guarantees a monotonous decrease in $\|x_k - x^*\|$ for all k. The following result was proved by Poljak [27] for a special case.

Theorem 5.4. *Suppose* f *has a unique minimizer* x^*, *the optimal value* $f(x^*)$ *is known and with some* $\ell > 0$

$$f(x) - f(x^*) \geq \ell\|x - x^*\| \text{ for all } x.$$

Then for every starting point x_0 *iteration* (5.3) *with stepsize rule*

(5.12) *provides a sequence converging to* x^* *with geometric convergence rate.*

The function

$$f(x) := \max\{f_1(x),\ldots,f_m(x),0\},$$

whose minimization is equivalent to the solution of the inequality system (compare (2.5))

$$\textit{find } x \textit{ such that } f_i(x) \leq 0 \quad (1 \leq i \leq m),$$

can serve as an example where one knows $f(x^*)$ (=0) without knowing x^*. It proves that for linear f_i's iteration (5.3) with stepsize rule (5.12) is closely connected to the relaxation method for solving the above inequality system; see Goffin [7].

Another interesting t_k-choice which does not require the a-priori knowledge of $f(x^*)$, but nevertheless enforces geometric convergence (however to a possibly nonoptimal point), was proposed by Shor [31] and studied in detail by Goffin [8]. Put in (5.3)

$$t_k := M\rho^k \textit{ with } M > 0 \textit{ and } 0 < \rho < 1. \tag{5.13}$$

Obviously, $\Lambda := \sum_{j=o}^{\infty} t_j = M/(1-\rho) < \infty$ and thus the Convergence Theorem 5.2 does not apply any more. The above choice, however, ensures that all iterates lie in a ball with radius Λ around the starting point x_o; compare (5.8). By a compactness argument the x_k-sequence has a cluster point $\bar{x}$, i. e., the x_k (or a suitable subsequence) converge to $\bar{x}$. Now

$$\begin{aligned}\|x_k - x_{k+m}\| &\leq \|x_k - x_{k+1}\| + \ldots + \|x_{k+m-1} - x_{k+m}\| \\ &= M\rho^k + \ldots + M\rho^{k+m-1} = M\rho^k(1-\rho^m)/(1-\rho),\end{aligned}$$

hence for $m \to \infty$

$$\|x_k - \bar{x}\| \leq M\rho^k/(1-\rho),$$

i. e., we have convergence to $\bar{x}$ with geometric convergence rate. However, there is no guarantee that $\bar{x}$ is an optimal point. Goffin showed how to choose M and ρ in (5.13) (hence the radius of the ball around x_o in which all iterates lie) to ensure the limit point $\bar{x}$ to be optimal. Unfortunately this proposal by Goffin is not implementable, since it requires information about f which is not disposable; see [8] for more details.

5.3. In case we neither know the optimal function value $f(x^*)$ nor do we want to risk convergence to a nonoptimal $\bar{x}$, then there remains only one way to accelerate the convergence of the Subgradient iteration: we have to give up the Markov nature of iteration (5.3) by adding information obtained in previous steps. Shor made some proposals along this line; see e. g. [32], [33]. One basic idea: suppose, starting from x_o, we have found x_1 by a move along $-g_o$ and we have selected g_1 as the new subgradient at x_1. Then it makes sense to shorten that part of $-g_1$ which is colinear to the just used direction $-g_o$. This dilatation of the space means a multiplication of g_1 with some matrix H_1. If we repeat this at every step and write it in a condensed form (due to Shokov [30]), then we obtain the following iteration in which the dilatation matrices H_k are updated in a very convenient way:

$$\begin{aligned} x_{k+1} &:= x_k - t_k H_k g_k \Big/ \sqrt{g_k^T H_k g_k} \quad \textit{with } g_k \in \partial f(x_k) \\ H_{k+1} &:= \alpha_k (H_k - \beta_k d_k d_k^T); \end{aligned} \tag{5.14a}$$

here d_k stands for $H_k g_k \Big/ \sqrt{g_k^T H_k g_k}$; α_k, β_k and the steplength t_k are suitable positive parameters and the starting matrix is a positive multiple of the unit matrix, $H_o = \alpha I$. The above iteration shows a strong analogy to (5.2) if we choose in (5.2) as update scheme the DFP or the BFGS formula. For $\alpha = 1$, $\alpha_k \equiv 1$ and $\beta_k \equiv 0$ we get back (5.3).

One parameter choice proves to be especially interesting. Let n be the dimension of the space and choose α_k, β_k and t_k in (5.14a) as constants:

$$\alpha_k := n^2/(n^2-1), \quad \beta_k := 2/(n+1), \quad t_k := 1/(n+1) \\ \textit{for } k = 0,1,2,\ldots \tag{5.14b}$$

Then the following convergence result holds; see Shor [34] and for more details Goffin [9]:

Theorem 5.5. *Suppose the starting point* x_o *and the starting matrix* $H_o = \alpha I$ *are such that* $\|x_o - x^*\|^2 \leq \alpha$ *for some optimal* x^*. *Then there exists* $M > 0$ *and* $q < 1$ *such that for the sequence generated by* (5.14a,b):

$$\min_{0 \leq j \leq k} [f(x_j) - f(x^*)] \leq M q^k \quad \textit{for all } k. \tag{5.15}$$

(5.15) guarantees that a subsequence of the function values $f(x_k)$ converges to $f(x^*)$ with geometric convergence rate. A closer analysis shows that $q = [n^2/(n^2-1)]^{1/2}[1-2/(n+1)]^{1/2n}$ which is approximately

$1-1/2n^2$. Hence, unfortunately, q is close to 1, the closer the greater the dimension.

We supplement Theorem 5.5 by some *remarks*. The first one concerns another proposal for the minimization of a convex functional on $\mathbb{R}^n$, the *method of centers of gravity* due to Levin [21]. This idea exploits cutting plane techniques. After k steps we have built up a polyhedron with center x_k as intersection of k halfspaces. By construction this polyhedron contains at least one optimal point x^*. In case x_k is not yet optimal then we intersect with another halfspace determined by a hyperplane through x_k with normal vector $-g_k$, where g_k is any subgradient at x_k. By construction x^* is also a point of the smaller polyhedron with center x_{k+1}; compare the geometric meaning of $-g_k$ in Figure 2.

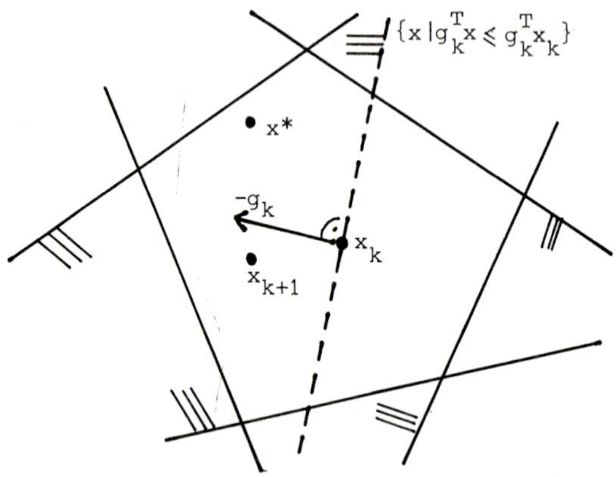

Figure 4: Method of centers of gravity

It can be shown that a convergence statement like (5.15) holds, but now with q close to $1-\frac{1}{n}$, i. e., the convergence factor is somewhat better than in (5.15). Levin's idea does not seem to be implementable in its pure form; however one can implement weakened versions. Such a modification is the so-called *Ellipsoid method* where one replaces the above polyhedron by an ellipsoid whose center is easier to compute. Surprisingly it proves that the analytic form of the ellipsoid method coincides with the above Subgradient iteration (5.14a,b). Hence, very different approaches (namely, cutting plane ideas on one side and generalized gradient techniques on the other side) lead to just the same algorithm. This correlation has a remarkable consequence. A result due to Yudin/Nemorovskii [38] roughly says that no minimization method, using the information given by an oracle, which returns a subgradient at each iterate (just like in a Subgradient method or in Levin's idea), can improve significantly on Levin's method. Hence geometric convergence as in (5.15) with $q \approx 1 - \frac{1}{n}$ (which is close to the q in (5.15)) is the "best" we can expect. There is no hope for a substantial improvement as long as we are content in (5.3) and (5.14), respectively, with any g_k from $\partial f(x_k)$; this random choice excludes a line search, there is no implementable stopping criterion etc.. In the next chapter we will see how the Bundle idea overcomes these essential drawbacks of the Subgradient idea, without requiring more information at x_k than one random subgradient.

We close this chapter by mentioning another proposal due to Shor [26], where the dilatation matrix H_k in (5.14) aims at shortening $g_k - g_{k-1}$. This is justified by the fact that for quadratic f the ideal direction $x^* - x_k$ is perpendicular to $g_k - g_{k-1}$, i. e.,

$$\langle x^* - x_k, g_k - g_{k-1} \rangle = 0.$$

It is said that this algorithm behaves better in practice (provided the t_k are chosen in an appropriate way); see e. g. Lemaréchal [17]. However, there is no convergence result like Theorem 5.2, and there are even (counter)examples for which this method does not converge.

6. BUNDLE METHODS

6.1. A conceptual first order ε-descent method. Suppose for the moment that we know the whole subdifferential at the current iterate x_k; in Section 6.2 we will drop this unrealistic assumption. Then, instead of performing a step in the direction of a randomly chosen subgradient (as in a Subgradient method), let us select that g_k in $\partial f(x_k)$ which provides the steepest descent. The directional derivative $f'(x_k;d)$ is a measure for the descent we can expect (in a first order sense and for small t) along $x_k + td$, $t > 0$. This leads to the direction finding subproblem

$$\min_{\|d\| \leqslant 1} f'(x_k;d). \tag{6.1}$$

The additional constraint $\|d\| \leqslant 1$ in (6.1) becomes necessary, since $f'(x_k;\cdot)$ is positively homogeneous. In case f is smooth one has $f'(x_k;d) = f'(x_k)d$, i. e., (6.1) is just the minimization of the classical *first order model* (3.1). We use (1.7) to rewrite (6.1) in the form

$$\min_{\|d\|\leq 1} \max_{g\in\partial f(x_k)} g^T d. \tag{6.2}$$

The unit ball and $\partial f(x_k)$ are nonempty convex compact sets. Hence, by a well-known Minimax Theorem, (6.2) is equivalent to

$$\max_{g\in\partial f(x_k)} \min_{\|d\|\leq 1} g^T d. \tag{6.3}$$

For given g the minimizing d of length 1 is $-g/\|g\|$ and (6.3) becomes

$$\max_{g\in\partial f(x_k)} \min_{\|d\|\leq 1} g^T d = \max_{g\in\partial f(x_k)} g^T(-g/\|g\|) = -\min_{g\in\partial f(x_k)} \|g\|.$$

Hence, for solving (6.1), we have to study the minimum-norm problem (which is uniquely solvable since $\partial f(x_k)$ is a nonempty closed convex set)

$$\min_{g\in\partial f(x_k)} \|g\|. \tag{6.4}$$

If and only if g_k is the subgradient at x_k, closest to the origin, then $d_k := -g_k/\|g_k\|$ minimizes $f'(x_k;\cdot)$ and one has $f'(x_k;d_k) = -\|g_k\|$. (1.9) shows that $g_k \neq 0$ for nonoptimal x_k and thus $f'(x_k;d_k) < 0$, i. e., (6.1) provides a descent direction as long as we are not yet optimal. We are led to the iteration scheme:

$$x_{k+1} := x_k + t_k d_k$$

$$\textit{where } d_k := -g_k/\|g_k\| \textit{ with } \|g_k\| = \min_{g\in\partial f(x_k)} \|g\| \tag{6.5}$$

$$\textit{and } f(x_k + t_k d_k) = \min_{t\geq 0} f(x_k + t d_k).$$

In case the gradient exists at x_k, the subdifferential shrinks to $\nabla f(x_k)$ and (6.5) reduces to the classical steepest descent method (with exact line search). Hence the example discussed in Chapter 3.1 warns us that for nonsmooth f the above iteration may collapse close to a kink. It is important to understand the reason for this failure and to develop a feeling what could serve as safeguard in (6.5) against shortcomings in the nonsmooth situation. For this purpose consider once more the function (3.3) and the path shown in Figure 1. Going back to definition (1.4) we realize that the limits

$$\begin{pmatrix} 9 \\ \pm 16 \end{pmatrix} = \lim_{\pm z_2 \downarrow 0} \nabla f(0,z_2) \text{ and } \begin{pmatrix} 15 \\ 0 \end{pmatrix} = \lim_{z_1 \downarrow 0} \nabla f(z_1,0)$$

belong to the subdifferential of f at the origin $\bar{x} := (0,0)^T$. A more detailed (but straightforward) analysis shows that $\partial f(\bar{x})$ is part of an ellipsoid:

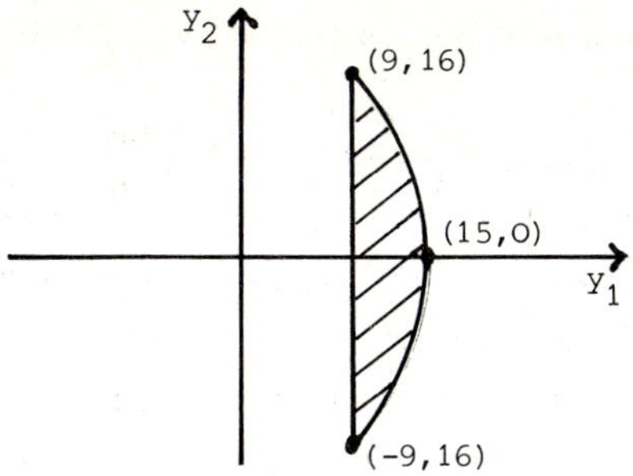

Figure 5: Subdifferential of f at $\bar{x} = (0,0)^T$

Consequently iteration (6.5) would use at $x_k := \bar{x}$ the direction

$$\bar{d} := -(9,0)^T/\|(9,0)\| = -(1,0)^T$$

and in one step we would reach the minimum, i. e., $x_{k+1} = x^*$. However, Figure 1 shows that we are trapped. We walk in shorter and shorter steps on a zigzagging path along the two directions

$$\begin{aligned} d' &:= d_k = -\nabla f(x_k)/\|\nabla f(x_k)\| \quad \textit{for } k = 0,2,4,\ldots \\ d'' &:= d_k = -\nabla f(x_k)/\|\nabla f(x_k)\| \quad \textit{for } k = 1,3,5,\ldots, \end{aligned} \tag{6.6}$$

however, without ever reaching $\bar{x}$. In a strictly local sense d' and d" provide the steepest descent at x_k, whereas in a more global sense d' and d" are less and less profitable. The more we approach $\bar{x}$, the poorer the (sub)gradient information becomes due to the nondifferentiability at $\bar{x}$. This suggests to enrich the information at x_k and to replace $\partial f(x_k)$ in (6.5) by $\bigcup_{y \in U} \partial f(y)$, where U is a suitable neighbourhood of x_k. Then for x_k close to $\bar{x}$, the enlarged set $\bigcup_{y \in U} \partial f(y)$ will contain $\partial f(\bar{x})$, i. e., the above ideal direction $\bar{d}$ will be a candidate for the line search and we will escape. Luckily we have at hand a device in NSO which helps to realize what we have in mind. Fix some small $\varepsilon > 0$ and add this ε in the inequality, defining the subdifferential (we attach ε as a subscript),

$$\partial_\varepsilon f(x) := \{g \in \mathbb{R}^n \mid g^T(z-x) \leq f(z) - f(x) + \varepsilon \textit{ for all } z \in \mathbb{R}^n\}. \tag{6.7}$$

This so-called *ε-subdifferential* is again a nonempty convex compact set (same proof as for $\varepsilon = 0$) and the basic auxiliary result holds:

<u>Lemma 6.1</u>. *For every* x *and every neighbourhood* U *of* x *there exists*

some $\bar{\varepsilon} > 0$ *such that*

$$\partial_\varepsilon f(x) \supset \bigcup_{y \in U} \partial f(y) \quad \textit{for } \varepsilon \geq \bar{\varepsilon}.$$

<u>Proof</u>. Let x and U be given. Further, let B_ρ be a ball of radius ρ around x which contains U in its interior and let L be the Lipschitz constant of f with respect to B_ρ. Then for every $y \in U$ and $g \in \partial f(y)$ one has with suitable $v \in B_\rho$

$$\|g\| \|v-y\| = g^T(v-y) \leq f(v) - f(y) \leq L\|v-y\|;$$

hence,

$$\bar{\varepsilon} := \max_{y \in U} |f(x) - f(y)| + \max_{\substack{g \in \partial f(y) \\ y \in U}} \|g\| \cdot \max_{y \in U} \|y-x\| \quad (\leq 2L\rho)$$

is a well-defined positive real number.

Now let $y \in U$ and $g \in \partial f(y)$ be given. Then for arbitrary z

$$\begin{aligned} g^T(z-x) &= g^T(z-y) + g^T(y-x) \\ &\leq f(z) - f(y) + g^T(y-x) \\ &= f(z) - f(x) + [f(x) - f(y) + g^T(y-x)] \\ &\leq f(z) - f(x) + \bar{\varepsilon}, \end{aligned}$$

i. e., $g \in \partial_\varepsilon f(x)$ for $\varepsilon \geq \bar{\varepsilon}$. □

Lemma 6.1 implies that the ε-subdifferential realizes what we have in mind: the set $\partial_\varepsilon f(x_k)$ contains in a condensed form the subgradient information from a whole neighbourhood of x_k. As we will see, this will help to overcome the shortcomings caused by nonsmoothness.

To become more precise, fix some small $\varepsilon > 0$ throughout the following and replace $\partial f(x)$ by $\partial_\varepsilon f(x)$. If we add the same ε in the definition of the directional derivative (for convex f the operation lim in (1.2) can be replaced by inf)

$$f'_\varepsilon(x;d) := \inf_{t>0} \frac{1}{t} [f(x+td) - f(x) + \varepsilon], \tag{6.8}$$

then this ε-*directional derivative* again is the support function of the ε-subdifferential (same proof as for $\varepsilon = 0$):

$$f'(x;d) = \max_{g \in \partial_\varepsilon f(x)} g^T d.$$

The arguments used at the beginning of this chapter show that also the ε-modified problems

$$\min_{\|d\|\leqslant 1} f'_\varepsilon(x_k;d) \tag{6.9}$$

and

$$\min_{g\in\partial_\varepsilon f(x_k)} \|g\| \tag{6.10}$$

correspond to each other. In words: if g_k solves (6.10), then $d_k := -g_k/\|g_k\|$ minimizes $f'_\varepsilon(x_k;\cdot)$ and $f'_\varepsilon(x_k;d_k) = -\|g_k\|$. As long as $0 \notin \partial_\varepsilon f(x_k)$ the minimal g_k in (6.10) is not the zero-vector and thus $f'_\varepsilon(x_k;d_k) < 0$. Definition (6.8) implies that a move along d_k guarantees a decrease of at least ε, i. e.,

$$0 \notin \partial_\varepsilon f(x_k) \implies \begin{array}{l} f(x_k + td_k) < f(x_k) - \varepsilon \\ \textit{for suitable } t > 0. \end{array} \tag{6.11}$$

Now suppose $0 \in \partial_\varepsilon f(x_k)$. Put $g = 0$ in definition (6.7) to realize that x_k is already "almost" optimal; more precisely:

$$0 \in \partial_\varepsilon f(x_k) \implies \begin{array}{l} f(x_k) \leqslant f(x) + \varepsilon \\ \textit{for all } x. \end{array} \tag{6.12}$$

(6.11) together with (6.12) motivates the following ε-modification of iteration (6.5):

$$\begin{array}{l} x_{k+1} := x_k + t_k d_k \\ \textit{where } d_k := -g_k/\|g_k\| \textit{ with } \|g_k\| = \displaystyle\min_{g\in\partial_\varepsilon f(x_k)} \|g\| \\ \textit{and } f(x_k + t_k d_k) = \displaystyle\min_{t\geqslant 0} f(x_k + td_k). \end{array} \tag{6.13}$$

Obviously (6.11) and (6.12) exclude a pathological behaviour as in Figure 1. In particular, if $\inf f(x) > -\infty$, then in finitely many steps one must reach some x_k such that $0 \in \partial_\varepsilon f(x_k)$, i. e., x_k is ε-*optimal*. Hence $0 \in \partial_\varepsilon f(x_k)$ serves as a stopping criterion. We summarize:

Theorem 6.2. *Let* $f^* := \inf f(x)$ *and let* $\{x_k\}$ *be the sequence generated by* (6 13) *for arbitrary starting point* x_0.

(a) *If* $f^* = -\infty$ *then* $\lim_{k\to\infty} f(x_k) = -\infty$.

(b) *If* $f^* > -\infty$ *then there exists* $k \in \mathbb{N}$ *such that* $f(x_k) \leqslant f^* + \varepsilon$.

Before we discuss an implementation of the above idea let us give another explanation for the change which is caused in the convergence behaviour by the transition from (6.5) to (6.13). The disaster in Figure 1 stems from the discontinuity of the subproblem in (6.5):

$$x \to d(x) := -g(x)/\|g(x)\| \; where \; \|g(x)\| = \min_{g\in\partial f(x)} \|g\|.$$

Although the x_k converge to $\bar{x}$ the directions (compare (6.6))

$$d(x_k) = d' \; for \; 2|k \; and \; d(x_k) = d'' \; for \; 2\nmid k$$

do not converge to $d(\bar{x}) = (-1,0)^T$. Using a result due to Nurminskii [23] (see also [13]) about the Lipschitz continuity of the point-to-set mapping $x \to \partial_\varepsilon f(x)$, one can easily verify that the slightly modified mapping in (6.13)

$$x \to d(x) := -g(x)/\|g(x)\| \; where \; \|g(x)\| = \min_{g\in\partial_\varepsilon f(x)} \|g\|$$

becomes continuous. Thus for x_k close to $\bar{x}$ the directions $d(x_k)$, used in (6.13), will be sufficiently close to the ideal direction $d(\bar{x}) = (-1,0)^T$ and we will make a move away from x_k in the direction of x^*.

6.2. Implementation. Let us come back to the general situation (1.10) where at x_k we know but one subgradient g_k. Despite this minimal information, the Bundle idea realizes (6.13) in a rather sophisticated way. The basic idea of a *bundle-type-algorithm* consists in replacing $\partial_\varepsilon f(x_k)$ by some inner approximating polytope P (*forming of the bundle*) and to solve (6.13) with $\partial_\varepsilon f(x_k)$ replaced by P. Provided P is a sufficiently good approximation, then we will find a direction along which a line search provides some x_{k+1} with a decrease of almost ε. In case P is a bad approximation (the line search will let us know this), then we stay at x_k and try to improve the approximating P by adding a further subgradient (so-called *nullstep*). These two items are the crucial ingredients of all Bundle methods, which were initiated by Wolfe [36] (conjugate subgradient method) and developed further on by Lemaréchal [15], Mifflin [22] and others. We will try to describe in very short terms the two mentioned basic steps by avoiding all technical datails (which, of course, are very important for an efficient implementation); in particular we will not say anything about the line search which has to take care of the nondifferentiability. For further technical details see Lemaréchal/Strodiot/Bihain [19].

Let g_j, $j = k-1,k-2,\ldots$, be the subgradients computed at the previous steps and assume, for simplicity, that $g_j \notin \partial_\varepsilon f(x_k)$ for all j. It is a trivial exercise to show that

$$\tilde{g}_j := (1-\varepsilon/\varepsilon_j)g_k + \varepsilon/\varepsilon_j g_j \in \partial_\varepsilon f(x_k)$$
$$for \; \varepsilon_j := f(x_k) - f(x_j) - g_j^T(x_k - x_j).$$

Hence the bundle of the subgradients already computed provides us with an inner approximation

$$P := \text{conv}\,(\{g_k\} \cup \{\tilde{g}_j \mid j = k-1, k-2, \ldots\})$$

of $\partial_\varepsilon f(x_k)$ (*forming of the bundle*). Replacing $\partial_\varepsilon f(x_k)$ by P in (6.13) the determination of a search direction reduces to a quadratic programming problem (we put $\tilde{g}_k := g_k$)

$$\min \|\sum_{j\leq k} \lambda_j \tilde{g}_j\| \quad \text{s. t.} \quad \sum_{j\leq k} \lambda_j = 1,\ \lambda_j \geq 0. \tag{6.14}$$

In case P provides a good approximation of $\partial_\varepsilon f(x_k)$, then we will find a direction d yielding a decrease of almost ε and we are done. Now suppose that, on the contrary, P is a bad approximation. Then it may happen that the substitute problem (6.14) provides a direction d which is not even a descent direction; recall that not every negative subgradient guarantees a decrease. Then we compute a further subgradient g at $x_k + td$ for some small $t > 0$. By Lemma 6.1 g is an ε-subgradient at x_k. A series of technical arguments shows g to lie below a hyperplane through some $\hat{x}$ with normal vector d; here $\hat{x}$ is a point on the line from 0 to the solution of (6.14) and dividing this distance by a fixed ratio determined by some line search parameters. We stay at x_k and improve the approximating polytope (*nullstep*):

$$\tilde{P} := \text{conv}(P \cup \{g\}).$$

Iteration (6.13) with $\tilde{P}$ instead of P will yield a better $\tilde{d}$.

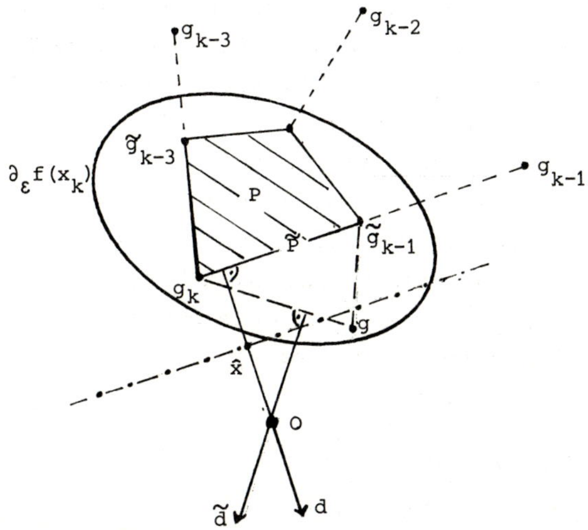

Figure 6: Forming of the Bundle and a Nullstep

One can prove that only finitely many nullsteps are needed to build up an approximation of $\partial_\varepsilon f(x_k)$ which provides a direction yielding a sufficient decrease. Practical experience shows that, in general, less than one nullstep is needed to proceed from x_k to x_{k+1}. As soon as one has found some x_k with $0 \in \partial_\varepsilon f(x_k)$ one replaces ε by a smaller ε, say $\varepsilon/10$, and repeats the above procedure. In [19] an implementation of these ideas is discussed in details. For this implementation a convergence result holds which is almost a copy of Theorem 6.2.

7. NUMERICAL COMPARISON

The Table below gives a short numerical comparison of the BFGS method, of Shor's proposal (5.14a,b) (i. e. the Ellipsoid method) and of the Bundle method, as discussed in 6.2, on a small number of rather academic examples. We did not try to incorporate any sophisticated programming techniques when implementing the methods. Hence the number of steps, reported below, should not be considered as "optimal"; nevertheless the results allow a rather fair comparison of the methods, since all methods were implemented "on the same level". Our results are confirmed by experiments reported by Lemaréchal [17].

In the implementation of the Bundle idea we used parts of a code written by Strodiot. The same line search was incorporated in BFGS and Bundle; see [16], [36]. The data for the indirect approaches were taken from literature; see Charalambous and Conn [2] for Example 1 and El Attar et al. [4] for Example 3. The columns show the total amount of function and (sub)gradient evaluations needed to reach the $f(x_k)$-value on the right side; the correct figures are underlined. BFGS was stopped when (3.4) was satisfied or when the line search did not yield any further decrease within 100 line search steps. In Bundle we iterated till $0 \in \partial_\varepsilon f(x_k)$. Shor was stopped after the minimum was reached (the minimum is known in the examples!) or a after a prescribed number of iterations.

Example 1 is the Rosen/Suzuki problem (see [2])

$$\min f_o(x) \text{ s. t. } g_i(x) \leqslant 0 \qquad (i = 1,2,3). \tag{7.1}$$

Here

$$f_o(x) := x_1^2 + x_2^2 + 2x_3^2 + x_4^2 - 5x_1 - 5x_2 - 21x_3 + 7x_4$$
$$g_1(x) := x_1^2 + x_2^2 + x_3^2 + x_4^2 + x_1 - x_2 + x_3 - x_4 - 8$$
$$g_2(x) := x_1^2 + 2x_2^2 + x_3^2 + 2x_4^2 - x_1 - x_4 - 10$$
$$g_3(x) := x_1^2 + x_2^2 + x_3^2 + 2x_1 - x_2 - x_4 - 5.$$

In order to solve (7.1) one considers the equivalent minimax problem

$$\min_{x} \max_{0 \leq i \leq 3} f_i(x)$$

with $f_i(x) := f_o(x) + 10 g_i(x)$ *for* $i = 1,2,3.$

The dimension is 4, the optimal value is $x^* = (0,1,2,1)^T$ with $f(x^*) = -44$. As starting point we used $x_o = (0,0,0,0)^T$.

Example 2 is a minimax problem with five quadratic functions on $\mathbb{R}^{10}$ (compare Lemaréchal [17]):

$$\min_{x} \max_{1 \leq k \leq 5} x^T A_k x - b_k^T x;$$

here

$$a_k^{(i,j)} := \begin{cases} e^{i/j} \cos(i \cdot j) \sin k & \textit{for } i < j \textit{ and } i > j \\ \frac{i}{10} \sin k + \sum_{i \neq j} a_k^{(i,j)} & \textit{for } i = j \end{cases}$$

and

$$b_k^{(i)} := e^{i/k} \sin(i \cdot k).$$

The solution is $f(x^*) = -0.8414$; the starting point was $x_o = (1, \dots, 1)^T$.

Example 3 is an ℓ_1-approximation problem in $\mathbb{R}^6$ (compare El Attar et al. [4]). Given are equidistant points $t_i := i/10$, $i = 0,1,\dots,50$, on the real line and a function

$$h(t) := \frac{1}{2}e^{-t} - e^{-2t} + \frac{1}{2}e^{-3t} + \frac{3}{2}e^{-\frac{3}{2}t} \sin 7t + e^{-\frac{5}{2}t} \sin 5t.$$

The aim is to determine the parameters $x_1, \dots, x_6$ in

$$k(x_1, \dots, x_6; t) := x_1 e^{-x_2 t} \cos(x_3 t + x_4) + x_5 e^{-x_6 t}$$

such that

$$f(x) := \sum_{i=0}^{50} |h(t_i) - k(x; t_i)|$$

becomes minimal. The solution is $f(x^*) = 0.5598\dots$; as starting point $x_o = (2,2,7,0,-2,1)^T$ was used.

Example 4 is the dual of a transportation problem (see Held [11] and Lemaréchal [17] for more details):

$$\min f(x) := -\{\sum_{i=1}^{48} s_i x_i + \sum_{j=1}^{48} d_j \min_{1 \leq i \leq 48} (a_{ij} - x_i)\}.$$

The dimension is $n = 48$, the solution $f(x^*) \approx -638500$ and $x_o = (0, \dots, 0)^T$.

Example 5 is a problem in $\mathbb{R}^2$

$$\min f(x) \text{ s. t. } g_i(x) \leq 0 \quad (i = 1,2) \qquad (7.2)$$

with

$$f(x) := (x_1-2)^2 + (x_2-1)^2$$

$$g_1(x) := x_1^2 - x_2$$

$$g_2(x) := x_1 + x_2 - 2.$$

(7.2) is replaced by the equivalent exact penalty problem

$$\min f(x) + 100(\max\{0,g_1(x)\} + \max\{0,g_2(x)\}).$$

The solution is $x^* = (1,1)^T$ and $f(x^*) = 1$; as starting point $x_o = (2,2)^T$ was used.

Example 6 is the minimization of the (smooth!) Rosenbrock function (see Himmelblau [12])

$$\min 100(x_2-x_1^2)^2 + (1-x_1)^2.$$

The optimum is $x^* = (1,1)^T$ with $f(x^*) = 0$; the starting point was $x_o = (-1.2,1)^T$.

Example 7 is again a smooth (but extremely ill-conditioned) problem. We minimize a quadratic function on $\mathbb{R}^{50}$ (compare [17]):

$$\min \tfrac{1}{2}x^TAx - b^Tx;$$

A is the 50 × 50 Hilbertmatrix and b is chosen such that $x^* = (1,\dots,1)^T$ is optimal. The Table shows the truncation error in x_k after stopping (due to line search difficulties).

All examples were run on a CYBER 173.

Example	BFGS		INDIRECT APPROACH		SHOR (ELLIPSOID)		BUNDLE	
	f/g-eval.	$f(x_k)$	f/g-eval.	$f(x_k)$	f/g-eval.	$f(x_k)$	f/g-eval.	$f(x_k)$
1	122	-43.8615	66	-44.0000	354	-43.9999	63	-44.0000
2	612	- 0.8414	--	--	1446	- 0.8414	93	- 0.8414
3	251	0.5567	166	0.5598	1000	0.5962	352	0.5598
4	1655	- 638460	--	--	3000	- 636103	528	- 638463
5	107	1.0000	--	--	112	1.0000	33	1.0000
6	92	0.0000	--	--	100	0.1719	58	0.0000
7	28	$\Vert x_k-x^*\Vert^2 = 0.0484$	--	--	100	$\Vert x_k-x^*\Vert^2 = 0.0558$	32	$\Vert x_k-x^*\Vert^2 = 0.0103$

Table 2

The above results clearly demonstrate the superiority of the Bundle algorithm in a nonsmooth context. Only for Example 3 the indirect approach (hence a highly specialized method) proves to be more efficient. It is remarkable that even for the smooth (but stiff) Examples 6 and 7 the Bundle can compete with BFGS. The nonsmooth examples show that BFGS (without additional safeguards) will break down before reaching the minimum. As discussed before this is due to the use of a not appropriate model. Further, the numerical results confirm the inferiority of Shor's method. Note that in addition there is no implementable stopping criterion for Subgradient ideas.

8. A SECOND ORDER MODEL

For smooth f and $\varepsilon \to 0$ iteration (6.13) reduces to the classical steepest descent method which is not more than linearly convergent. To obtain faster convergence one has to replace the first order model (3.1) behind the steepest descent idea by the second order model (3.2). Minimization of (3.2) leads to (Quasi-)Newton methods of type (5.2) which are at least superlinearly convergent and consequently behave much better in practice. There arises the natural and challenging question in NSO: Can we develop a "second order" model for nonsmooth f, based on the ε-modifications of standard definitions from convex analysis introduced in Chapter 6, i. e., a model which for smooth f and $\varepsilon \to 0$ reduces to (3.2)? The minimization of such a model should lead to a much better convergence behaviour.

We start with the following proposal for a *second directional derivative* at x in direction d (provided the limit exists)

$$f''(x;d) := \lim_{t \downarrow 0} \frac{1}{t}[f'(x+td;d) - f'(x;d)]. \tag{8.1}$$

For C^2-f the Mean-Value-Theorem implies

$$\frac{1}{t}[f'(x+td;d) - f'(x;d)] = \frac{1}{t}[f''(x+\Theta td)td]^T d \textit{ with } \Theta \in (0,1);$$

taking the limit we see that $f''(x;d)$ reproduces the Hessian $f''(x)$:

$$f''(x;d) = f''(x)(d,d) \tag{8.2}$$

We remark that, contrary to $f'(x;d)$, the above limit does not exist for every convex f; the function $f(x) := |x|^{\alpha}$ with $\alpha \in (1,2)$ and $x = 0$ may serve as a counterexample.

Provided $f''(x;d)$ exists we can define the model

$$f'(x;d) + \frac{1}{2}f''(x;d). \tag{8.3}$$

For C^2-f (8.3) is identical with (3.2) and thus minimization of (8.3) leads to Newton's method. This looks rather promising. However, we still feel uneasy. Firstly, an ε is missing somewhere in (8.3). As we know from Chapter 6.1 such ε helps to avoid shortcomings due to nonsmoothness. Secondly, we do not dispose of a max-expression of type (1.7) for $f''(x;d)$. Such a relation, however, is important for an eventual implementation of (8.3); compare Chapter 6.2. The following idea proves to be helpful. Consider another limit (provided it exists)

$$c(x;d) := \lim_{\varepsilon \downarrow 0} \frac{f'_\varepsilon(x;d) - f'(x;d)}{\varepsilon^{1/2}}. \tag{8.4}$$

A straightforward geometric argument shows that the infimum $f'_\varepsilon(x;d)$ is the slope of the line through $\binom{x}{f(x)-\varepsilon}$ which supports the graph of $f(x+\alpha d)$, $\alpha \geqslant 0$, from below. Hence in (8.4) we compare the difference of this slope and of the slope of the tangent at $\binom{x}{f(x)}$ to the *vertical* increment $\varepsilon^{1/2}$. In (8.1) the tangents in direction d at x + td and x are compared to the *horizontal* increment t. At the first glance there is no relation between these two limits. A closer analysis shows however that thanks to convexity they are closely linked together.

Theorem 8.1. *Let* x *and* d *be given. Suppose one of the limits* (8.1) *and* (8.4), *respectively, exists and is positive. Then the other one exists as well and*

$$f''(x;d) = \frac{1}{2}c(x;d)^2.$$

The proof of the above statement is very technical. It can be found for a more general version than above in [20].

Provided $f''(x;d)$ exists and is positive, then, because of Theorem 8.1, we may write $c(x;d)^2/4$ instead of $f''(x;d)/2$ in (8.3). Further, we will not cause a disaster if we replace $c(x;d)$ by $[f'_\sigma(x;d) - f'(x;d)]/\sigma^{1/2}$ for fixed small positive σ. This leads to the modification of (8.3):

$$M_\sigma(x;d) := f'(x;d) + \frac{1}{4\sigma}[f'_\sigma(x;d) - f'(x;d)]^2. \tag{8.5}$$

This $M_\sigma(x;\cdot)$ is the model at x which we will discuss in the following; x is the current iterate which is kept fixed (we have omitted the index k) and d is the variable. It follows from Theorem 8.1 and (8.2) that (8.5) coincides with (3.2) for C^2-f and $\sigma \to 0$; hence (8.5) can be considered as a generalized *second order model* for nonsmooth convex f. We remark that, differently from (8.3), the model (8.5) is well-defined even if $f''(x;d)$ does not exist. Further note that the two disturbing points in connection with model (8.3) are no longer present: All quantities in (8.5) can be expressed via suitable max-terms. Moreover the σ in (8.5) will prevent us from shortcomings due to nonsmoothness; see the end of this chapter.

Let us consider the minimization of model (8.5). In case f is linear along the halfline $x+td$, $t \geq 0$, then $f'_\sigma(x;d) = f'(x;d)$ and (8.5) reduces to a pure first order model. To exclude this trivial situation we assume for the following that $f'_\sigma(x;d) > f'(x;d)$ for $d \neq 0$. Further suppose that the current iterate x is not yet σ-optimal, i. e., $0 \notin \partial_\sigma f(x)$. For simplicity we restrict the discussion to the special case where the gradient exists at x; the general situation is studied in full detail in [20]. Then $f'(x;d) = \nabla f(x)^T d$ and thus $M_\sigma(x;\cdot)$ is a convex function; hence

$$d_\sigma \textit{ minimizes } M_\sigma(x;\cdot) \iff 0 \in \partial M_\sigma(x;d_\sigma). \tag{8.6}$$

Let us compute the subdifferential of $M_\sigma(x;\cdot)$. Using repeatedly a chain rule (see [1]) we obtain

$$\partial M_\sigma(x;d) = \nabla f(x) + \frac{1}{2\sigma}[f'_\sigma(x;d) - f'(x;d)]\cdot(\partial f'_\sigma(x;d) - \nabla f(x))$$

and thus

$$0 \in \partial M_\sigma(x;d) \iff \frac{\lambda(d)-1}{\lambda(d)} \nabla f(x) \in \partial f'_\sigma(x;d)$$
$$\textit{with } \lambda(d) := \frac{1}{2\sigma}[f'_\sigma(x;d) - f'(x;d)].$$

This together with (8.6) and some technical arguments on $\lambda := \lambda(d_\sigma)$ shows that

$$d_\sigma \textit{ minimizes } M_\sigma(x;\cdot) \iff \lambda > 1 \textit{ and } \frac{\lambda-1}{\lambda}\nabla f(x) \in \partial f'_\sigma(x;d_\sigma). \tag{8.7}$$

A well-known formula for the subdifferential of a max-function (see e. g. Pshenichnyi [28]) yields for $f'_\sigma(x;d) = \max_{g\in\partial_\sigma f(x)} g^T d$:

$$\partial f'_\sigma(x;d) = \{h \in \partial_\sigma f(x) \mid h^T d \geq g^T d \textit{ for all } g \in \partial_\sigma f(x)\}. \tag{8.8}$$

Hence $\partial f'_\sigma(x;d_\sigma)$ is a subset of the boundary of $\partial_\sigma f(x)$, i. e., λ in (8.7) must be such that $\frac{\lambda-1}{\lambda}\nabla f(x)$ lies on the boundary of $\partial_\sigma f(x)$. (8.8) implies moreover that d_σ in (8.7) is the normal vector of a hyperplane which supports $\partial_\sigma f(x)$ at $\frac{\lambda-1}{\lambda}\nabla f(x)$. The following picture shows d_σ and the direction d_ε provided by the first order model, discussed in Chapter 6.1. Further, d_{st} denotes the classical steepest descent direction.

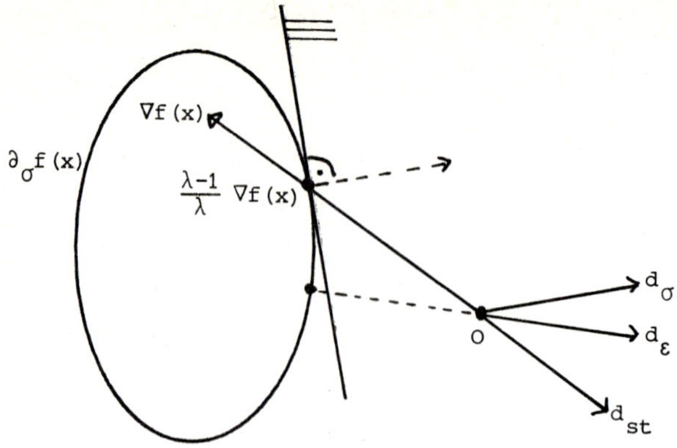

Figure 7: Minimization of the second order model

To justify the above model let us show that the direction d_σ provided by (8.5) is as least as good as the direction d_ε which we use in the corresponding first order iteration (6.13). We proceed in several steps from the smooth to the nonsmooth situation.

Consider a quadratic function $f(x) := \frac{1}{2}x^TAx + b^Tx$ with symmetric positive definite matrix A. Straightforward computation of the inf in (6.8) gives

$$f'_\sigma(x;d) = f'(x;d) + \sqrt{2\sigma d^TAd},$$

hence

$$M_\sigma(x;d) = x^TAd + b^Td + \frac{1}{2}d^TAd.$$

We obtain as minimizing d_σ for every $\sigma > 0$ the Newton-direction:

$$d_\sigma = -A^{-1}(Ax+b) = -f''(x)^{-1}\nabla f(x).$$

This direction is optimal, in one step we reach the minimum. Obviously there is no reason for d_ε to be close to the Newton-direction.

In case f is C^2 then for small σ our model $M_\sigma(x;d)$ is close to the second order model (3.2). Hence we can expect that, at least for small σ, the direction d_σ will be close once more to the Newton direction, whereas for small ε the first order direction d_ε will be close to the steepest descent direction.

We skip the C^1-situation for which we can prove a convergence result comparable to the one for the steepest descent method and smooth f; see [20]. Let us come to the case $f \notin C^1$. If f is differentiable at the current iterate x then the above Figure implies

$$f'_\sigma(x;d_\sigma) = \max_{g\in\partial_\sigma f(x)} g^T d_\sigma = \frac{\lambda-1}{\lambda}\nabla f(x)^T d_\sigma < 0. \quad (8.9)$$

In [20] it is shown that $f'_\sigma(x;d_\sigma) < 0$ holds in the general case. Hence a move in the direction d_σ guarantees a decrease of at least σ. We summarize: the above model combines the advantage of Newton's model in the region where f is smooth with a guaranteed decrease of at least σ in every step.

The above model is still waiting for an implementation. Some progress in this direction is reported in a paper by Lemaréchal and Strodiot [18].

REFERENCES

(the following list contains only papers quoted in the text; a more complete reference list on papers in NSO can be found in [24])

[1] A. Ben-Tal and J. Zowe, "Necessary and sufficient optimality conditions for a class of nonsmooth minimization problems" *Mathematical Programming* 24 (1982) 70 - 91.

[2] C. Charalambous and A. R. Conn, "An efficient method to solve the minimax problem directly", *SIAM Journal on Numerical Analysis* 15 (1982) 162 - 187.

[3] A. R. Conn, "An efficient second order method to solve the (constrained) minimax problem", University of Waterloo, Report CORR-79-5.

[4] El-Attar, R. Vidyasagar and S. R. Dutta, "An algorithm for ℓ_1-norm minimization with application to nonlinear ℓ_1-approximation", *SIAM Journal on Numerical Analysis* 16 (1979) 70 - 86.

[5] Yu. M. Ermoliev, "Methods of solution of nonlinear extremal problems", *Cybernetics* 2, 4, 1 - 16.

[6] R. Fletcher, *Practical methods of optimization II* (John Wiley & Sons, New York, 1981).

[7] J. L. Goffin, "Nondifferentiable optimization and the relaxation method" in: C. Lemaréchal and R. Mifflin (eds.) *Nonsmooth Optimization* (Pergamon Press, New York, 1978).

[8] J. L. Goffin, "On convergence rate of subgradient optimization methods", *Mathematical Programming* 13 (1977) 329 - 347.

[9] J. L. Goffin, "Convergence results in a class of variable metric subgradient methods" in: O. L. Mangasarian, R. R. Meyer and S. M. Robinson (eds.) *Nonlinear Programming 4* (Academic Press, New York, 1981).

[10] J. L. Goffin, "Convergence rates of the ellipsoid method on general convex functions", *Mathematics of Operations Research* 8 (1983) 135 - 149.

[11] M. Held, P. Wolfe and H. Crowder, "Validation of subgradient optimization", *Mathematical Programming* 6 (1974) 62 - 88.

[12] D. M. Himmelblau, *Applied Nonlinear Programming* (McGraw-Hill, 1972).

[13] J. B. Hiriart-Urruty, "Lipschitz r-continuity of the approximate subdifferential of a convex function", *Mathematica Scandinavica* 47 (1980) 123 - 134.

[14] R. Hornung, "An efficient algorithm for solving the discrete minisum problems, *Operations Research Letters* 2 (1983) 115 - 118.

[15] C. Lemaréchal, "Nondifferentiable optimization", in: Dixon, Spedicato and Szegö (eds.) *Nonlinear Optimization, Theory and Algorithms* (Birkhäuser, Boston, 1980).

[16] C. Lemaréchal, "A view of line searches" in: W. Oettli and J. Stoer (eds.) *Optimization and Optimal Control*, Lecture Notes in Control and Information, (Springer, 1981).

[17] C. Lemaréchal, "Numerical experiments in nonsmooth optimization" in: E. A. Nurminskii (eds.) *Progress in Nondifferentiable Optimization* (IIASA Proceedings, Laxenburg, 1982).

[18] C. Lemaréchal and J. J. Strodiot, "Bundle methods, cutting plane algorithms and σ-Newton directions", Preprint, University of Namur, 1984.

[19] C. Lemaréchal, J. J. Strodiot and A. Bihain, "On a Bundle algorithm for nonsmooth optimization" in: O. L. Mangasarian, R. R. Meyer and S. M. Robinson (eds.), *Nonlinear Programming 4* (Academic Press, New York, 1981).

[20] C. Lemaréchal and J. Zowe, "Some remarks on the construction of higher order algorithms in convex optimization", *Journal of Applied Mathematics and Optimization* 10 (1983) 51 - 61.

[21] A. Y. Levin, "On an algorithm for the minimization of convex functions", *Soviet Mathematical Doklady* 6, 286 - 290.

[22] R. Mifflin, "An algorithm for constrained optimization with semismooth functions", *Mathematics of Operations Research* 2 (1977) 191 - 207.

[23] E. A. Nurminskii, "On ε-differential mappings and their applications in nondifferentiable optimization", Working paper 78 - 58, IIASA, 1978.

[24] E. A. Nurminskii (ed.), *Progress in nondifferentiable optimization* (IIASA Proceedings, Laxenburg, 1982).

[25] T. Pietrzykowski, "An exact penalty method for constrained maxima", *SIAM Journal on Numerical Analysis* 6 (1969) 299 - 304.

[26] B. T. Poljak, "A general method of solving extremal problems", *Soviet Mathematical Doklady* 8, 593 - 597.

[27] B. T. Poljak, "Minimization of unsmooth functionals", *USSR Computational Mathematics and Mathematical Physics* 9, 14 - 29.

[28] B. N. Pshenichnyi, *Necessary conditions for an extremum* (Marcel Dekker, New York, 1971).

[29] R. T. Rockafellar, *Convex Analysis* (Princeton University Press, 1972).

[30] V. A. Shokov, "Note on minimization methods using space dilation", *Cybernetics* 10, 4, 689 - 692.

[31] N. Z. Shor, "The rate of convergence of the generalized gradient descent method", *Cybernetics* 4, 3, 79 - 80.

[32] N. Z. Shor, "Utilization of the operation of space dilation in the minimization of convex functions", *Cybernetics* 6, 2, 102 - 108.

[33] N. Z. Shor, "Convergence of a gradient method with space dilation in the directions of the difference between two successive gradients", *Cybernetics* 11, 4, 564 - 570.

[34] N. Z. Shor, "Cut-off method with space extension in convex programming problems", *Cybernetics* 13, 94 - 96.

[35] N. Z. Shor, "Generalized gradient methods of nondifferentiable optimization employing space dilation operations" in: A. Bachem, M. Grötschel and B. Korte (eds.) *Mathematical programming, the state of the art* (Springer, Berlin, 1983).

[36] P. Wolfe, "A method of conjugate subgradients for minimizing nondifferentiable functions", *Mathematical Programming Study* 3 (1975) 145 - 173.

[37] R. S. Womersley, "Numerical methods for structured problems in nonsmooth optimization", University of Dundee, 1981.

[38] D. B. Yudin and A. S. Nemorovskii, "Estimation of the informational complexity of mathematical programming problems, *Matekon* 13, 2 - 25, 25 - 45.

PARALLEL COMPUTING IN OPTIMIZATION

Robert B. Schnabel*

Department of Computer Science
University of Colorado
Boulder, Colorado 80309

September, 1984

*This research supported by ARO contract DAAG 29-84-K-0140

Presented at the NATO Advanced Study Institute on Computational Mathematical Programming, Bad Windsheim, Germany F.R., July 1984

NATO ASI Series, Vol. F15
Computational Mathematical Programming
Edited by K. Schittkowski

Abstract

One of the major developments in computing in recent years has been the introduction of a variety of parallel computers, and the development of algorithms that effectively utilize their capabilities. Very little of this parallel algorithm development, however, has been in numerical optimization. Nevertheless, significant opportunities exist for the utilization of parallelism in optimization, especially on computers that support independent concurrent processes. This paper first gives a very brief survey of parallel architectures and general characteristics of parallel algorithms. Next we indicate what we see as the leading opportunities for the utilization of parallelism in optimization. Then we survey the small amount of existing research in parallel optimization; most of this has been conducted at The Hatfield Polytechnic. Finally we discuss some recently initiated research at the University of Colorado concerned with solving optimization problems by parallel algorithms suitable for implementation on a local area network of computers; we focus on a new parallel algorithm for global optimization.

1. Introduction

One of the major developments in computing in recent years has been the introduction of a variety of parallel computers, and the development of algorithms that effectively utilize their capabilities. By parallel computer, informally we mean any computer capable of performing two or more computations simultaneously or concurrently. Algorithms that are designed to use simultaneous or concurrent operations will be called "parallel algorithms". In contrast, we will call traditional computers that are designed to execute one instruction at a time "sequential computers", and algorithms designed for such computers "sequential algorithms".

The development of parallel computation is motivated by the same objectives as most progress in computing : the desire to solve an ever-increasing range of problems, and the desire to solve problems as quickly or cheaply as possible. In some cases, the main goal of parallel computation is raw speed, enabling the solution of problems that heretofore were too time-consuming to be solved in a reasonable amount of time, or allowing the solution of problems within the time constraints of real-time systems. In other cases, the main goal of parallel computation is to solve problems more cheaply than with sequential computers.

While in recent years there has been considerable development and implementation of parallel numerical algorithms, very little of this activity has been in numerical optimization. (By numerical optimization we mean the wide category of problems involving the minimization of a linear or nonlinear objective function, perhaps subject to linear or nonlinear constraints. In this paper, we only consider problems in real, as opposed to discrete, variables. For a general description of numerical optimization problems, see for example Gill, Murray, and Wright [1981] or Dennis and Schnabel [1983].) Instead, parallel numerical computation has centered mainly on the solution of partial differential equations, and associated areas such as numerical linear algebra. The reasons for this are simple and clear. First, the most obvious candidates for high speed computation are large scale problems, and these arise far more commonly in the solution of differential equations than in optimization. Secondly, the parallel computers that have been most widely available have been vector computers such as the Cray-1 and Cyber-205, and as we will discuss, these are far better suited to solving partial differential equations and linear algebra problems than optimization problems.

There are many opportunities for the use of parallelism in optimization, however, and good reasons to take advantage of these opportunities. It is well known that many optimization problems are very expensive, for one or more of the following reasons : the size of the problem (number of variables and constraints) is large; the objective function or constraints are expensive to evaluate; many iterations or function evaluations are required to solve the problem; a large number of variations of the same problem must be solved. Indeed, often optimization problems are not solved, or are only "solved" in a very approximate manner, because the cost of solving them with existing sequential computers and algorithms is prohibitive or does not fit within real-time constraints. The availability of faster or cheaper solutions to optimization problems is likely to cause a considerable increase in the number of optimization problems that are formulated and solved.

The main purpose of this paper is to discuss the prospects for the application of parallel computation to numerical optimization problems, and to survey the small amount of research that has been conducted in this area. While a self-contained treatment of this subject also would require a thorough discussion of parallel computer architectures, it is not within the scope of this paper to discuss this subject in detail. Therefore in Section 2 we provide a thumbnail sketch of the main types of parallel architectures, stressing those characteristics relevant to our discussion of parallel optimization. Section 3 provides a brief overview of important general characteristics of parallel algorithms and their relation to the types of architectures discussed in Section 2. Together, Sections 2 and 3 provide a minimal background in parallel computation; two excellent and thorough references on this subject are the books by Hockney and Jesshope [1981] and Hwang and Briggs [1984].

Sections 4-7 are concerned with parallel optimization. In Section 4 we summarize what we see as the main opportunities for the utilization of parallelism in optimization. Section 5 surveys the small amount of research that has been conducted to date in parallel optimization; much of this work has been conducted at The Hatfield Polytechnic. We and our colleagues at the University of Colorado have recently begun a large project in distributed computation that includes a significant component in parallel optimization; we discuss this work in Section 6. Section 7 contains some concluding remarks on the future of parallel optimization.

2. Brief overview of parallel computer architectures

The development of parallel computer architectures and parallel algorithms are concurrent activities, each motivated partially by the capabilities and needs of the other. This paper mainly discusses parallel algorithms suitable for several broad classes of parallel computer architectures. In this section we provide a minimal description of these classes of parallel computers. Parallel computer architectures are discussed in many research papers and several books; especially recommended are the books by Hockney and Jesshope [1981] (for pipeline and processor array (SIMD) computers), Hwang and Briggs [1984] (pipeline, processor array, and multiprocessor (MIMD) computers), and Lampson, Paul, and Siegert [1981] (distributed systems).

The four types of parallel architectures we will refer to are pipeline computers (specifically its subclass, vector computers), processor arrays (SIMD computers), multiprocessors (MIMD computers), and local area networks of computers. The terms SIMD and MIMD are from the well-known taxonomy by Flynn [1966] and are explained below; Flynn's taxonomy is not adequate, however, to differentiate the four classes mentioned above.

Pipeline / vector computers

Pipeline computers are machines that overlap computations by subdividing and interleaving them in assembly line fashion. That is, roughly speaking, computer instructions are subdivided into m portions, and the pipeline machine executes part 1 of instruction k, part 2 of instruction k+1, ..., and part m of instruction of instruction k+m-1 simultaneously. Vector computers are a subclass of pipeline machines that use this technique to rapidly compute arithmetic operations on vectors. For example, by dividing the addition instruction into several (typically 6-10) subparts, a vector machine can perform the vector instruction

$$z(i) = x(i) + y(i), \qquad i = 1, \cdots, n$$

rapidly by interleaving the successive additions. Vector computers are intended to perform the above operation, as well as the componentwise subtraction and multiplication of vectors, and the "saxpy" operation

$$z(i) = a * x(i) + y(i), \qquad i = 1, \cdots, n$$

rapidly as long as n is moderately large. Thus they are best suited to large scale problems where these vector computations are dominant, such as solving large systems of linear equations and many algorithms for solving systems of partial differential equations. The primary vector computers in use today include the Cray-1 and Cyber-205. The designs of these machines differ in important ways that effect, for example, the length (n) at which vector processing becomes efficient, but these considerations are beyond the scope of this paper.

Processor arrays (SIMD computers)

Processor arrays are computers capable of executing a given instruction simultaneously on multiple data. Hence in Flynn's taxonomy they are referred to as "single instruction multiple data" (SIMD) computers. A processor array thus can execute a given program segment simultaneously (in "lockstep") on multiple data sets so long as the program segment contains no data dependent branches. Processor arrays typically consist of hundreds or thousands of processing elements, each with a small amount of local memory, connected to a single control processor from which instructions are issued. Usually an interconnection network permits data to be exchanged between various processing elements; sometimes data is shared between neighboring processors. Examples of processor arrays are the Illiac-IV, the Burroughs Scientific Processor, the Goodyear Massively Parallel Processor, and the ICL Distributed Array Processor (DAP) of Queen Mary College, London.

Multiprocessors (MIMD computers)

Multiprocessors are computers capable of executing multiple instructions on multiple data independently and concurrently. Hence in Flynn's taxonomy they are referred to as "multiple instruction multiple data" (MIMD) computers. Multiprocessors thus are versatile parallel computers that can be used to concurrently perform similar or dissimilar tasks. Typically, multiprocessors consist of a small number (2-50) of processors, usually with their own local memory, which share access to common memory modules and are controlled by a single operating system. Several multiprocessors have been developed in experimental environments, including the C.mmp and Cm* of Carnegie Mellon University, the NEPTUNE of Loughborough University, the EMPRESS of ETH Zurich, and the Erlangen General Purpose Array. Commercially available multiprocessors include the Denelcor Heterogeneous Element Processor (HEP) and the Cray X-MP. (The HEP achieves MIMD performance through the use of multiple pipelined functional units.)

Local area networks of computers

Local area networks of computers are multiple computers, within fairly close physical proximity, connected by a high speed communication network. While such networks usually are not primarily intended for parallel computation, they may be used for parallel computation by distributing various portions of a concurrent algorithm among various computers on the network. (Considerable operating systems support is required to make this convenient.) Thus local area networks of computers can be used to achieve concurrent independent processing similar to that available on a multiprocessor, but communication between processors is substantially slower than between multiple processors on a single computer. The use of local area networks for concurrent computation is currently being explored in various university and industrial research projects; projects considering the use of such networks for concurrent numerical computation include the Crystal project of the University of Wisconsin, and the ENCOMP project of the University of Colorado that is discussed in Section 6.

The above list of parallel computer architecture categories is sufficient for the purposes of this paper, but it is by no means exhaustive. Another important category of parallel machines for numerical computation are "multi-microprocessors", computers consisting of substantial numbers of simple processing elements. These may be configured as SIMD or MIMD computers and thus could also be considered as further examples of processor arrays or multiprocessors. Examples include the NASA Langley finite element machine, the University of Texas reconfigurable array processor, and the University of Maryland ZMOB. Additional parallel architectures include "data-flow machines" and "systolic arrays"; machines of this type are just beginning to appear in practice and have hardly been used for numerical computation.

3. General characteristics of parallel algorithms

Before discussing parallel optimization algorithms, it is helpful to mention some general characteristics of parallel algorithms, and their relation to the classes of parallel computers discussed in the previous section. In this section we briefly discuss five attributes that are emerging as important characteristics of parallel algorithms in general and parallel optimization algorithms in particular. These attributes are : the level of parallelism; communication requirements; uniformity of operations; generation of processes; and control of processes. We also discuss how the suitability of various parallel architectures to a particular parallel algorithm depends on these characteristics. The attributes we consider constitute a primitive taxonomy of parallel algorithms, something now just starting to emerge. Indeed, this section draws upon the ideas discussed at the Taxonomy of Parallel Algorithms Workshop held in Santa Fe in 1983, especially the paper by Siegel [1983]. An interesting discussion of characteristics of parallel algorithms for multiprocessors is given in Kung [1976].

To aid our considerations, it is helpful to consider a simple optimization task that is very conducive to parallelism, the calculation of a finite difference gradient. That is, given a function $f(x)$ of a vector x of n variables, approximate ∇f at a specific point z by g, where

$$g_i = (f(x+h_i e_i) - f(x)) \,/\, h_i \,, \qquad i=1, \cdots, n \,.$$

Here e_i is the ith unit vector and h_i is an appropriate stepsize. The most obvious way to parallelize this calculation is to have various processors concurrently compute various components of g. Another alternative is to parallelize each computation of $f(x)$ in some manner. Now we will relate these possibilities to the attributes of parallel algorithms given above.

The level of parallelism refers to the size and complexity of the portions of the algorithm that are executed concurrently (or alternatively, the position of the concurrent segments in a top down description of the algorithm). In a high-level parallel algorithm, major portions of the algorithm are separated and executed concurrently. In a low-level parallel algorithm, individual instructions or small groups of instructions are executed concurrently. For example, the first parallel finite difference gradient algorithm mentioned above, distributing the calculation of various components of g to various processors, is a high-level subdivision of this algorithm. Such a subdivision is clearly well suited to a multiprocessor. It would be suited to a processor array only if the calculation of $f(x)$ contained no branches dependent upon x. A vector computer would not be appropriate for this calculation. In general, high-level parallel algorithms are best suited to multiprocessors, and may also be suited to local area networks of computers if relatively little inter-processor communication is required. Returning to the finite difference example, suppose the calculation of $f(x)$ consists mainly of large scale matrix-vector operations. An algorithm that calculates the components of g sequentially but vectorizes these portions of each calculation of $f(x)$ makes a low-level parallelization of the algorithm. As this example suggests, vector computers are best suited to low-level parallelism.

Communication requirements refer to the amount of data that must be shared between concurrent processes, as well as the dependence of the execution of some processes upon the receipt of information from other processes. In the finite difference example, processes computing various components of g would not have to communicate with each other at all, but presumably would pass their value of g_i back to a master process. In contrast, a Gaussian elimination algorithm where each processor holds one row of the matrix will require considerable communication between processors. In general, multiprocessors are usually well suited to parallel algorithms with high communication requirements, while networks of computers are only suited to algorithms with low communication requirements. Processor arrays often are well suited to fairly high communication requirements, especially between nearby processing elements.

Uniformity of operations refers to whether identical or different operations are performed in parallel. Clearly, vector computers require the highest uniformity, multiple applications of the same arithmetic operation to vectors of data, while processor arrays require the next highest uniformity, execution of the same sequence of instructions to multiple data in lockstep. Only multiprocessors and networks of computers can perform different instructions concurrently.

By generation of processes we refer to the determination of when and where processes are generated. Parallel algorithms may generate processes statically, meaning that the schedule for generating processes is known before execution of the program begins, or they may generate processes dynamically, meaning that the number and distribution of processes is determined during execution and usually varies as the program progresses. Most of the algorithms discussed in Section 5 generate processes statically, while the algorithm discussed in Section 6 generate processes dynamically.

Control of processes refers to the dependence of the execution of processes upon the progress of other processes. Parallel algorithms that require the start of some processes to wait for the completion of other processes are called synchronous algorithms, algorithms without this requirement are called asynchronous algorithms. Clearly, algorithms that run on processor arrays must be highly synchronous. Multiprocessors and local area networks are used for both synchronous and asynchronous algorithms, and examples of both types of optimization algorithms are given in this paper. Asynchronous algorithms usually are more difficult to design than synchronous algorithms, but often offer the greatest chance of near-optimal usage of independent multiple processors.

Finally, we need to briefly discuss how one measures the performance of algorithms on parallel computers. For vector computers, performance usually is measured in the average number of floating point operations performed per second. For the remaining parallel architectures, the issue is much less clear cut. Traditional sequential measures such as arithmetic complexity, number of iterations, and number of function evaluations, often are not appropriate for assessing parallel algorithms. Instead, one often wants some measure of how fully the parallel algorithm utilizes the computational power of the parallel computer, and how much advantage is gained over sequential algorithms. This measure will be influenced not only by the number of arithmetic operations required, but also by delays caused by synchronization, communication, and contention for shared data. The term most often used to characterize this measure of performance is "speedup"; however, no uniform definition of speedup exists. Speedup often is defined as the time required to solve a problem on a parallel machine using a given p-processor algorithm divided by the time to solve the same problem on the same machine using the best 1-processor algorithm; in this case, the speedup should be between 1 and p. Sometimes, however, speedup is defined as the time to solve a problem by a given parallel algorithm on a given parallel machine divided by the time to solve the same problem by a sequential algorithm on a sequential machine; in this case, wider variation is possible. The results mentioned in Section 5 use both definitions.

4. Opportunities for parallelism in optimization

The primary objective of parallel computation is to solve expensive problems more quickly or more cheaply. Since many optimization problems are expensive to solve, it is worthwhile to consider whether parallel algorithms can aid in their solution. In this section, we discuss four leading reasons why optimization problems are expensive, and for each, summarize the opportunities for parallelism that are suggested and the types of parallel architectures most likely to be suitable. The four causes of expense we consider are : the size of the problem is large; the objective function or constraints are expensive to evaluate; many iterations or function evaluations are required to solve the problem; many variations of the problem must be solved. Another view of the opportunities for parallelism in optimization is presented by Dixon, Patel, and Ducksbury [1983].

By large problems we mean problems where the number of variables or constraints is large, say greater than 100. The solution of such problems is likely to require large scale linear algebra calculations at each iteration; in addition, the evaluation of the objective function, constraints, and their derivatives may be expensive due to the number of variable and constraints. Each of these areas presents opportunities for parallelism. The most obvious use for parallelism for large problems is to speed the linear algebra calculations. Thus vector computers are excellent candidates. Note, however, that the vector algorithms required rarely are particular to optimization, and have mainly been developed. Since it is also possible to construct efficient parallel linear algebra algorithms for processor arrays and multiprocessors, these computers are also candidates for large problems. If the evaluation of the functions or derivatives also involves large scale linear algebra calculations, vector computers or processor arrays again are suggested, whereas a multiprocessor could be used to evaluate a number of constraints concurrently. Yet another possibility is to evaluate or approximate derivative components concurrently; multiprocessors or processor arrays are most likely to be suitable. In summary, large scale optimization problems present opportunities for virtually all parallel architectures, but many of the parallel computations they seem to suggest are not particular to optimization.

Problems where the objective function or constraints are expensive to evaluate suggest two types of opportunities for parallelism. The first is to perform each individual function evaluation by a parallel algorithm. For example, if the objective function is itself a system of partial differential equations, a parallel partial differential equation algorithm and a computer appropriate for this algorithm may be suggested. Again, the development of this routine usually is outside the realm of optimization. The second possibility is to perform multiple evaluations of the objective function concurrently. The most obvious use of concurrent function evaluations in optimization algorithms is in concurrent finite difference derivative algorithms such as the one mentioned in the previous section. The development of other effective optimization algorithms that utilize concurrent function evaluations is one of the fundamental challenges in the development of parallel optimization algorithms; some possibilities are mentioned in Sections 5 and 6. Concurrent function evaluation algorithms are best suited to multiprocessors or networks of computers, although processor arrays also can be used if the function has no data dependent branches.

Optimization problems often require many iterations (and function evaluations) when the starting guess is far from a minimizer and the objective function or constraints are highly nonlinear. Another important class of problems that usually require many function evaluations are "global optimization" problems, where multiple local minima exist and the lowest of these is required. In general, these problems present two obvious opportunities for parallelism : concurrent function evaluations, or concurrent iterations. The use of concurrent iterations in the search for a single ("local") minimizer is an interesting possibility that has hardly been explored. The use of concurrent function evaluations or iterations in solving the global optimization problems appears to be one of the most fruitful areas for the application of parallelism to optimization, and is discussed in both Sections 5 and 6. Again, multiprocessors or networks of computers are appropriate parallel architectures for these types of concurrent algorithms, while processor arrays might sometimes be appropriate.

Finally, optimization applications are often expensive because many versions of the same problem must be solved. A common example is a parametric study where the objective function depends not only on the variables x but also on some parameters y, and the minimizer with respect to x is required for various values of y. Here the most obvious use of parallelism is to run multiple problems concurrently; this should be very effective and requires no new algorithm development. In some cases, each run depends upon information from previous runs; in this case, it may be necessary to speed individual runs instead. Techniques for this case are likely to be application dependent, and we are not aware of any existing applications of parallelism to this area.

There is one important case when the use of parallelism may be important even though the optimization problem being solved is not particularly expensive. This is the case of "real-time" problems that must be solved within an allotted time limit. Clearly, the best ways to speed the solution of real-time problems may be application dependent. Since the predominant cost of most optimization algorithms is either linear algebra, expense per function evaluation, or number of function evaluations and iterations, however, the approaches indicated above for expensive problems should often be appropriate for real-time problems as well.

5. Survey of parallel optimization research

In recent years a considerable amount of research activity has been devoted to developing and testing parallel numerical algorithms. Very little of this activity, however, has been in numerical optimization. The reasons for this become clear if one considers the opportunities for parallel optimization mentioned in the previous section together with the availability of parallel computers so far. From the previous section, it appears that most of the opportunities for parallelism in optimization involve independent concurrent processes, and are therefore best implemented on multiprocessors or local area networks of computers, and sometimes may be implemented on processor arrays. There does not appear to be much opportunity for the exploitation of vector processing that is particular to optimization. The availability of parallel computers up to this time has been the opposite of these needs, that is, good access to vector computers but very little availability of processor arrays, multiprocessors, or networks of computers. This is an important reason why parallel optimization has been relatively slow to develop.

There has been a limited amount of work in parallel optimization, however, as well as work in parallel linear algebra that is especially relevant to optimization. In this section we survey this research. First we give references to some relevant parallel linear algebra research, especially conjugate gradient methods for solving systems of linear equations. The predominant research in parallel optimization that we are aware of has been conducted by a group at The Hatfield Polytechnic. We summarize their research in parallel algorithms for four applications : a modified Newton's method for unconstrained optimization, a nonlinear conjugate gradient method, global optimization, and nonlinear least squares. We conclude the section with references to a few other research projects in parallel optimization.

Much of the research to date in parallel numerical computation has concerned the solution of systems of linear equations. Algorithms have been developed for full, sparse, tridiagonal, and banded matrices. Most work has concentrated on algorithms for vector computers but algorithms for processor arrays and multiprocessors have also been developed. Many algorithms are adaptations of the standard sequential algorithms such as Gaussian elimination, QR factorization, and Cholesky decomposition, some are more specially developed for parallel computation. Many of the implementations have achieved efficient speed or speedup. Some representative references include : Heller [1978] and Sameh [1977] (surveys of algorithms), Duff [1983], Fong and Jordan [1977], Jordan [1979], and Rodrigue [1982] (all mainly concerned with vector computation), Hockney and Jesshope [1981] (vector computers and processor arrays), and Dongarra and Hiromoto [1983], Kapur and Browne [1981], and Lord, Kowalik, and Kumar [1980] (multiprocessors).

Of special interest to optimization are parallel algorithms for solving linear systems by conjugate gradient methods. Linear conjugate gradient algorithms consist mainly of matrix-vector multiplications and inner products, so they lend themselves nicely to vectorization or concurrent computation. Much of the interesting research has concerned the effective use of preconditioners in these algorithms. Some references are Adams [1983], Adams and Ortega [1982], Kowalik and Kumar [1982], and van der Vorst [1982]. These techniques are of special interest to optimization because conjugate gradient algorithms for nonlinear optimization (see e.g. Gill, Murray, and Wright [1981]) are closely related to the linear algorithm. One parallel nonlinear conjugate gradient algorithm is described below.

A final parallel linear equations algorithm with applications to optimization is the parallel relaxation algorithm implemented by Baudet [1978] on the C.mmp multiprocessor. Baudet achieved high speedup by implementing a "chaotic relaxation" algorithm with asynchronous concurrent processes. An application of Baudet's approach to networks of computers and nonlinear optimization is mentioned at the end of the next section.

Turning to nonlinear optimization, most of the work on parallel algorithms that we are aware of has been conducted by a group at The Hatfield Polytechnic led by L. Dixon. Most of their research has been conducted on the ICL DAP, a processor array with 4096 bit serial processing elements that can compute in lockstep. Each processing element has 16K bits of local storage. The processing elements are connected in a 64 x 64 grid with each processor having access to the four neighboring elements and their storage.

The DAP is connected to an ICL 2980 computer. While we mentioned in Section 4 that processor arrays are not as well suited to parallel numerical optimization as are multiprocessors, they are reasonably well suited to *experimental* studies in parallel optimization. This is because any algorithm requiring parallel function evaluations can be tested on a processor array as long as the test functions contain no data dependent branches, a condition often satisfied by test (as opposed to real-world) objective functions. Processor arrays may not be used, however, to test algorithms that involve concurrent execution of different program segments. For this reason, most of the algorithms tested by the Hatfield group involve concurrent function evaluation or parallel linear algebra computations. A small amount of their work has been conducted on the Loughborough NEPTUNE computer, a multiprocessor with four processors.

The parallel optimization research at Hatfield has concentrated on four types of algorithms, modified Newton methods for unconstrained optimization, nonlinear conjugate gradient methods, global optimization methods, and methods for nonlinear least squares. It is reported in a number of survey papers, including Dixon and Patel [1982], Dixon, Patel, and Ducksbury [1983], and Patel [1982a], as well as many papers mentioned below.

The parallel unconstrained optimization methods discussed in Dixon [1981] and Patel [1982b] are adaptations to an processor array of a finite difference Newton's method - line search algorithm. Parallelism is achieved in three ways : by computing the function evaluations involved in calculating the finite difference gradient and Hessian in parallel; by using a parallel algorithm to solve the system of linear equations required to calculate the Newton step; and by conducting 1, 2, and 4 dimensional line searches which evaluate the function at up to 4096 points simultaneously. (The four possible search directions are the Newton direction, the steepest descent direction, and the directions to the previous past two iterates.) In all of the test results reported by the Hatfield group, the times required by the parallel algorithms on the DAP are compared to the times required by a sequential algorithm on a faster sequential machine (a DEC 1091). Since the relative speeds of the DAP and the sequential machine that are given as a benchmark are highly variable, it is difficult to calibrate the gains that are achieved. It is clear, however, that on test problems such as those reported with $n = 64$, the parallel algorithm is substantially faster than the sequential Newton algorithm or a sequential variable metric method. This is largely because the finite difference calculations and the solution of linear equations dominate the computing cost, and are done considerably faster by the parallel machine than by the sequential machine. The intriguing idea of concurrent multiple dimension, multiple step length line searches is not shown to produce any great advantage; the computational results show no clear cut advantage in going from one to two or four dimensions, while the data reported do not permit an assessment of the effect of multiple step lengths.

Experience in running parallel linear and nonlinear conjugate gradient algorithms on the DAP is reported in Dixon, Ducksbury, and Singh [1982]. The algorithms are designed specifically for finite element problems, with each processor handling one finite element. For the linear problem, the matrix-vector products and inner products that constitute the bulk of the algorithm are computed in parallel by having each processor concurrently compute the contribution of its finite element to the desired result. The nearest neighbor connections of the DAP are very helpful here. For the nonlinear problem, the gradient is also computed in parallel by combining the contribution of each finite element. Test results for both algorithms are reported for linear problems only; for problems with enough nodes to utilize most of the processors on the DAP, the parallel algorithm clearly is much faster than a sequential algorithm since up to 4096 computations are performed concurrently during much of the parallel algorithm. Neither preconditioning, nor the interesting question of how to handle problems with more nodes than the number of processors, are addressed.

Parallel global optimization algorithms designed for the DAP and the NEPTUNE multiprocessor are considered in Dixon and Patel [1981] and Ducksbury [1982]. As we stated in Section 4, the global optimization problem is perhaps the most obvious candidate for parallel optimization, because the solution of these problems usually requires many function evaluations, and because the problem does not appear to require as inherently sequential algorithms as the calculation of a single local minimizer.

Dixon, Patel, and Ducksbury investigate how a primitive global optimization algorithm, the algorithm of Price [1981] that is largely intended for mini-computers, can be implemented on parallel computers. Price's algorithm consists of the random sampling of $f(x)$ at a number of points, followed by the repeated replacement of a high point by a new lower point generated by a reflection operation through some of the current points. No derivative information or local minimization techniques are used. The version developed by the Hatfield group for the DAP samples $3n$ groups of 4096 points concurrently, resulting in $3n$ points at each processor; then it repeatedly generates new points at each processor concurrently. The total number of function evaluations required by the parallel algorithm on test problems is usually 10 to 100 times that required by the sequential algorithm, but since 4096 test function evaluations are performed concurrently on the DAP, the parallel algorithm requires less time than the sequential algorithm on the faster sequential machine, by factors of 3 to 68. The multiprocessor algorithm is a different adaptation of Price's algorithm where the ith processor, $i = 0, \cdots, 3$, repeatly attempts to replace the ith worst remaining point by reflection. The algorithm is asynchronous with the processors accessing a global store of the sampled points. Speedups in the range 2 to 3.8 are reported when testing the four-processor algorithm against a similar one-processor algorithm on the NEPTUNE.

While the results achieved by the Hatfield group show considerable speedup of Price's algorithm on processor arrays and multiprocessors, it must be acknowledged that Price's algorithm is not a competitive global optimization algorithm for standard sequential machines. For example, in tests by Rinnooy Kan and Timmer [1983], a more sophisticated version of Price's algorithm still usually requires an order of magnitude more function evaluations than a state-of-the-art stochastic method. Parallel global optimization that use more sophisticated optimization techniques still need to be developed. A brief discussion of a more sophisticated parallel global optimization algorithm is included in Dixon and Patel [1982]; another alternative is presented in Section 6.

The final type of parallel optimization research reported by the Hatfield group is a brief study by McKoewn [1979] of two-processor nonlinear least squares algorithms run on an earlier, two-processor version of the NEPTUNE multiprocessor. Two techniques are considered for achieving parallelism. The first is the obvious idea of evaluating different portions of the sum of squares function concurrently. Clearly for a data fitting problem where each residual function requires a similar amount of time to evaluate, this should lead to good speedup on a multiprocessor, and McKeown's computational results support this contention. The second parallel algorithm tested is an interesting asynchronous algorithm where one processor computes the search directions while the other does line searches using the latest available information. Test results showed that the total number of function evaluations required may degrade very little (the desired result) or significantly over the comparable sequential algorithm, depending on seemingly unrelated factors, such as amount of input/output, that effect the timing. Speedup information is not reported.

In summary, it should be noted that most of the parallel optimization algorithms developed by the Hatfield group are closely related to existing sequential optimization algorithms. While these are important, in fact groundbreaking, contributions, it is our opinion that future developers of parallel optimization algorithms will also need to consider new algorithmic approaches that are not necessarily best or even suitable for sequential computation. This principle has already been demonstrated many times in the development of parallel algorithms; see for example the discussion of recurrences and tridiagonal linear systems in Hockney and Jesshope [1981]. We believe that the design of fundamentally parallel optimization algorithms offers many new and exciting possibilities; hopefully the material in the next section provides an indication of approaches geared specifically to parallel computation.

A few other projects related to parallel optimization have been reported. Straeter [1973], Straeter and Markos [1975], Housos and Wing [1980], and van Laarhoven [1984] have proposed parallel versions of a variety of unconstrained optimization algorithms, including variable metric and conjugate direction methods. Mohan [1982] has developed a parallel traveling salesman algorithm and tested it on the Cm*. Feijoo and Meyer [1984] currently are testing parallel nonlinear network optimization algorithms on the local area network of computers developed under the Crystal project at the University of Wisconsin.

6. Parallel optimization at the University of Colorado

We and several colleagues at the University of Colorado have recently begun research into the development of parallel optimization algorithms suitable for implementation on a local area network of computers. This research is part of a larger project at the Computer Science Department of the University of Colorado , the ENCOMP project, investigating the use of a local area network of computers for parallel computation. In this section we first give a very brief description of the ENCOMP project. Then we discuss some of the approaches to parallel optimization we are investigating, focusing on an approach to global optimization that is suitable to multiprocessors as well as networks of computers.

The aim of the ENCOMP project is to develop and test parallel numerical algorithms designed specifically for a local area network of computers. This computing environment is of interest for three main reasons. First, it is becoming increasingly common in practice. Second, as will be indicated below, the areas of numerical computation we are interested in, such as optimization and VLSI design, lend themselves naturally to solution by concurrent algorithms that require little inter-process communication. Thus they are well suited to parallel solution on a local area network of computers, where the communication speed between processors is relatively slow compared to the computing speed of each processor. Third, there now exist commercially available operating systems that support inter-process communication, thereby enabling one to readily develop and test concurrent programs on networks of computers. In particular, we are using a network of Sun workstations and VAX computers with each node running the Berkeley Unix 4.2 operating system, and the nodes connected on an Ethernet, and we have been able to develop and run concurrent programs almost immediately upon installation of this network. Naturally, additional support software is still desirable to make the development, debugging, testing, and evaluation of concurrent programs more convenient.

The main class of optimization problems we are considering share the characteristic that they can solved by partitioning the original problem, often dynamically, into a number of subproblems, and then solving each subproblem, with limited communication required between the subproblems. In many cases this leads to a quite different algorithm than has been considered for sequential computation. The foremost optimization problem we propose to solve in this manner is the global optimization problem,

$$\underset{x \in D}{\text{minimize}} \; f : R^n \to R \; .$$

Other problems that can be approached in this way include determining the feasibility of a system of constraints,

$$\text{given} \quad D \in R^n \quad \text{and} \quad c_i : R^n \to R, \quad i=1, \cdots, m$$

$$\text{determine whether there exists } x \in D \text{ for which } c_i(x) \le 0, \quad i=1, \cdots, m \; ,$$

the nonlinear mini-max problem,

$$\underset{x \in D}{\text{minimize}} \; \alpha$$

$$\text{subject to} \quad c_i(x) \le \alpha, \quad i=1, \cdots, m$$

and the global solution to a system of nonlinear equations.

The global optimization algorithm we are developing is most closely related to the stochastic optimization method of Boender et al [1982], to our knowledge the most efficient known sequential stochastic optimization method. Our algorithm basically consists of adaptively partitioning the variable space into subregions likely to contain one local minimizer each, and then simultaneously finding these local minimizers using separate processors, recurring this procedure as necessary. Additional efficiency is attained by the early termination of subregions where the function is high. Below we give a high level, oversimplified description of the algorithm that is executed by each process invoked by the concurrent algorithm. Variable and procedure names are italicized and inter-process communication is in boldface.

Global (*Function*, *Subregion*, *Lowest-anywhere-so-far* (* shared variable *))

1. evaluate *Function* at several points in *Subregion* ;
   throw away very high points and take a steepest descent step from the rest
      (* for initial (entire) region, this step can be executed in parallel,

and *lowest-anywhere-so-far* is initialized to the lowest function value found *)

2. cluster points with the aim of forming one cluster for all points in the region of attraction of each local minimizer ;
   partition *Subregion* into smaller subregions containing these clusters
3. IF (number of subregions identified at step 3) > 1
   THEN (attempt to) **create a process for each new subregion**
   (* subsequent instantiations of *Global* begin at step 4 *)
4. Run a local minimization algorithm in *Subregion* for at most a predetermined number of steps
5. IF (lowest function value found in *Subregion*) < *Lowest-anywhere-so-far*)
   THEN **broadcast this new value of** *Lowest-anywhere-so-far* **to all other processes**
6. IF (local minimizer found at step 4) THEN
   perform steps 1-2
   IF (number of clusters found) > 1
   THEN go to step 3
   ELSE terminate this process
   (* global minimizer in *Subregion* has been found *)
7. IF (lowest function value found in *Subregion*) = *lowest-anywhere-so-far*
   THEN go to step 4
   ELSE
   (* try to determine whether this subregion can be discarded because *Function* has high values throughout *)
   estimate a lower bound on *Function* in *Subregion*
   (* may involve further sampling *)
   IF (lower bound) < *lowest-anywhere-so-far*
   THEN go to step 4
   ELSE terminate this process

The parallelism in this algorithm is at the highest level; each process is an embellished local optimization algorithm. The communication requirements between processes, namely initiating and terminating a small number of processes and broadcasting and receiving values of the single shared variable *lowest-anywhere-so-far*, are very small in comparison to the computational requirements of each process, since each evaluation of *Function* typically takes many milliseconds or seconds. The "shared variable" actually can be maintained by each process and updated asynchronously. Thus the entire algorithm can be implemented as an asynchronous concurrent algorithm on a loosely or tightly coupled multiprocessor or on a local area network of computers.

The parallel global optimization algorithm contains several features which represent significant departures from existing sequential algorithms, and will require substantial research. While sequential algorithms such as Boender et al [1982] include clustering, the clustering at step 2 is different in that it tries to locate convex regions, and not just proximity as do most existing clustering algorithms. The partitioning required at the end of step 2 may be accomplished by formulating the partitioning problem as a linear program or by various heuristic approaches such as the perceptron algorithm; these approaches must be compared and new ones may be investigated. Of course the partitioning problem may be infeasible, in which case the clusters must be modified. Perhaps the most interesting aspect of the global optimization algorithm is the lower bound required at step 7. This step is related in its objective to the termination step of existing sequential global optimization algorithms, but better procedures should be possible because the objective function should often be convex inside the subregion. One possible way to calculate a lower bound is to calculate a convex function that interpolates or underestimates all the points sampled in the subregion so far, and take the minimum of this underestimating function as the lower bound.

We have not yet implemented the above parallel global optimization algorithm, so we can only speculate on its efficiency. It would appear that on problems where a sequential algorithm must identify even a handful of local minimizers before declaring that it has found the global minimizer, the parallel algorithm has a good chance of doing this work concurrently and hence achieving a speedup close to the number of processors. In fact, even greater speedup over existing sequential algorithms may be possible from the early termination of subregions, if this allows the parallel algorithm to skip finding local minimizers that the sequential algorithm would find. Only experimentation will show if these hopes are realized. Experimentation is also required to determine whether the communication requirements are, in fact, insignificant in relation to the total amount of computation.

The feasibility of nonlinear constraints and nonlinear mini-max problems also lend themselves nicely to concurrent solution techniques based on partitioning. First, both problems are themselves global optimization problems, so it should be useful to partition the feasible region into subregions as discussed above. In addition, the constraints may be partitioned into subsets. Combining these two types of partitioning leads to interesting algorithmic possibilities. Consider for example the feasibility of constraints problem with two constraints. Suppose we divide the region D into two subregions $D1$ and $D2$, and then determine whether each constraint is feasible in each subregion. If, for example, c_1 is infeasible in $D1$ and c_2 is infeasible in $D2$, then the entire problem is infeasible. If both constraints are feasible only in $D1$, then the algorithm can restrict its attention to this subregion. These types of approaches again seem to offer the possibility of effective use of multiple processors with small interprocessor communication requirements, but experimentation is required to determine whether this goal is achieved.

Our experience in implementing and testing concurrent optimization algorithms so far is very limited, because the Berkeley Unix 4.2 operating system which we require to run concurrent algorithms on a network of computers has only been available for a few months. So far we have shown the feasibility of using our local area network for concurrent computing by implementing a finite difference gradient algorithm of the type discussed in Section 3, and a version of the chaotic relaxation algorithm of Baudet [1978] for solving linear systems. We plan next to develop and test a chaotic relaxation method for solving systems of nonlinear equations. This is another natural parallel optimization method, but it may require a higher rate of inter-process communication than the algorithms discussed above, and hence be better suited to multiprocessors than to networks of computers.

7. Concluding remarks

The development of parallel algorithms for nonlinear optimization clearly is in its infancy. Hopefully, the previous three sections have indicated that parallel optimization is a fruitful area for future research, especially as multiprocessors and local area networks of computers become available. We have already summarized, in Section 4, the opportunities we see for parallelism in optimization, grouped according to the main factors that make optimization problems expensive. We conclude by focusing on a few of the main types of optimization problems that seem most conducive to parallelization. Our discussion is related to that in Dixon, Patel, and Ducksbury [1983] and in many other papers by the Hatfield group.

We have already stated that the global optimization problem may be the most obvious candidate for significant gains from parallelism in optimization. The stochastic methods discussed in Sections 5 and 6 are just two of a myriad of possibilities for parallel global optimization. Many sequential approaches to solving global optimization problems exist, including various types of stochastic algorithms (see e.g. Rinnooy Kan and Timmer [1983] for further references), deterministic methods such as the tunneling algorithm of Levy et al [1981], and a large class of methods for constrained global optimization problems with concave objective functions (see e.g. Rosen [1983]). All of these approaches seem to suggest excellent possibilities for parallel algorithms suitable to multiprocessors, networks of computers, and in some cases, processor arrays.

While numerical optimization problems are not as inherently large scale as, say, the solution of many differential equations problems, there do exist a number of important optimization problems that are inherently large. These include semi-infinite programming problems, optimal control problems, and network optimization problems. Many of the approaches to solving these problems, including conjugate gradient methods, relaxation methods, branch and bound methods, and the partially separable methods introduced recently by Griewank and Toint [1982], appear well suited to parallelization. Vector computers may be advantageous in some of these cases.

For local (as opposed to global) optimization problems where function evaluation is expensive, but the optimization problem itself is not too difficult to solve, the best approach may often be to parallelize the function evaluation itself. Barring this, the most obvious utilization of parallelism is the concurrent calculation of components of finite difference derivatives. But there remain many other interesting possibilities, mostly unexplored, for the utilization of concurrent function evaluations.

Real-time optimization problems occur in a variety of applications including on-line process control. A far wider range of possibilities may be admissible for solving real-time problems than for standard off-line problems, since sub-optimal use of computing resources may be far more tolerable if it leads to satisfaction of a time bound. The solution of real-time optimization problems by parallel algorithms is likely to be application dependent, however.

The above list is by no means exhaustive. As in any new research area, the best problems may not yet have been identified, and some of the best solutions almost certainly have not yet been developed. In fact, the above list falls somewhat into the trap of suggesting parallel versions of sequential algorithms. We close by reiterating the need to also look at fundamentally new algorithms specifically designed for parallel computers.

8. References

L. Adams [1983], "An M-step preconditioned conjugate gradient method for parallel computation", *Proceedings of the 1983 International Conference on Parallel Processing*, pp. 36-43.

L. Adams and J. Ortega [1982], "A multi-color SOR method for parallel computation", *Proceedings of the 1982 International Conference on Parallel Processing*, pp. 53-56.

G. Baudet [1978], "Asynchronous iterative methods for multiprocessors", *Journal of the Association for Computing Machinery* 25, pp. 226-244.

C. G. E. Boender, A. H. G. Rinnooy Kan, L. Stougie, and G. T. Timmer [1982], "A stochastic method for global optimization", *Mathematical Programming* 22, pp. 125-140.

J. E. Dennis Jr. and R. B. Schnabel [1983], *Numerical Methods for Nonlinear Equations and Unconstrained Optimization*, Prentice-Hall, Englewood Cliffs, New Jersey.

L. C. W. Dixon [1981], "The place of parallel computation in numerical optimization I, the local problem", Technical Report No. 118, Numerical Optimisation Centre, The Hatfield Polytechnic.

L. C. W. Dixon and K. D. Patel [1981], "The place of parallel computation in numerical optimization II, the multiextremal global optimisation problem", Technical Report No. 119, Numerical Optimisation Centre, The Hatfield Polytechnic.

L. C. W. Dixon and K. D. Patel [1982], "The place of parallel computation in numerical optimization IV, parallel algorithms for nonlinear optimisation", Technical Report No. 125, Numerical Optimisation Centre, The Hatfield Polytechnic.

L. C. W. Dixon, K. D. Patel, and P. G. Ducksbury [1983], "Experience running optimisation algorithms on parallel processing systems", Technical Report No. 138, Numerical Optimisation Centre, The Hatfield Polytechnic.

L. C. W. Dixon, P. G. Ducksbury, and P. Singh [1982], "A parallel version of the conjugate gradient algorithm for finite element problems", Technical Report No. 132, Numerical Optimisation Centre, The Hatfield Polytechnic.

J. J. Dongarra and R. E. Hiromoto [1983], "A collection of parallel linear equations routines for the Denelcor HEP", Technical Report ANL/MCS-TM-15, Mathematics and Computer Science Divsion, Argonne National Laboratory.

P. G. Ducksbury [1982], "The implementation of a parallel version of Price's (CRS) algorithm on an ICL DAP", Technical Report No. 127, Numerical Optimisation Centre, The Hatfield Polytechnic.

I. S. Duff [1983], "The solution of sparse linear systems on the Cray-1", Technical Report CSS 125 (revised), Computer Science and Systems Divsion, AERE Harwell.

B. Feijoo and R. R. Meyer [1984], "Piecewise-linear approximation methods for non-separable convex optimization", Technical Report No. 521, Computer Sciences Department, University of Wisconsin - Madison (to appear).

M. J. Flynn [1966], "Very high-speed computing systems", *Proceedings of the IEEE* 54, pp. 1901-1909.

K. W. Fong and T. L. Jordan [1977], "Some linear algebraic algorithms and their performance on the Cray-1", Report LA-6774, Los Alamos National Laboratory.

P. E. Gill, W. Murray, and M. H. Wright [1981], *Practical Optimization*, Academic Press, London.

A. O Griewank and Ph. L. Toint [1982], "On the unconstrained optimization of partially separable functions", in *Nonlinear Optimization 1981*, M. J. D. Powell ed., Academic Press, London, pp. 301-312.

D. Heller [1978], "A survey of parallel algorithms in numerical linear algebra", *SIAM Review* 20, pp. 740-777. pp. 409-436.

R. W. Hockney and C. R. Jesshope [1981], *Parallel Computers*, Adam-Hilger Ltd., Bristol, England.

E. C. Housos and O. Wing [1980], "Parallel nonlinear minimization by conjugate directions", *Proceedings of the 1980 International Conference on Parallel Processing*, pp. 157-158.

K. Hwang and F. A. Briggs [1984], *Computer Architecture and Parallel Processing*, McGraw-Hill, New York.

T. L. Jordan [1979], "A performance evaluation of linear algebra software in parallel architectures", in *Performance Evaluation of Numerical Software*, L. D. Fosdick, ed., North-Holland, Amsterdam, pp. 59-76.

R. Kapur and J. Browne [1981], "Block tridiagonal system solution on reconfigurable array computers", *Proceedings of the 1981 International Conference on Parallel Processing*, pp. 92-99.

J. Kowalik and S. P. Kumar [1982], "An efficient parallel block conjugate gradient method for linear equations", *Proceedings of the 1982 International Conference on Parallel Processing*, pp. 47-52.

H. T. Kung [976], "Synchronized and asynchronous parallel algorithms for multiprocessors", in *Algorithms and Complexity : Recent Results and New Directions*, J. E. Traub, ed., Addison-Wesley, pp. 153-200.

B. W. Lampson, M. Paul, and H. J. Siegert [1981], eds., *Distributed Systems - Architecture and Implementation*, Springer-Verlag, Berlin.

A. V. Levy, A. Montalvo, S. Gomez, and A. Calderon [1981], "Topics in global optimization", in *Proceedings of the Third IIMAS Workshop*, Cocoyoc, Mexico, January 1981, J. P. Hennart, ed.

R. Lord, J. Kowalik, and S. Kumar [1980], "Solving linear algebraic equations on a MIMD computer", *Proceedings of the 1980 International Conference on Parallel Processing*, pp. 205-210.

J. J. McKeown [1979], "Experiments in implementing a nonlinear least-squares algorithm on a dual-processor computer", Technical Report No. 102, Numerical Optimisation Centre, The Hatfield Polytechnic.

J. Mohan [1982], "A study in parallel computation -- the traveling salesman problem", Technical Report CMU-CS-82-136, Department of Computer Science, Carnegie-Mellon University.

K. D. Patel [1982b], "Implementation of a parallel (SIMD) modified Newton method on the ICL DAP", Technical Report No. 131, Numerical Optimisation Centre, The Hatfield Polytechnic.

K. D. Patel [1982a], "Parallel Computation and Numerical Optimisation", Technical Report No. 129, Numerical Optimisation Centre, The Hatfield Polytechnic.

W. L. Price [1981], "A new version of the controlled random search procedure for global optimisation", Technical Report, Engineering Department, University of Leicester, England.

A. H. G. Rinnooy Kan and G. T. Timmer [1983], "Stochastic methods for global optimization", Report 8317/0, Econometric Institute, Erasmus University, Rotterdam.

G. Rodrigue [1982], ed., *Parallel Computations*, Academic Press, New York.

J. B. Rosen [1983], "Global minimization of a linearly constrained concave function by partition of feasible domain", *Mathematics of Operations Research* 8, pp.

A. Sameh [1977], "Numerical parallel algorithms - a survey", in *High Speed Computer and Algorithm Organization*, D. Kuck, D. Lawrie, and A. Sameh, eds., Academic Press, pp. 207-228.

L. J. Siegel [1983], "Characteristics of Parallel Algorithms", presented at the Taxonomy of Parallel Algorithms Workshop, Santa Fe, New Mexico, December 1983.

T. A. Straeter [1973], "A parallel variable metric optimization algorithm," NASA Technical Note D-7329, Langley Reseach Center, Hampton, Virginia.

T. A. Straeter and A. T. Markos [1975], "A parallel Jacobson-Oksman optimization algorithm," NASA Technical Note D-8020, Langley Reseach Center, Hampton, Virginia.

H. van der Vorst [1982], "A vectorizable variant of some ICCG methods", *SIAM Journal on Scientific and Statistical Computing* 3, pp. 350-356.

P. van Laarhoven [1984], "Parallel algorithms for unconstrained optimization," *Mathematical Programming*, to appear.

SOFTWARE FOR MATHEMATICAL PROGRAMMING

K. Schittkowski

Institut für Informatik
Universität Stuttgart
Azenbergstr. 12
D-7000 Stuttgart 1, Germany F.R.

The NATO Advanced Study Institute on Computational Mathematical Programming was accomplished by several activities of a Software Fair. The most important one was the display of information material on existing optimization software. More than 50 code descriptions were submitted by the authors to describe characteristic features of their programs. In all cases, the codes are well tested and documented, are currently used for solving practical optimization optimization problems, and are available on request either in form of a listing or on a magnetic tape. In most cases, the programming language is FORTRAN.

The purpose of this paper is to publish the code descriptions, and thus to present some information on available mathematical programming software. By no means it is claimed that a complete survey will be given. But the underlying mathematical problems cover a wide range in optimization and the given information might support a first decision.

Each code is described in a fixed format, pointing out the specific mathematical problem to be solved, some organizational details, and the conditions of availability. More precisely, the following information will be given:

1. Name of the code.
2. Name of the author or firm.
3. Mathematical problem.
4. Domain of applications.

NATO ASI Series, Vol. F15
Computational Mathematical Programming
Edited by K. Schittkowski

5. Mathematical method.
6. Programming language.
7. Computer systems where the code was implemented.
8. Special program features.
9. Practical application problems solved by the code.
10. Conditions of availability.
11. Charge.
12. Address from where the code can be ordered.
13. References.

To facilitate the search for a special problem solver, some condensed information will be summarized in form of a table. It contains the names of the codes, the names of the authors or distributing firms, the mathematical problem types and some specific features which could restrict the application area. The following abbreviations are used to classify the mathematical programming problem:

LP - Linear programming.
IP - Integer (linear) programming.
MG - Matrix generator for linear programming.
SP - Stochastic programming
NO - Network optimization.
NLP - General constrained nonlinear programming.
LSNLP - Large scale nonlinear programming.
NLS - Nonlinear least squares programming.
UNLP - Unconstrained nonlinear programming.
GP - Geometric programming.
QP - Quadratic programming.
NSO - Non-smooth optimization.
OC - Optimal control.

Subsequently, the reader will find the code descriptions submitted by the program authors. Only some editorial changes are made to stay within the predetermined format. The descriptions are arranged in the same order as used in the table.

Code	Author(s)	Problem	Special Features
ILLIP-2	M.-H. Young e.al.	IP	0-1 variables only
MINI LP	SCICON Ltd.	LP	
MGG	SCICON Ltd.	MG	
SCICONIC	SCICON Ltd.	LP/IP	
WHIZARD	KETRON Inc.	LP	also network models
MSPIII	KETRON Inc.	LP	
MIPIII	KETRON Inc.	LP/IP	
VARIFORM GUB	KETRON Inc.	LP	
DATAFORM	KETRON Inc.	MG	
MAGIC	R. Day	MG	
ZOOM/XMP	R. Marsten	LP/IP	0-1 integer variables
XMP	R. Marsten	LP	
NDSP	J. Birge	SP	
STO	A. Gaivoronski	SP	
LFGM	A. King	SP	
STOCH	A. Boehme, K. Marti	SP	
SPORT	L. Nazareth	SP	
FIM	Q. Liqun	SP	
STOSUB	A. Ruszczynski	SP	
PCSP	T. Szántai	SP	
STOCHNET	S.W. Wallace	SP	
GRAPHLIB	S. Pappe, T. Rubach	NO	various graph problems
NETLIB	S. Pappe, T. Rubach	NO	
NLPNET	R.S. Dembo	NO/LSNLP	nonlinear objective function, linear network constraints
PSPMIN	P. Toint	LSNLP	bound constraints only
MINOS	B. Murtagh, M.A. Saunders	LSNLP	
OPT2	S.H. Wang, K.M. Raysdell	LSNLP	
SEQ	A. Dourado, J.L. Calvet	LSNLP	serially connected systems
NLPQL	K. Schittkowski	NLP	
NLPSYS	K. Schittkowski	NLP	model building

Table: List of code descriptions (continued).

Code	Author(s)	Problem	Special Features
WIEM	R.T. Wijmenga, G.van der Hoek	NLP	
TWOFAS	A.S. Louter, G. van der Hoek	NLP	
NPSOL	P.E. Gill e.al.	NLP	
ALRQP	M.C. Bartholomew-Biggs	NLP	
IDESIGN	J.S. Arora	NLP	
GRG2	L. Lasdon, A. Waren	NLP	
OPT	G.A. Gabriele, K.M. Ragsdell	NLP	
BIAS	R.R. Root, K.M. Ragsdell	NLP	
TRUOPT	S.H. Wang, K.M. Ragsdell	NLP	structural mechanical optimization
OPTLIB	G.A. Gabriele, K.M. Ragsdell	NLP	education system
SYSOPT	M. Rijckaert, E. Walraven	NLP	education system
EA3	M. Kupferschmid, J.E. Ecker	NLP	inequality constraints only
UNT	A. Zilinskas	NLP	global optimization
ENLSIP/ ELSUNC	P. Lindström, P.-A. Wedin	NLS	
MULTEX/ISAAC	H.G. Bock. J. Schlöder	NLS	
PQ1	R. Mifflin	NLP	one-dimensional problems
M1FC1	C. Lemarechal	UNLP	
UNCMIN	R.B. Schnabel e.al.	UNLP	
QNBD	J.F. Bonnans	UNLP	bound constraints
GQOPT3	R.E. Quandt e.al.	UNLP	
GPKTC	M. Rijckaert, E. Walraven	GP	
QPSOL	P.E. Gill e.al.	QP	
ZQPCVX	M.J.D. Powell	QP	convex objective function
DUAL QP	A. Idnani	QP	convex objective function
CONQUA,START	W.J. Kribbe	QP	

Table: List of code descriptions (continued).

Code	Author(s)	Problem	Special Features
L1NONL	J. Hald	NSO	L1-function, linear constraints
MMLA1Q	J. Hald	NSO	min-max function, linear constraints
TOMP	D. Kraft	OC	
PARFIT	H.G. Bock	OC	parameter identification
MUSCOD	H.G. Bock, K.J. Plitt	OC	
OPCON	H.G. Bock		dynamic processes
FAUST	H.G. Bock, P. Krämer-Eis	OC	feedback control

Table: List of code descriptions

NAME OF THE CODE: ILLIP-2
AUTHOR(S): M.-H. Young, T.-K. Liu, C.R. Baugh, and S. Muroga
MATHEMATICAL PROBLEM: 0-1 variable linear programming
DOMAIN OF APPLICATIONS: General linear programming problems where variables assume values, 0 or 1.
MATHEMATICAL METHOD: Implicit enumeration method with "pseudo-underlining" and many other speedup techniques
PROGRAMMING LANGUAGE: FORTRAN IV and PL/1
COMPUTER SYSTEMS WHERE THE CODE HAS BEEN IMPLEMENTED: IBM 360/75I
SPECIAL PROGRAM FEATURES: This program is designed for large-scale 0-1 variable linear programming problems. The size is currently limited to 399 variables and 999 inqualities. The program is faster by an order of magnitude for many problems than commercial integer-programming packages which are not specialized to 0-1 variable linear programming.
PRACTICAL APPLICATION PROBLEMS SOLVED BY THE CODE: Factory production planning, Design problem of logic networks, Construction scheduling, General covering problem.
CONDITIONS OF AVAILABILITY: This belongs to public domain
CHARGE: Small charge to cover material and handling.
ADDRESS FROM WHERE THE CODE CAN BE ORDERED: Prof. Saburo MUROGA, Dept. of Computer Science, University of Illinois, 222 Digital Computer Laboratory, 1304 West Springfield Avenue, Urbana, Illinois 61801
REFERENCES: T. Ibaraki, T.-K. Liu, C.R. Baugh, and S. Muroga, "An implicit enumeration program for zero-one programming," International Journal of Computer and Information Sciences, Vol. 1, No. 1, March 1972, pp. 75-92. M.-H. Young, T.-K. Liu, C.R. Baugh, and S. Muroga, "A code for zero-one integer programming ILLIP-2 (a programming manual for ILLIP-2)," Dept. of Computer Science, Univ. of Ill. at Urbana-Champaign, UIUCDCS-R-77-858, April 1977, 175 pp. (Available from NTIS, U.S. Department of Commerce.)

NAME OF THE CODE: MINI LP

AUTHOR(S): Scicon Ltd.

MATHEMATICAL PROBLEM: Version 1.12. MINI LP is designed for the in-core solution of linear programming problems on minicomputers

DOMAIN OF APPLICATIONS:

MATHEMATICAL METHOD:

PROGRAMMING LANGUAGE: Fortran

COMPUTER SYSTEMS WHERE THE CODE HAS BEEN IMPLEMENTED:
DEC PDP11, RSX-11M
DEC VAX, VMS
Data General Nova, RDOS
Data General Eclipe, AOS
Data General MV8000, AOS/VS
Norsk Data ND-100 series, SINTRAN III
Wang VS-80
Prime, Primos

SPECIAL PROGRAM FEATURES: MPS format matrix input
Basis input/output
Right-hand side and cost ranging
Output files for access by report writers

PRACTICAL APPLICATION PROBLEMS SOLVED BY THE CODE:

CONDITIONS OF AVAILABILITY:

CHARGE: Perpetual Licence Fee (PLF) is £5,000
Annual Maintenance is 12% p.a. of PLF
Annual Lease (including maintenance) is 40% p.a. of PLF
Software purchased by PLF carries 90 day warranty

ADDRESS FROM WHERE THE CODE CAN BE ORDERED: Dieter Giesen, SCS GmbH, Oehleckerring 40, 2000 Hamburg 62, West Germany.
Pieter Klapwijk, ORES, Konninginneweg 83, 1075 CJ Amsterdam, Netherlands.
Paul Bryan-Tatham, Scicon Limited, Brick Close, Kiln Farm, Milton Keynes, MK11 3EJ. United Kingdom.

REFERENCES:

NAME OF THE CODE: MGG

AUTHOR(S): Scicon Ltd.

MATHEMATICAL PROBLEM: Matrix generator generator

DOMAIN OF APPLICATIONS:

MATHEMATICAL METHOD: Version 1.04. MGG is designed to accept the mathematical programming formulation of a linear programming problem as input and to produce a Fortran matrix generator program as output. MGG imposes a logical structure on the formulation and this, together with the consistency checks carried out by MGG, means that MGG not only elminates programming errors but can also prevent errors in the problem formulation. In addition to the matrix generator, MGG also produces code to read the LP solution file and to set the results up for report writing. The generated matrix generator produces the matrix in standard MPS format. Versions can be supplied to handle solution files from packages such as SCICONIC/VM and MINI LP.

PROGRAMMING LANGUAGE: Fortran

COMPUTER SYSTEMS WHERE THE CODE HAS BEEN IMPLEMENTED:
DEC PDP11 series, RSX-11M
DEC VAX series, VMS
Systime, VMS
Data General MV8000, AOS/VS
Norsk Data ND-500 series, SINTRAN III
ICL 2900 series, VME
Prime series, Primos
IBM 370 series, DOS/VSE
IBM 3030, 3080, 4300 series, MVS/CMS

SPECIAL PROGRAM FEATURES: Batch mode is possible, but no interactive mode.

PRACTICAL APPLICATION PROBLEMS SOLVED BY THE CODE:

CONDITIONS OF AVAILABILITY:

CHARGE: Perpetual Licence Fee (PLF) is £6,000
Annual Maintenance is 12% p.a. of PLF
Annual Lease (including maintenance) is 40% p.a. of PLF
Software purchased by PLF carries 90 day warranty

ADDRESS FROM WHERE THE CODE CAN BE ORDERED: Dieter Giesen, SCS GmbH, Oehleckerring 40, 2000 Hamburg 62, West Germany.
Pieter Klapwijk, ORES, Konninginneweg 83, 1075 CJ Amsterdam, Netherlands.
Paul Bryan-Tatham, Scicon Limited, Brick Close, Kiln Farm, Milton Keynes, MK11 3EJ. United Kingdom.

REFERENCES:

NAME OF THE CODE: SCICONIC
AUTHOR(S): Scicon Ltd.
MATHEMATICAL PROBLEM: Mathematical programming
DOMAIN OF APPLICATIONS:
MATHEMATICAL METHOD: Version 3. This is an interactive mathematical programming system. SCICONIC provides efficient linear problem solving techniques and handles binary, integer, semi-continuous variables and special ordered sets.
PROGRAMMING LANGUAGE: Fortran
COMPUTER SYSTEMS WHERE THE CODE HAS BEEN IMPLEMENTED: IBM 370 series, DOS/VSE IBM 3030, 3080, 4300 series, MVS/CMS Univac 1100 series, OS/1100 DEC VAX series, VMS Systime, VMS Prime 50 series, Primos Data General MV series, AOS/VS Norsk Data ND-500 series, SINTRAN III ICL 2900 series, VME CDC 180 series, NOS/VE
SPECIAL PROGRAM FEATURES: SCICONIC conforms to the MPS Industry Standard format for Linear Programming.
PRACTICAL APPLICATION PROBLEMS SOLVED BY THE CODE: SCICONIC is used across all industry, commerce and administration in both problem solving and "what-if" simulation exercises. Typical applications are production planning, transport and distribution, mixing and blending, tender evaluation, manpower planning, facilities location, investment planning, inventory control and scheduling.
CONDITIONS OF AVAILABILITY:
CHARGE: Perpetual Licence Free (PLF) is £15,000 Annual Maintenance is 12% p.a. of PLF Annual Lease (including maintenance) is 40% p.a. of PLF Software purchased by PLF carries 90 day warranty Supplied by request
ADDRESS FROM WHERE THE CODE CAN BE ORDERED: Dieter Giesen, SCS GmbH, Oehleckerring 40, 2000 Hamburg 62, West Germany. Pieter Klapwijk, ORES, Konninginneweg 83, 1075 CJ Amsterdam, Netherlands. Paul Bryan-Tatham, Scicon Limited, Brick Close, Kiln Farm, Milton Keynes, MK11 3EJ, United Kingdom.
REFERENCES:

NAME OF THE CODE: WHIZARD	

NAME OF THE CODE: WHIZARD

AUTHOR(S): KETRON, INC.

MATHEMATICAL PROBLEM: Linear Programming

DOMAIN OF APPLICATIONS: General

MATHEMATICAL METHOD: Revised simplex - primal, dual network

PROGRAMMING LANGUAGE: Assembly

COMPUTER SYSTEMS WHERE THE CODE HAS BEEN IMPLEMENTED:
IBM Mainframe (360, 370, 40xx, 30xx or compatible) MVT, MVS, VW/CMS

SPECAIL PROGRAM FEATURES: WHIZARD is an in-core mathematical programming optimizer. It has the reputation for being the fastest, most reliable optimizer available for IBM mainframe processors. WHIZARD repacks the LP matrix data into a minimum amount of memory in order to get it all in core. The benefits are a virtual elimination of programmed I/O and enhancement of solution algorithms by taking advantage of core residence of the matrix. Three different algorithms are included: primal, dual, and pure network. WHIZARD's speed is enhanced by presolve and postsolve facilities that automatically reduce the model prior to optimization and mathematically extend the reduced basis after optimization. The ability to conveniently process large (up to 16,383 constraints) and complex models is aided by automatic adjustment of tolerances and selection of algorithm elements to math model requirements. WHIZARD runs in either the MPSIII and MPSX/370 basic system enviroment.

PRACTICAL APPLICATION PROBLEMS SOLVED BY THE CODE: Full range of LP and network models. Routinely used to solve models up to 16,000 constraints, 100,000 variables.

CONDITIONS OF AVAILABILITY: Mounthly or paid-up license agreement

CHARGE: $550/month, requires MPSIII or MPSX/370 Basic System

ADDRESS FROM WHERE THE CODE CAN BE ORDERED: Ketron, Inc. 1700 N. Moore Street Arlington, VA 22209 USA, Telephone (703) 558-8701 TWX 710 955 0219

REFERENCES:

NAME OF THE CODE: MPSIII	

AUTHOR(S): KETRON, INC.

MATHEMATICAL PROBLEM: Linear programming

DOMAIN OF APPLICATIONS: General

MATHEMATICAL METHOD: Revised simplex - primal and dual

PROGRAMMING LANGUAGE: Assembly

COMPUTER SYSTEMS WHERE THE CODE HAS BEEN IMPLEMENTED:
IBM Mainframe (360, 370, 40xx, 30xx or compatible) MVT, MSV, VM/CMS

SPECIAL PROGRAM FEATURES: MPSIII is the most comprehensive mathematical programming system available. The MPSIII Basic System comprises all features expected to be in a linear programming system, plus many more: The Primal and Dual optimizers employ, as appropriate, variable multiple pricing, sectional pricing, semi-flags, and composite rhs and obj. to solve models with up to 16,383 constraints. The P^4 matrix inversion technique has set the standard for speed and accuracy. Standard MPS format is used for model input and revision; optional DATAFORM input can be used. Solution results and postoptimal and parametric analysis is output to printer or to machine readable files. Basis preservation and restoration is handled in several different ways to meet varying application requirements. Vector subsetting is done via standard selection lists or unique bit map facility. Both the batch (MVS) version and the interactive (CMS) version provide host environments for the optional subsystems MIPIII, GUB, WHIZARD, and DATAFORM.

PRACTICAL APPLICATION PROBLEMS SOLVED BY THE CODE:
Full range of LP applications; in commercial use for over 12 years.

CONDITIONS OF AVAILABILITY: Monthly or paid-up license agreement

CHARGE: $300/month (batch), $500/month (conversational).

ADDRESS FORM WHERE THE CODE CAN BE ORDERED: Ketron, Inc., 1700 N. Moore Street Arlington, VA 22209 USA, Telephone (703) 558-8701 TWX 710 955 0219

REFERENCES:

NAME OF THE CODE: MIPIII
AUTHOR(S): KETRON, INC.
MATHEMATICAL PROBLEM: Mixed integer and separable programming
DOMAIN OF APPLICATIONS: General
MATHEMATICAL METHOD: Branch and bound (Uses WHIZARD for submodels)
PROGRAMMING LANGUAGE: DATAFORM
COMPUTER SYSTEMS WHERE THE CODE HAS BEEN IMPLEMENTED: IBM Mainframe (360, 370, 40xx, 30xx or compatible) MVT, MVS, VM/CMS
SPECIAL PROGRAM FEATURES: MIPIII is the optional mixed integer programming (MIP) subsystem of MPSIII. It uses the branch and bound integer node tree search procedure on models with as many as six different kinds of discrete variables: zero/one variables model yes/no decisions; bivalent variables may be zero or some discrete (non-one) value; a semicontinous variable may be either zero or have a value in a given range; general integer variables may assume only integer values; special ordered sets of degree one model multiple choice; and special ordered sets of degree two model non linear functions in the fashion of lambda separable programming. Execution control for MIPIII runs is via simple control tables instead of the traditional control program (the control program is available for complex applications). The analyst's knowledge of the model can be conveyed to MIPIII via priority tables and forced-first-solution tables. The LP subproblems are optimized by WHIZARD, thus making available primal, dual, or network algorithms as appropriate.
PRACTICAL APPLICATION PROBLEMS SOLVED BY THE CODE: Successfully used on large LP models with hundreds of integer decision variables.
CONDITIONS OF AVAILABILITY: Monthly or paid-up license agreement
CHARGE: $550/month, requires MPSIII Basic System
ADDRESS FROM WHERE THE CODE CAN BE ORDERED: Ketron, Inc., 1700 N. Moore Street, Arlington, VA 22209 USA, Telephone (703) 558-8701 TWX 710 955 0219
REFERENCES:

NAME OF THE CODE: VARIFORM GUB	

NAME OF THE CODE: VARIFORM GUB

AUTHOR(S): KETRON, INC.

MATHEMATICAL PROBLEM: LP and Lambda Separable Programming

DOMAIN OF APPLICATIONS: General

MATHEMATICAL METHOD: Revised simplex

PROGRAMMING LANGUAGE: Assembly

COMPUTER SYSTEMS WHERE THE CODE HAS BEEN IMPLEMENTED:
IBM Mainframe (360, 370, 40xx, 30xx or compatible) MVT, MVS, VM/CMS

SPECIAL PROGRAM FEATURES: The Generalized Upper Bounding of MPSIII provides the ability to optimize certain mathematical programming models that are far too large for standard optimizers. A model is a candidate for GUB processing if a significant percentage of its constraints can be expressed as only contiguous ones with a non-zero rhs. If so, the relevant row is removed from the matrix and replaced by a vector group marker. Much of the resultant speed improvement is derived from the smaller matrix row size. Models of any size with a large GUB component (90 % GUB) benefit from GUB processing; more important is the ability to solve models with 20,000 to 50,000 constraints. GUB honors composite rhs and obj and handles four group types: equality, less than or equal, balance (positive and negative variables sum to zero), and function (equivalent to lambda separable sets). GUB is an optional component of MPSIII.

PRACTICAL APPLICATION PROBLEMS SOLVED BY THE CODE: Used on wide range of problems. Largest recorded has 50,000 constraints (mostly GUB groups) and 250,000 variables. Routinely used to solve model with 15,000 constraints and 200,000 variables.

CONDITIONS OF AVAILABILITY: Montly or paid-up license agreement

CHARGE: $1,000/month, requires MPSIII Basic System

ADDRESS FROM WHERE THE CODE CAN BE ORDERED: Ketron, Inc., 1700 N. Moore Street, Arlington, VA 22209 USA, Telephone (703 558-8701 TWX 710 955 0219

REFERENCES:

NAME OF THE CODE: DATAFORM
AUTHOR(S): KETRON, INC.
MATHEMATICAL PROBLEM: Programming language for model and data management.
DOMAIN OF APPLICATION: Mathematical programming modeling support
MATHEMATICAL METHOD: Successive linear programming
PROGRAMMING LANGUAGE: FORTRAN, Assembly
COMPUTER SYSTEMS WHERE THE CODE HAS BEEN IMPLEMENTED: IBM Mainfram (360, 370, 40xx, 30xx or compatible) MVT, MVS, VM/CMS
SPECIAL PROGRAM FEATURES: DATAFORM is the mathematical programming model management subsystem of MPSIII which comprises a model data-base manager and a data manipulation language specifically oriented to MP modeling. The model database contains all model oriented data: data tables used for model generation and report generation; the LP model matrix; the optimal solution results and tableau; and saved bases. Language features include matrix generation and revision facilities which are easily learned and extremely powerful. DATAFORM applications are data driven and react to changes in table sizes and content. The report generation facilities include easy access to solution values and automatic wide paging and pagination control. Much of the power of DATAFORM is due to its integration with MPSIII; this provides full communication with other MPSIII procedures and is an excellent environment for successive LP. DATAFORM extensions include a host language interface; alien language subprograms; an interactive, full-screen interface; and a graphics generator.
PRACTICAL APPLICATION PROBLEMS SOLVED BY THE CODE: Matrix generation, report generation, model management for wide range of mathematical programming applications; used to implement MIPIII and other successive LP applications.
CONDITIONS OF AVAILABILITY: Monthly or paid-up license agreement
CHARGE: $950/month, requires MPSIII Basic System
ADDRESS FROM WHERE THE CODE CAN BE ORDERED: Ketron, Inc., 1700 N. Moore Street, Arlington, VA 22209 USA, Telephone (703) 558-8701, TWX 710 955 0219
REFERENCES:

NAME OF THE CODE: MAGIC
AUTHOR(S): R.Day
MATHEMATICAL PROBLEM: Takes Mathematical Definition of a Mathematical Programming problem (Σ notation, indexing etc.) and produces a file in MPS format
DOMAIN OF APPLICATIONS: Any Linear or Integer Programming Model.
MATHEMATICAL METHOD:
PROGRAMMING LANGUAGE: FORTRAN, FORTRAN 77
COMPUTER SYSTEMS WHERE THE CODE HAS BEEN IMPLEMENTED: ICL 2900 Series, IBM Large Mainframes
SPECIAL PROGRAM FEATURES: Interactive Syntax Checking
PRACTICAL APPLICATION PROBLEMS SOLVED BY THE CODE: All problems in "Model Building in Mathematical Programming" by H.P. Williams Contractor Allocation Problem (508x864), Power Station Scheduling (760x720), Dynamic Leontref Model (786x1170), Multiperiod Investment (946x945).
CONDITIONS OF AVAILABILITY: On Application
CHARGE: £1000
ADDRESS FROM WHERE THE CODE CAN BE ORDERED: EMS, 25 Warrender Park Rd, Edinburgh, U.K.
REFERENCES: MAGIC User's Manual from above address

NAME OF THE CODE: ZOOM/XMP	
AUTHOR(S): Roy Marsten	
MATHEMATICAL PROBLEM: Zero/One Mixed-Integer Programming	
DOMAIN OF APPLICATIONS: General zero/one mixed-integer programs with no special structure. Medium sized problems with up to about 200 zero/one variables.	
MATHEMATICAL METHODS: Branch-and bound with bounds computed by solving linear programs	
PROGRAMMING LANGUAGE: FORTRAN	
COMPUTER SYSTEMS WHERE THE CODE HAS BEEN IMPLEMENTED: IBM, DEC, Prime, Burroughs, Control Data	
SPECIAL PROGRAM FEATURES: 1) Uses the XMP linear programming library; 2) Uses Balas and Martin Pivot & Complement heuristic to find a good initial feasible solution; 3) Novel branching strategy includes Best Bound and LIFO as special cases; 4) A novel "cheating" strategy that deflates the LP bounds by an amount that depends on the depth in the search tree.	
PRACTICAL APPLICATION PROBLEMS SOLVED BY THE CODE: This is a new code, but has been used so far to schedule maintenance for electrical generators, to solve vehicle routing problems, and on investment sequencing problems. Companies using prototype versions of ZOOM/XMP include ALCOA, Energy Management Associates, Scientific Systems Inc., and Multiplications Inc.	
CONDITIONS OF AVAILABILITY: Annual or perpetual lease	
CHARGE: Depends on type of organization and type of application.	
ADDRESS FROM WHERE THE CODE CAN BE ORDERED: Prof. Roy Marsten, Dept. of Management Information Systems, University of Arizona, Tucson, AZ 85721, USA	
REFERENCES:	

NAME OF THE CODE: XMP

AUTHOR(S): Roy Marsten

MATHEMATICAL PROBLEM: Linear Programming

DOMAIN OF APPLICATIONS: General linear programming problems, including large problems (a few thousand constraints)

MATHEMATICAL METHODS: Primal and dual simplex methods with bounded variables, postoptimality analysis, GUB, triangular factorization of the basis matrix

PROGRAMMING LANGUAGE: FORTRAN

COMPUTER SYSTEMS WHERE THE CODE HAS BEEN IMPLEMENTED: IBM, DEC, Univac, Burroughs, Amdahl, Prime, HP, NCR, Hitachi, Perkin-Elmer, Control Data, Harris

SPECIAL PROGRAM FEATURES: 1) portable subroutine library; 2) can be used as a traditional stand-alone LP system; 3) can be used in conjunction with other techniques such as simulation or econometrics; 4) can be used as a "black box" inside some more complex algorithm or decision support system; 5) can be easily modified to work from any user specified data structure; 6) has solved problems with up to 6000 constraints on a VAX 11/780; 7) the SLP/XMP extension solves general nonlinear programs; 8) the ZOOM/XMP extension solves general zero/one mixed-integer programs.

PRACTICAL APPLICATION PROBLEMS SOLVED BY THE CODE: Over 120 installations, including about 20 in Europe. Companies using XMP include ALCOA, AMOCO, Booz, Allen, & Hamilton, Data Resources Inc., Dow Chemical, Flying Tiger, GTE Laboratories, Shell Oil

CONDITIONS OF AVAILABILITY: Annual or perpetual lease

CHARGE: Depends on type of organization and type of application.

ADDRESS FROM WHERE THE CODE CAN BE ORDERED: Prof. Roy Marsten, Dept. of Management Information Systems, University of Arizona, Tucson, AZ 85721, USA

REFERENCES:

NAME OF THE CODE: NDSP	

AUTHOR(S): J. Birge

MATHEMATICAL PROBLEM: Multi-Stage Stochastic Linear Programs

DOMAIN OF APPLICATIONS:

MATHEMATICAL METHOD: Nested Decomposition for Stochastic Programming based on van Slyke & Wets' L-Shaped Method

PROGRAMMING LANGUAGE: FORTRAN IV

COMPUTER SYSTEMS WHERE THE CODE HAS BEEN IMPLEMENTED:
VAX 11/780 (UNIX)

SPECIAL PROGRAM FEATURES:

PRACTICAL APPLICATION PROBLEMS SOLVED BY THE CODE:

CONDITIONS OF AVAILABILITY:
Send tape to address below

CHARGE:

ADDRESS FROM WHERE THE CODE CAN BE ORDERED:
Project Secretary, ADO/SDS, IIASA, A-2361 Laxenburg, Austria

REFERENCES:

R. Van Slyke & R. Wets, "L-SHAPED LINEAR PROGRAMS WITH APPLICATIONS TO OPTIMAL CONTROL AND STOCHASTIC LINEAR PROGRAMS," SIAM JOURNAL ON APPLIED MATHEMATICS, v. 17 pp, 638-663, 1969

J. Birge, "DECOMPOSITION AND PARTITIONING METHODS FOR MULTI-STAGE STOCHASTIC LINEAR PROGRAMS," TECHNICAL REPORT 82-6, DEPT. OF INDUSTRIAL AND OPERATIONS ENGINEERING, UNIVERSITY OF MICHIGAN, 1982

J. Birge, "ANL-SHAPED METHOD COMPUTER CODE FOR MULTI-STAGE STOCHASTIC LINEAR PROGRAMS", IN NUMERICAL METHODS FOR STOCHASTIC OPTIMIZATION, Yu.Ermol'ev & R. Wets (Eds), To appear.

NAME OF THE CODE: STO
AUTHOR(S): Gaivoronski, A.
MATHEMATICAL PROBLEM: Stochastic linear programs
DOMAIN OF APPLICATIONS:
MATHEMATICAL METHOD: Ermol'ev & Gaivoronski's quasigradient method
PROGRAMMING LANGUAGE: FORTRAN 77 (?)
COMPUTER SYSTEMS WHERE THE CODE HAS BEEN IMPLEMENTED: VAX 11/780 (UNIX)
SPECIAL PROGRAM FEATURES:
PRACTICAL APPLICATION PROBLEMS SOLVED BY THE CODE:
CONDITIONS OF AVAILABILITY: Send tape to address below
CHARGE:
ADDRESS FROM WHERE THE CODE CAN BE ORDERED: Project Secretary, ADO/SDS, IIASA, A-2361 Laxenburg, Austria
REFERENCES: Yu. Ermol'ev and A. Gaivoronski, "Stochastic quasigradient methods and their implementations", IIASA Working Paper WP-84-55, IIASA, Laxenburg, 1984

NAME OF THE CODE: LFGM	
AUTHOR(S): A. King	
MATHEMATICAL PROBLEM: Stochastic programs with simple recourse	
DOMAIN OF APPLICATIONS:	
MATHEMATICAL METHOD: Rockafellar & Wets' Lagrangian finite generation method	
PROGRAMMING LANGUAGE: FORTRAN 77	
COMPUTER SYSTEMS WHERE THE CODE HAS BEEN IMPLEMENTED: VAX 11/780 (UNIX)	
SPECIAL PROGRAM FEATURES:	
PRACTICAL APPLICATION PROBLEMS SOLVED BY THE CODE:	
CONDITIONS OF AVAILABILITY: Send tape to address below	
CHARGE:	
ADDRESS FROM WHERE THE CODE CAN BE ORDERED: Project Secretary, ADO/SDS, IIASA, A-2361 Laxenburg, Austria	
REFERENCES: R.T. Rockafellar & R. Wets, "A Lagrangian finite generation technique for solving linear-quadratic problems in stochastic programming," IIASA Working-Paper WP-84-25, IIASA, Laxenburg, 1984 A. King, "An implementation of the LFGM," in Numerical Methods for Stochastic Optimization, Yu. Ermol'ev & R. Wets (Eds.), to appear	

NAME OF THE CODE: STOCH
AUTHOR(S): B. Boehme and K. Marti
MATHEMATICAL PROBLEM: Stochastic linear programs with recourse
DOMAIN OF APPLICATIONS:
MATHEMATICAL METHOD: Marti's descent stochastic quasigradient method
PROGRAMMING LANGUAGE: FORTRAN 77 (?)
COMPUTER SYSTEMS WHERE THE CODE HAS BEEN IMPLEMENTED: VAX 11/780 (UNIX)
SPECIAL PROGRAM FEATURES:
PRACTICAL APPLICATION PROBLEMS SOLVED BY THE CODE:
CONDITIONS OF AVAILABILITY: Send tape to address below
CHARGE:
ADDRESS FROM WHERE THE CODE CAN BE ORDERED: Project Secretary, ADO/SDS, IIASA, A-2361 Laxenburg, Austria
REFERENCES: K. Marti, "Descent stochastic quasigradient methods," HSBw München, 1984

NAME OF THE CODE: SPORT (Version 1.1)	
AUTHOR(S): L. Nazareth	
MATHEMATICAL PROBLEM: Stochastic programs with recourse	
DOMAIN OF APPLICATIONS:	
MATHEMATICAL METHOD: Nazareth and Wets' inner linearization method and a bounded variable method	
PROGRAMMING LANGUAGE: FORTRAN IV	
COMPUTER SYSTEMS WHERE THE CODE HAS BEEN IMPLEMENTED: VAX 11/780 (UNIX)	
SPECIAL PROGRAM FEATURES:	
PRACTICAL APPLICATION PROBLEMS SOLVED BY THE CODE:	
CONDITIONS OF AVAILABILITY: Send tape to address below	
CHARGE:	
ADDRESS FROM WHERE THE CODE CAN BE ORDERED: Project Secretary, ADO/SDS, IIASA, A-2361 Laxenburg, Austria	
REFERENCES: J.L. Nazareth & R. Wets, "Stochastic programs with recourse: Algorithms and Implementations," IIASA Working Paper, Forthcoming. J.L. Nazareth, "Algorithms based upon generalized linear programming for stochastic programs with recourse," IIASA Working Paper WP-84-81, IIASA, Laxenburg, 1984 J.L. Nazareth & R. Wets, "Algorithms for stochastic programs: The case of nonstochastic tenders," IIASA Working Paper WP-83-5, IIASA, Laxenburg 1983	

NAME OF THE CODE: FIM	
AUTHOR(S): Q. Liqun	
MATHEMATICAL PROBLEM: Stochastic transportation problems	
DOMAIN OF APPLICATIONS:	
MATHEMATICAL METHOD: Qi's forest iteration method	
PROGRAMMING LANGUAGE: FORTRAN 77	
COMPUTER SYSTEMS WHERE THE CODE HAS BEEN IMPLEMENTED: VAX 11/780 (UNIX)	
SPECIAL PROGRAM FEATURES:	
PRACTICAL APPLICATION PROBLEMS SOLVED BY THE CODE:	
CONDITIONS OF AVAILABILITY: Send tape to address below	
CHARGE:	
ADDRESS FROM WHERE THE CODE CAN BE ORDERED: Project Secretary, ADO/SDS, IIASA, A-2361 Laxenburg, Austria	
REFERENCES: L.Qi, "Forest iteration method for stochastic transportation problems," to appear in Math. Prog. Study	

NAME OF THE CODE: STOSUB

AUTHOR(S): A. Ruszczynski

MATHEMATICAL PROBLEM: Stochastic linear programs

DOMAIN OF APPLICATIONS:

MATHEMATICAL METHOD: Ruszczynski & Syski's stochastic subgradients

PROGRAMMING LANGUAGE: FORTRAN 77

COMPUTER SYSTEMS WHERE THE CODE HAS BEEN IMPLEMENTED:
VAX 11/780 (UNIX)

SPECIAL PROGRAM FEATURES: STOSUB is a collection of subroutines and requires a driver

PRACTICAL APPLICATION PROBLEMS SOLVED BY THE CODE:

CONDITIONS OF AVAILABILITY: Send tape to address below

CHARGE:

ADDRESS FROM WHERE THE CODE CAN BE ORDERED:
Project Secretary, ADO/SDS, IIASA, A-2361 Laxenburg, Austria

REFERENCES:
A. Ruszcyzynski & W. Syski, "A method of aggregate stochastic subgradients with on-line stepsize rues for convex stochastic programming problems," Report R.I. 21/83, Instytut Automatyki, Politechnika Warszawska, 1983 (to appear in Math. Prog. Study)
A. Ruszczynski, "STOSUB: A package for solving stochastic programming problems - user's manual," Manuscript from the Institut für Operations Research der Universität Zürich, October 1984

NAME OF THE CODE: PCSP	
AUTHOR(S): T. Szántai	
MATHEMATICAL PROBLEM: Probabilistic constrained stochastic programs	
DOMAIN OF APPLICATIONS:	
MATHEMATICAL METHOD: Veinott's supporting hyperplane algorithm	
PROGRAMMING LANGUAGE: FORTRAN 77	
COMPUTER SYSTEMS WHERE THE CODE HAS BEEN IMPLEMENTED: VAX 11/780 (UNIX)	
SPECIAL PROGRAM FEATURES:	
PRACTICAL APPLICATION PROBLEMS SOLVED BY THE CODE:	
CONDITIONS OF AVAILABILITY: Send tape to address below	
CHARGE:	
ADDRESS FROM WHERE THE CODE CAN BE ORDERED: Project Secretary, ADO/SDS, IIASA, A-2361 Laxenburg, Austria	
REFERENCES: A.F. Veinott, "Supporting Hyperplane Method," Operations Research, v. 15 pp. 147-152, 1967	

NAME OF THE CODE: STOCHNET	

AUTHOR(S): S.W. Wallace

MATHEMATICAL PROBLEM: Stochastic programs with pure network recourse

DOMAIN OF APPLICATIONS:

MATHEMATICAL METHOD: Based on Van-Slyke & Wets' L-shaped method (uses Schur complements for pure networks to update the basis without a matrix inversion)

PROGRAMMING LANGUAGE: FORTRAN 77

COMPUTER SYSTEMS WHERE THE CODE HAS BEEN IMPLEMENTED:
VAX 11/780 (UNIX)

SPECIAL PROGRAM FEATURES:

PRACTICAL APPLICATION PROBLEMS SOLVED BY THE CODE:

CONDITIONS OF AVAILABILITY: Send tape to address below

CHARGE:

ADDRESS FROM WHERE THE CODE CAN BE ORDERED:
Project Secretary, ADO/SDS, IIASA, A-2361 Laxenburg, Austria

REFERENCES:
R. van Slyke & R. Wets, "L-shaped linear programs with applications to optimal control and stochastic linear programs," SIAM Journal on Applied Mathematics, v. 17 pp. 638-663, 1969
S.W. Wallace, "On network structured stochastic optimization problems," Report No. 842555-8, Chr. Michelsen Institute, 1984
J. Birge & R. Wets, "Designing Approximation schemes for stochastic optimization problems, in particular for stochastic programs with recourse," IIASA Working Paper WP-83-111, IIASA, Laxenburg, 1983

NAME OF THE CODE: GRAPHLIB

AUTHOR(S): Stephan Pappe, Thomas Rubach

MATHEMATICAL PROBLEM: graph algorithms
- reachability and connectivity
- minimum spanningtrees and shortest paths
- network flows

DOMAIN OF APPLICATIONS: shortest route problems
transportation and assignment problems
etc.

MATHEMATICAL METHOD: Bellman's method, Prim's method, Dijkstra's method, Ford's method, Floyd-Warshall method, Ford-Fulkerson method, Busacker-Gowen method, out-of-kilter method

PROGRAMMING LANGUAGE: Pascal, Z80-Assembler

COMPUTER SYSTEMS WHERE THE CODE HAS BEEN IMPLEMENTED:
-ALTOS ACS 8000, Z80-based CP/M-Micro (also MP/M)
-Hewlett-Packard Micro/1000 (A600) with RTE-A (available soon)

SPECIAL PROGRAM FEATURES:
-timecritical subroutines are written in assembler
-the library includes an interactive networkgenerater, an interactive networkeditor, and a network spooler.

PRACTICAL APPLICATION PROBLEMS SOLVED BY THE CODE: none

CONDITIONS OF AVAILABILITY: Open files on 8'SSSD and 8'SSDD floppy disks for CP/M and MP/M - Systems, on 3.5'SSSD Microfloppy for HP/1000 with RTE-A

CHARGE: material, post & packing (for non-profit users)

ADDRESS FROM WHERE THE CODE CAN BE ORDERED: Institut für Wirtschaftstheorie und Operations Research, Universität Karlsruhe, Kaiserstr. 12, D-7500 Karlsruhe 1

REFERENCES:

Lawler, E.L.: Combinatorial Optimization. Networks and Matroids. Holt, Rinehart and Winston, New York 1976.

Neumann, K.: Graphen und Netzwerke. Report WIOR-202. Institut für Wirtschaftstheroie und Operations Research der Universität Karlsruhe, 1983.

Rubach, T.: Algorithmen für Graphen und Netzwerke: Beschreibungen - Programme - Dokumentation. Report WIOR-195. Institut für Wirtschaftstheorie und Operations Research der Universität Karlsruhe, 1983.

NAME OF THE CODE: NETLIB

AUTHOR(S): Stephan Pappe, Thomas Rubach

MATHEMATICAL PROBLEM: network algorithms

DOMAIN OF APPLICATIONS: projekt planning by network models in the sense of time planning: CPM, PERT, MPM and in the sense of cost optimization.

MATHEMATICAL METHOD: Bellman's method
Ford's method
Kelley method

PROGRAMMING LANGUAGE: Pascal, Z80-Assembler

COMPUTER SYSTEMS WHERE THE CODE HAS BEEN IMPLEMENTED:
- ALTOS ACS 8000, Z80-based CP/M-Micro (also MP/M)
- Hewlett-Packard Micro/1000 (A600) with RTE-A (available soon)

SPECIAL PROGRAM FEATURES:
- timecritical subroutines are written in assembler
- the library includes an interactive networkgenerater, an interactive networkeditor, and a network spooler

PRACTICAL APPLICATION PROBLEMS SOLVED BY THE CODE: none

CONDITIONS OF AVAILABILITY: open files on 8'SSSD and 8'SSDD floppy disks for CP/M and MP/M - Systems, on 3.5'SSSD Microfloppy for HP/1000 with RTE-A

CHARGE: material, post & packing (for non-profit users)

ADDRESS FROM WHERE THE CODE CAN BE ORDERED: Institut für Wirtschaftstheorie und Operations Research, Universität Karlsruhe, Kaiserstrasse 12, D-7500 Karlsruhe 1

REFERENCES:
Neumann, K.: Netzplantechnik, Report WIOR-209, Institut für Wirtschaftstheorie und Operations Research der Universität Karlsruhe, 1983
Delivorias, P., Rubach, T.: Algorithmen zur Netzplantechnik: Beschreibungen - Programme, Dokumentation. Report WIOR-212, Institut für Wirtschaftstheorie und Operations Research der Universität Karlsruhe, 1984

NAME OF THE CODE: NLPNET

AUTHOR(S): R.S. Dembo

MATHEMATICAL PROBLEM:

MIN $f(x)$

S.t. $Ax = b$

$l \leq x \leq u$,

where A is a node arc incidence matrix of a network

DOMAIN OF APPLICATIONS: Nonlinear network optimization
Large scale

MATHEMATICAL METHOD: Primal Truncated-Newton Method

PROGRAMMING LANGUAGE: FORTRAN

COMPUTER SYSTEMS WHERE THE CODE HAS BEEN IMPLEMENTED:
IBM, DEC 20/60, VAX

SPECIAL PROGRAM FEATURES:

PRACTICAL APPLICATION PROBLEMS SOLVED BY THE CODE:
Matrix balancing problems (2300 variable)
Water distribution models

CONDITIONS OF AVAILABILITY: negotiable

CHARGE: negotiable

ADDRESS FROM WHERE THE CODE CAN BE ORDERED: R.S. Dembo, Yale School of Management Box 1A, New Haven, G.06520, USA

REFERENCES:
R.S. Dembo, "A Primal Truncated Newton Method for Large-Scale Linearly-Contrained Optimization", SOM Working Paper 72, Yale University (1983)
R.S. Dembo, "NLPNET - A Nonlinear Network Optimizer - Users Guide", SOM Working Paper 70 (1983)

NAME OF THE CODE: PSPMIN (VE08)
AUTHOR(S): Ph.Toint
MATHEMATICAL PROBLEM: Unconstrained or bound constraint optimization of nonlinear partially separable functions
DOMAIN OF APPLICATIONS: Finite elements, spline approximation, speech recognition nonlinear least squares, minimum surface
MATHEMATICAL METHOD: Partitioned quasi-Newton method
PROGRAMMING LANGUAGE: FORTRAN 4 (PFORT)
COMPUTER SYSTEMS WHERE THE CODE HAS BEEN IMPLEMENTED: DEC 2050, IBM 3033, VAX 11
SPECIAL PROGRAM FEATURES: Subroutine with two levels of use: 1) a simple argument list 2) an extended argument list allowing more user control
PRACTICAL APPLICATION PROBLEMS SOLVED BY THE CODE: Torsional mode in turbo alternator shafts and other engineering applications
CONDITIONS OF AVAILABILITY: At costs from the author, or in the Harwell Library. Contacts have been taken with NAG and IMSL.
CHARGE:
ADDRESS FROM WHERE THE CODE CAN BE ORDERED: Ph. Toint, Dept. of Mathematics, FNDP, 61, rue de Bruxelles, B-5000 Namur (Belgium)
REFERENCES: A. Griewank and Ph.L.Toint, "Numerical experiments with partially separable optimization problems", to appear in "Numerical Analysis, Dundee 1983", ed. D. Griffiths, Springer Verlag.

NAME OF THE CODE: MINOS Version 5.0

AUTHOR(S): Bruce Murtagh, Michael A. Saunders

MATHEMATICAL PROBLEM:

$$\text{minimize} \quad F(x)$$

$$\text{subject to} \quad \ell \leq \begin{Bmatrix} x \\ Ax \\ c(x) \end{Bmatrix} \leq u, \text{ where } F(x) \text{ and } c(x) \text{ are smooth.}$$

DOMAIN OF APPLICATIONS: In-core solution of sparse linear and nonlinear programming problems.

MATHEMATICAL METHOD:
Linear programs: the simplex method.
Nonlinear objective: a reduced-gradient/quasi-Newton method.
Nonlinear constraints: a projected-Langrangian algorithm (related to Robinson's method).

PROGRAMMING LANGUAGE: ANSI Fortran 66. Compatible with most Fortran 77 compilers.

COMPUTER SYSTEMS WHERE THE CODE HAS BEEN IMPLEMENTED: All major US/European mainframe/super-mini computers.

SPECIAL PROGRAM FEATURES:
(1) Stable numerical methods, including sparse LU factors for the basis matrix.
(2) Automatic scaling of linear constraints.
(3) Automatic estimation of some or all gradients of F(x) and c(x).
(4) Stand-alone system. File formats compatible with industry MPS standard.
(5) Requires at least 500K bytes of storage for code and data. Best suited to machines with virtual storage.

PRACTICAL APPLICATION PROBLEMS SOLVED BY THE CODE: The code has been distributed to 250 sites world wide. It has been applied to numerous models in economics, engineering, agriculture, geophysics, etc.

CONDITIONS OF AVAILABILITY: Copyright is held by Stanford University. A simple license agreement (limiting redistribution) must be signed.

CHARGE: At cost to universities/nonprofit organizations (typically $300). $5000 for single-site commercial licenses.

ADDRESS FROM WHERE THE CODE CAN BE ORDERED: Office of Technology Licensing, Room 105 Encina Hall, Stanford University, Stanford, California 94305, USA

REFERENCES: B.A. Murtagh and M.A. Saunders (1983). "MINOS 5.0 User's Guide," Report SOL 83-20, Department of Operations Research, Stanford University, California.

NAME OF THE CODE: OPT 2
AUTHOR(S): S.H. Wang & K.M. Ragsdell
MATHEMATICAL PROBLEM: Min: $f(x);\ x \in R^N$ Subject to: $g_j(x) \geq 0,\ j = 1,2,3,\ldots,J$ $h_k(x) = 0,\ k = 1,2,3,\ldots,K$ With $x_i(l) \leq x_i \leq x_i(u)\ \ i = 1,2,3,\ldots,N$
DOMAIN OF APPLICATIONS: Nonlinear programming problems with nonlinear equality & inequality constraints; especially suited for large problems with sparse constraint Jacobian.
MATHEMATICAL METHOD: Reduced Gradient Method with sparse matrix algebra.
PROGRAMMING LANGUAGE: Fortran
COMPUTER SYSTEMS WHERE THE CODE HAS BEEN IMPLEMENTED: DEC, VAX-11/780; CDC: Cyber-170; D.G.: MV10000; IBM370, IBM PC
SPECIAL PROGRAM FEATURES: OPT2 uses dynamic storage allocation, which implies a minimum amount of core storage
PRACTICAL APPLICATION PROBLEMS SOLVED BY THE CODE: Mechanical and Structural Design
CONDITIONS OF AVAILABILITY: Available to members of the Design Optimization Laboratory Software Users Group by legal agreement and annual fee .
CHARGE: Varies according to member grade-Regular, Subscribing or University
ADDRESS FROM WHERE THE CODE CAN BE ORDERED: Mechanical and Aerospace Engineering, University of Missouri, Columbia, Missouri 65211
REFERENCES: Wang, S.H. and Ragsdell, K.M., "OPT2 Users Manual", DOL, University of Arizona, 1983

NAME OF THE CODE: SEQ	
AUTHOR(S): A. Dourado and J.L. Calvet	
MATHEMATICAL PROBLEM: Static optimization of a large scale nonlinear system, serially connected.	
DOMAIN OF APPLICATIONS: Optimization of a cascade reactor for ethanol fermentation.	
MATHEMATICAL METHOD: Hierarchical method of sequential decomposition	
PROGRAMMING LANGUAGE: FORTRAN IV	
COMPUTER SYSTEMS WHERE THE CODE HAS BEEN IMPLEMENTED: IBM, MIWI 6 (H.B.)	
SPECIAL PROGRAM FEATURES: Good convergence and acceptable computational volume.	
PRACTICAL APPLICATION PROBLEMS SOLVED BY THE CODE: Optimization of a cascade of several reactors serially connected for ethanol fermentation.	
CONDITIONS OF AVAILABILITY: By demand	
CHARGE:	
ADDRESS FROM WHERE THE CODE CAN BE ORDERED: LAAS du CNRS, 7 Avenue du Colonel Roche, f 3/400 Toulouse France	
REFERENCES: A. Dourado, Production d'ethanol dans une cascade de fermenteurs; contribution à l'étude de la modelisation, de l'identification et de l'optimisation. Thèse de Docteur-Ingénieur, Décembre 1983, LAAS du CWRS, 7 Avenue du Colonel Roche, f3/400 Toulouse, France.	

NAME OF THE CODE: NLPQL	
AUTHOR(S): K. Schittkowski	
MATHEMATICAL PROBLEM: Solves the general constrained nonlinear programming problem, i.e. minimizing a nonlinear objective function subject to nonlinear equality and inequality constraints.	
DOMAIN OF APPLICATIONS: The problem functions must be continuously differentiable. The code solves problems up to 100 variables quite efficiently.	
MATHEMATICAL METHOD: The algorithm formulates a quadratic programming subproblem in each iteration-step by defining a quadratic approximation of the Lagrange function and by linearizing the constraints. A subsequent line search with an augmented Lagrangian merit function guarantees convergence. The Hesse matrix of the Lagrangian is approximated by BFGS-quasi Newton method.	
PROGRAMMING LANGUAGE: FORTRAN	
COMPUTER SYSTEMS WHERE THE CODE HAS BEEN IMPLEMENTED: Telefunken TR 440, IBM 370, VAX 11/780, by the author and most US/European main frame computers by other users	
SPECIAL PROGRAM FEATURES: Lower and upper bounds of the variables can be handled separately. Additional features are: Easy-to-use versions, reverse communications, automatic scaling facilities, restart capabilities, numerical differentiation, etc.	
PRACTICAL APPLICATION PROBLEMS SOLVED BY THE CODE: The code is in use at about 50 different sites world wide. Typical application problems are structural optimization problems in mechanical engineering.	
CONDITIONS OF AVAILABILITY: The code will be delivered on request on a magnetic tape.	
CHARGE: $ 100	
ADDRESS FROM WHERE THE CODE CAN BE ORDERED: K. Schittkowski, Institut für Informatik, Universität Stuttgart, Azenbergstr. 12, 7000 Stuttgart 1, Germany F.R.	
REFERENCES: NLPQL: A FORTRAN subroutine solving constrained nonlinear programming problems, Report, Institut für Informatik, Universität Stuttgart, 1984 (submitted for publication) User's guide for the nonlinear programming code NLPQL, Report, Institut für Informatik, Universität Stuttgart, 1984. On the convergence of a sequential quadratic programming method with an augmented Lagrangian line search function, Math. Operationsforschung u. Statistik, Ser, Optimization, Vol. 14 (1983), 197-216	

NAME OF THE CODE: NLPSYS

AUTHOR(S): K. Schittkowski

MATHEMATICAL PROBLEM: Interactive model building and solution of various nonlinear programming problems (e.g. general, least squares or quadratic objective function, linear of nonlinear constraints, parameter estimation).

DOMAIN OF APPLICATIONS: The user has to supply only problem parameters and some FORTRAN statements to define the problem functions. The main program for defining dimensioning statements, solution parameters or calling sequence of the chosen mathematical programming subroutine, and the subroutine necessary to link the problem functions are written by the system.

MATHEMATICAL METHOD: Problem parameters, problem functions, constant data and solution parameters are provided interactively and stored in form of a data base system. NLPSYS selects a suitable algorithm, writes the corresponding FORTRAN source code, and executes the program automatically. The basic code to solve the problem, is NLPQL or a certain variant to take advantage of special problem structures. Results are stored in form of a data base.

PROGRAMMING LANGUAGE: PASCAL

COMPUTER SYSTEMS WHERE THE CODE HAS BEEN IMPLEMENTED:
VAX 11/780

SPECIAL PROGRAM FEATURES: All actions are controlled by certain commands explained in form of menues. Names can be addressed to variables and functions. The data base system for all problems and results allows alteration of problem data, deletion of problems or results, detailed information and terminal or printer output of results in a way specified by the user. External files can be linked with the system.

PRACTICAL APPLICATION PROBLEMS SOLVED BY THE CODE:
Not yet available

CONDITIONS OF AVAILABILITY: Distributed on request by the author on a magnetic tape.

CHARGE: Contact author

ADDRESS FORM WHERE THE CODE CAN BE ORDERED: K. Schittkowski, Institut für Informatik, Universität Stuttgart, Azenbergstr. 12, D-7000 Stuttgart 1, Germany F.R.

REFERENCES:
NLPQL: A FORTRAN subroutine solving constrained nonlinear programming problems, Report, Institut für Informatik, Universität Stuttgart, 1984 (submitted for publication)
NLPSYS: A data base system to support interactively model building and numerical solution of nonlinear programming problems, Report, Mathematisches Institut, Universität Bayreuth, 1985.

NAME OF THE CODE: WIEM

AUTHOR(S): R.Th. Wijmenga and G. van der Hoek

MATHEMATICAL PROBLEM: The constrained nonlinear programming problem

$$\min \quad F(x)$$

s.t.

$$c_i(x) \begin{Bmatrix} \geq \\ = \end{Bmatrix} 0 \qquad i = 1,2,\ldots,m$$

with twice continuously differentiable problem functions

DOMAIN OF APPLICATIONS: Constrained nonlinear programming problems

MATHEMATICAL METHOD: WIEM is a research code which solves a constrained nonlinear programming problem via a sequence of Equality constrained Quadratic programming problems, WIEM is an implementation along the lines of Bigg's (1972) algorithm. The modifications concern the use of a self scaling update for the inverse Hessian approximation and a safeguarded inaccurate line search

PROGRAMMING LANGUAGE: FORTRAN

COMPUTER SYSTEMS WHERE THE CODE HAS BEEN IMPLEMENTED: DEC 20

SPECIAL PROGRAM FEATURES: Numerical differentiation, option for choosing several inverse Hessian update formulae

PRACTICAL APPLICATION PROBLEMS SOLVED BY THE CODE: no

CONDITIONS OF AVAILABILITY: Free of charge for research purposes on a user supplied tape

CHARGE: -

ADDRESS FROM WHERE THE CODE CAN BE ORDERED: R.Th. Wijmenga & G. van der Hoek, Econometric Institute, Erasmus University Rotterdam, P.O. Box 1738, 3000 DR Rotterdam, The Netherlands

REFERENCES:

1. M.C. Biggs, 1972, Constrained Minimization using Recursive Equality Quadratic Programming, in: Numerical Methods for Nonlinear Optimization, F.A. Lootsma (ed), Academic Press, London.
2. G. van der Hoek, 1980, Reduction Methods, in Nonlinear Programming, MC Tract 126, Mathematical Centre, (CWI), Amsterdam.
3. R.Th. Wijmenga and G. van der Hoek, 1982, User Manual to Subroutine Package WIEM for Constrained Nonlinear Programming, Report 8230/I, Econometric Institute, Erasmus University Rotterdam.

NAME OF THE CODE: TWOFAS

AUTHOR(S): A.S. Louter and G. van der Hoek

MATHEMATICAL PROBLEM: The constrained nonlinear programming problem:

$$\min \quad F(x)$$

$$\text{s.t. } c_i(x) \left\{ \begin{matrix} \geq \\ = \end{matrix} \right\} 0 \qquad i = 1,2,\ldots,m$$

with twice continuously differentiable problem functions

DOMAIN OF APPLICATIONS: Constrained nonlinear programming problems

MATHEMATICAL METHOD: TWOFAS consists of two phases. In phase I, a linearly constrained exterior penalty function is minimized to obtain an acceptable starting point.
In phase II linearly constrained reduced problems are solved, using a linear penalty function as discussed in Robinson (1972) and Van der Hoek (1980).

PROGRAMMING LANGUAGE: FORTRAN

COMPUTER SYSTEMS WHERE THE CODE HAS BEEN IMPLEMENTED:
DEC 20, VAX 11/750

SPECIAL PROGRAM FEATURES: Treatment of bounds, numerical differentiation

PRACTICAL APPLICATION PROBLEMS SOLVED BY THE CODE:
Technical problems, inventory problems

CONDITIONS OF AVAILABILITY: Free of charge for research purposes on a user supplied tape

CHARGE: -

ADDRESS FROM WHERE THE CODE CAN BE ORDERED: A.S. Louter & G. van der Hoek, Econometric Institute, Erasmus University, P.O. Box 1738, 3000 DR Rotterdam, The Netherlands

REFERENCES:

1. A.S. Louter and G. van der Hoek, 1984, TWOFAS: A 2-Phase Code for Constrained Nonlinear Programming, Econometric Institute, Erasmus University Rotterdam.
2. S.M. Robinson, 1972, A Quadratically Convergent Algorithm for General Nonlinear Programming Problems, Math. Progr., 3, 145-156.
3. G. van der Hoek, 1980, Reduction Methods in Nonlinear Programming, MC Tract 126, Mathematical Centre (CWI), Amsterdam.
4. G. van der Hoek, 1982, Asymptotic Properties of Reduction Methods Applying Linearly Constrained Reduced Problems, Math. Progr. 16, 162-189.

NAME OF THE CODE: NPSOL Version 2.0

AUTHOR(S): Philip E. Gill, Walter Murray, Michael A. Saunders, Margaret H. Wright

MATHEMATICAL PROBLEM:

$$\text{minimize} \quad F(x)$$

$$\text{subject to} \quad \ell \leqslant \begin{Bmatrix} x \\ Ax \\ c(x) \end{Bmatrix} \leqslant u.$$

DOMAIN OF APPLICATIONS: The code is designed to solve in-core nonlinear programming problems.

MATHEMATICAL METHOD: Version 2.0 is a sequential quadratic programming method with an augmented Lagrangian merit function. The quadratic subproblem is solved using Version 3.1 of QPSOL.

PROGRAMMING LANGUAGE: ANSI Fortran 66

COMPUTER SYSTEMS WHERE THE CODE HAS BEEN IMPLEMENTED: All major US/European mainframe/mini/super-mini computers

SPECIAL PROGRAM FEATURES:
1. The code takes advantage of linear constraints and bounds.
2. All iterates are feasible with respect to the linear constraints and bounds.
3. The code may be used for linearly constrained and unconstrained optimization.

PRACTICAL APPLICATION PROBLEMS SOLVED BY THE CODE: The code has been distributed to 75 sites world wide. The object code of Version 1.1. will be available at Mark 12 of the NAG library.

CONDITIONS OF AVAILABILITY: Copyright is held by Stanford University. A simple license agreement (limiting redistribution) must be signed.

CHARGE: At cost to universities/nonprofit organizations. $2000 for commerical licenses.

ADDRESS FROM WHERE THE CODE CAN BE ORDERED: Office of Technology Licensing, Room 105 Encina Hall, Stanford University, Stanford, California 94305, USA

REFERENCES: Gill, Murray, Saunders and Wrigth (1983). "User's Guide for SOL/NPSOL; A Fortran Package for Nonlinear Programming," Report SOL 83-12, Department of Operations Research, Stanford University, California

NAME OF THE CODE: ALRQP

AUTHOR(S): M.C. Bartholomew-Biggs

MATHEMATICAL PROBLEM:

Minimize F(x) s.t. $c_i(x) = 0$ $i=1,\ldots,m$; $c_i(x) \geq 0$ $i=m+1,\ldots,m'$
where F and the c_i are continuous and differentiable nonlinear functions.

DOMAIN OF APPLICATIONS: Small to medium sized nonlinear programming problems

MATHEMATICAL METHOD: Recursive Quadratic Programming. The method employs an equality constrained subproblem for finding the search directions which is derived from the Augmented Lagrangian (or Ideal penalty) function. The steplength procedure also uses the augmented Lagrangian as a line-search function.

PROGRAMMING LANGUAGE: FORTRAN 77

COMPUTER SYSTEMS WHERE THE CODE HAS BEEN IMPLEMENTED: DEC1091

SPECIAL PROGRAM FEATURES: Robustness in the presence of (near) dependence among the constraint normals. Simplicity of code due to the use of an equality constrained QP subproblem.

PRACTICAL APPLICATION PROBLEMS SOLVED BY THE CODE: Selection of the Hock & Schittkowski test problems

CONDITIONS OF AVAILABILITY: This prototype implementation is available to research workers in the area of nonlinear programming but ONLY FOR PURPOSES INVOLVED WITH THE STUDY AND DEVELOPMENT OF OPTIMIZATION ALGORITHMS.

CHARGE: F.O.C.

ADDRESS FROM WHERE THE CODE CAN BE ORDERED:
M.C. Bartholomew-Biggs; School of Information Sciences; The Hatfield Polytechnic; Hatfield; Hertfordshire; England

REFERENCES:
Technical Report 139, The Numerical Optimisation Centre, The Hatfield Polytechnic, Hatfield, Hertfordshire, England.

NAME OF THE CODE: IDESIGN

AUTHOR(S): J.S. Arora

MATHEMATICAL PROBLEM: Unconstrained, Linear, General Nonlinear Programming

min $f(x)$, $x \in S \subset R^n$

$S = \{x: h_i(x) = 0, i = 1, p; g_i(x) \leqslant 0, i = 1, m\}$

DOMAIN OF APPLICATIONS: Functions should be differentiable everywhere.

MATHEMATICAL METHOD:
Hybrid method: large step algorithm at start (cost function bounding, superlinearly convergent, RQP algorithm with active sets near the end.

PROGRAMMING LANGUAGE: FORTRAN 77

COMPUTER SYSTEMS WHERE THE CODE HAS BEEN IMPLEMENTED:
PRIME 750

SPECIAL PROGRAM FEATURES: User describes his problem in two FORTRAN subroutines USERF (function evaluations) and USERGR (gradients of active constraints). User can monitor as well as control progress of the iterative process interactively. Some graphics is also available.

PRACTICAL APPLICATION PROBLEMS SOLVED BY THE CODE: Many mathematical programming and structural optimization problems

CONDITIONS OF AVAILABILITY: Upon request (As is)

CHARGE: Cost of reproduction for academic research; Negotiable for commercial use.

ADDRESS FROM WHERE THE CODE CAN BE ORDERED: J.S. Arora, Director Applied-Optimal Design Laboratory, Department of Civil and Environmental Engineering, The University of Iowa, Iowa City, Iowa 52242, USA

REFERENCES:
1. Belsare, S.V. and Arora, J.S., "An Algorithm for Engineering Design Optimization," Int. J. for Num. Methods in Engrg., Vol. 19, 1983, pp. 841-858.
2. Belegundu, A.D. and Arora, J.S., "A Recursive Quadratic Programming Method for Structural Optimization with Active Set Strategy," Int. J. for Num. Meth. in Engrg., July 1984

NAME OF THE CODE: GRG2	

NAME OF THE CODE: GRG2

AUTHOR(S): Leon Lasdon, Allan Waren

MATHEMATICAL PROBLEM: General nonlinear program: unconstrained, linearly constrained, or nonlinear constraints.

DOMAIN OF APPLICATIONS: Does not exploit sparsity, so at most 100-200 constraints. Several hundred variables can be dealt with by conjugate gradient options.

MATHEMATICAL METHOD: Generalized Reduced Gradient Method

PROGRAMMING LANGUAGE: FORTRAN

COMPUTER SYSTEMS WHERE THE CODE HAS BEEN IMPLEMENTED:
Most mainframes, IBM PC

SPECIAL PROGRAM FEATURES: See documentation

PRACTICAL APPLICATION PROBLEMS SOLVED BY THE CODE:
Over 20 companies have used GRG2 in a wide variety of applications. It also has over 50 users as a module of IFPS/OPTIMUM.

CONDITIONS OF AVAILABILITY: Contact authors

CHARGE:

ADDRESS FROM WHERE THE CODE CAN BE ORDERED: Dept. of General Business, College of Business Administration, The University of Texas, Austin, Texas 78712, USA

REFERENCES: See documentation

NAME OF THE CODE: OPT

AUTHOR(S): G.A. Gabriele and K.M. Ragsdell

MATHEMATICAL PROBLEM:

Min: $f(x); x \in R^N$

Subject to: $g_j(x) \geqslant 0, j = 1,2,3,\ldots,J$

$h_k(x) \equiv 0, k = 1,2,3,\ldots,K$

With $x_i(l) \leqslant x_i \leqslant x_i(u) \; i = 1,2,3,\ldots,N$

DOMAIN OF APPLICATIONS: Nonlinear Programming problem with nonlinear equality and inequality constraints.

MATHEMATICAL METHOD: Generalized Reduced Gradient

PROGRAMMING LANGUAGE: FORTRAN

COMPUTER SYSTEMS WHERE THE CODE HAS BEEN IMPLEMENTED:
DEC, VAX-11/780; CDC: Cyber-170; D.G.: MV10000; IBM 370, IBM PC

SPECIAL PROGRAM FEATURES: OPT uses dynamic storage allocation, which implies a minimum amount of core storage

PRACTICAL APPLICATION PROBLEMS SOLVED BY THE CODE:
Mechanical and Structural Design

CONDITIONS OF AVAILABILITY: Available to members of the Design Optimization Laboratory Software Users Group by legal agreement and annual fee.

CHARGE: Varies according to member grade-Regular, Subscribing or University

ADDRESS FROM WHERE THE CODE CAN BE ORDERED: Mechanical and Aerospace Engineering, University of Missouri, Columbia, Missouri 65211

REFERENCES: Gabriele, G.A. and Ragsdell, K.M., "Opt Users Manual," Purdue University, 1976

NAME OF THE CODE: BIAS

AUTHOR(S): R.R. Root and K.M. Ragsdell

MATHEMATICAL PROBLEM:

Min: $f(x);\ x \in R^N$

Subject to: $g_j(x) \geq 0,\ j = 1,2,3,\ldots,J$

$h_k(x) \equiv 0,\ k = 1,2,3,\ldots,K$

With $x_i(l) \leq x_i \leq x_i(u)\ i = 1,2,3,\ldots,N$

DOMAIN OF APPLICATIONS: Nonlinear Programming problems with nonlinear equality and inequality constraints

MATHEMATICAL METHOD: Method of Multipliers

PROGRAMMING LANGUAGE: FORTRAN

COMPUTER SYSTEMS WHERE THE CODE HAS BEEN IMPLEMENTED:
DEC, VAX-11/780; CDC; Cyber-170; D.G.; MV10000; IBM 370, IBM PC

SPECIAL PROGRAM FEATURES: BIAS uses dynamic storage allocation, which implies a minimum amount of core storage.

PRACTICAL APPLICATION PROBLEMS SOLVED BY THE CODE:
Mechanical and Structural Design

CONDITIONS OF AVAILABILITY: Available to members of the Design Optimization Laboratory Software Users Group by legal agreement and annual fee.

CHARGE: Varies according to member grade-Regular, Subscribing or University

ADDRESS FROM WHERE THE CODE CAN BE ORDERED: Mechanical and Aerospace Engineering, University of Missouri, Columbia, Missouri 65211

REFERENCES: Root, R.R. and Ragsdell, K.M., "BIAS Users Manual" Purdue University, 1977

NAME OF THE CODE: TRUOPT

AUTHOR(S): S.H. Wang and K.M. Ragsdell

MATHEMATICAL PROBLEM:

Minimize: $F(A) = \sum_{i=1}^{N} \rho_i L_i A_i \quad , A \in R^N$ (weight)

Subject to: $g_j(A) \geq 0, \; j = 1,2,\ldots,J$

and $A_i^{(l)} \leq A_i \leq A_i^{(u)}, \; (= 1,2,\ldots,N)$

where g_j = displacement, stress, and buckling constraints

DOMAIN OF APPLICATIONS: Optimal design of three dimensional (space) trusses

MATHEMATICAL METHOD: Schmitt multilevel approximation philosophy and the generalized reduced gradient method.

PROGRAMMING LANGUAGE: FORTRAN

COMPUTER SYSTEMS WHERE THE CODE HAS BEEN IMPLEMENTED:
DEC-10, VAX-11/780; CDC; DG: MV10000; IBM 370, PC

SPECIAL PROGRAM FEATURES: Internally compatible in that lower level methods are options within the higher level methods. Variably dimensioned for a minimum of core storage use.

PRACTICAL APPLICATION PROBLEMS SOLVED BY THE CODE:
Structural Optimization

CONDITIONS OF AVAILABILITY: Available to members of the Design Optimization Laboratory Software Users Group by legal agreement and annual fee.

CHARGE: Varies according to member grade-Regular, Subscribing or University

ADDRESS FROM WHERE THE CODE CAN BE ORDERED: Mechanical and Aerospace Engineering, University of Missouri, Columbia, Missouri 65211

REFERENCES: Wang, S.H. and Ragsdell, K.M., "TRUOPT Users Manual", DOL, University of Arizona, 1983

NAME OF THE CODE: OPTLIB	
AUTHOR(S): G.A. Gabriele and K.M. Ragsdell	
MATHEMATICAL PROBLEM: Various Optimization problems	
DOMAIN OF APPLICATIONS: Optimization Education	
MATHEMATICAL METHOD: Library of modern Optimization methods including: constrained penalty methods, Box's complex, Griffith and Stewarts LP, reduced gradient; unconstrained methods: various gradient bases methods such as Davidon-Fletcher-Powell, Broyden-Fletcher-Shannon variable metric, pattern search, simplex search, Powell's conjugate direction, and several line search methods.	
PROGRAMMING LANGUAGE: FORTRAN	
COMPUTER SYSTEMS WHERE THE CODE HAS BEEN IMPLEMENTED: DEC-10, VAX-11/780; CDC; DG: MV10000; IBM 370, PC	
SPECIAL PROGRAM FEATURES: Internally compatible in that lower level methods are options within the higher level methods. Variably dimensioned for a minimum of core storage use.	
PRACTICAL APPLICATION PROBLEMS SOLVED BY THE CODE:	
CONDITIONS OF AVAILABILITY: Available to members of the Design Optimization Laboratory Software Users Group by legal agreement and annual fee.	
CHARGE: Varies according to member grade-Regular, Subscribing or University	
ADDRESS FROM WHERE THE CODE CAN BE ORDERED: Mechanical and Aerospace Engineering, University of Missouri, Columbia, Missouri 65211	
REFERENCES: Gabriele, Ragsdell, "OPTLIB Users Manual," Purdue University, 1977	

NAME OF THE CODE: SYSOPT	
AUTHOR(S): M. Rijckaert, E. Walraven	
MATHEMATICAL PROBLEM: Unconstrained and constrained optimization	
DOMAIN OF APPLICATIONS: To be used on APPLE-II - microcomputer	
MATHEMATICAL METHOD: A variety of methods to be selected by the user from several menu's	
PROGRAMMING LANGUAGE: BASIC	
COMPUTER SYSTEMS WHERE THE CODE HAS BEEN IMPLEMENTED: APPLE II	
SPECIAL PROGRAM FEATURES: Interactive and selection of method.	
PRACTICAL APPLICATION PROBLEMS SOLVED BY THE CODE: Many classical test problems.	
CONDITIONS OF AVAILABILITY: For academic purposes: special low price	
CHARGE:	
ADDRESS FROM WHERE THE CODE CAN BE ORDERED: Prof. Dr. M. Rijckaert, K.U. Leuven, B-3030 Heverlee, Belgium.	
REFERENCES:	

NAME OF THE CODE: EA3
AUTHOR(S): M. Kupferschmid and J.G. Ecker
MATHEMATICAL PROBLEM: minimize $f_o(x)$ subject to $f_i(x) \leqslant 0$, $i = 1,\ldots,m$, $x \in R^n$
DOMAIN OF APPLICATIONS: Convex and nonconvex inequality constrained minimizations involving $\leqslant$ 50 variables and $\leqslant$ 50 constraints
MATHEMATICAL METHOD: Ellipsoid algorithm with center cuts
PROGRAMMING LANGUAGE: Fortran IV
COMPUTER SYSTEMS WHERE THE CODE HAS BEEN IMPLEMENTED: IBM 3033U, IBM 3081D, IBM PC
SPECIAL PROGRAM FEATURES: Easy to use; no parameters to set; small size
PRACTICAL APPLICATION PROBLEMS SOLVED BY THE CODE: Cooling tower design feedback control gain optimization wastwater treatment plant design cogenerating power station design differential equation parameter estimation
CONDITIONS OF AVAILABILITY: Nondisclosure, as explained on the order form
CHARGE: Free for academic research, $100.00 for commercial users
ADDRESS FROM WHERE THE CODE CAN BE ORDERED: Michael Kupferschmid, Engineering Applications Consultant, Alan M. Voorhees Computing Center, Rensselear Polytechnic Institute, Troy, New York 12181, USA
REFERENCES: An Ellipsoid Algorithm for Nonlinear Programming, J.G. Ecker and M. Kupferschmid, Mathematical Programming 27, pages 83-106, 1983

NAME OF THE CODE: UNT	

NAME OF THE CODE: UNT

AUTHOR(S): A. Žilinskas

MATHEMATICAL PROBLEM: Global optimization of multimodal functions of several variables (number of variables $\leq$ 20, number of local essential minimas $\sim$ 20 - 30).

DOMAIN OF APPLICATIONS: Optimization of difficult (time consuming) multimodal functions occuring in computer aided design of technical equipments and systems.

MATHEMATICAL METHOD: Method is based on the use of statistical model of an multimodal objective function and axiomatically based rational search strategy

PROGRAMMING LANGUAGE: FORTRAN

COMPUTER SYSTEMS WHERE THE CODE HAS BEEN IMPLEMENTED: BESM-6, EC-1040, DEC-20, VAX-11/780.

SPECIAL PROGRAM FEATURES: Algorithm is robust with respect to errors (noise) of evaluating of objective function values.

PRACTICAL APPLICATION PROBLEMS SOLVED BY THE CODE: Design of magnetic deflection system of colour television tube, design of pigmental mixtures (colours) with given spectral and colour characteristics (number of variables was 10 - 15, time of evaluation of one objective function value was up to 17 sec. on BESM-6).

CONDITIONS OF AVAILABILITY: Individual agreement based mainly on some exchange of software.

CHARGE: no.

ADDRESS FROM WHERE THE CODE CAN BE ORDERED: A. Žilinskas, Inst. of Mathematics and Cybernetics, K. Pozelos 54, Vilnius, 232600, U.S.S.R.

REFERENCES:

1) A. Žilinskas. Axiomatic approach to statistical models and their use in multimodal optimization theory.- Mathematical Programming, 1982, v. 22, No. 1, p. 102-116.
2) A. Žilinskas. Results of application of multimodal optimization algorithms based on statistical models.- In: COMPSTAT 82, Wien: Physica Verlag, 1982, p. 457-462.

NAME OF THE CODE: ENLSIP and ELSUNC

AUTHOR(S): Per Lindström, Per-Ake Wedin

MATHEMATICAL PROBLEM: Nonlinear least squares problems.
ENLSIP for problems with nonlinear constraints.
ELSUNC for unconstrained problems.

DOMAIN OF APPLICATIONS: Parameter estimation.
Curve fitting.

MATHEMATICAL METHOD: A Gauss-Newton algorithm generalized to handle problems with constraints in the constrained case. If the algorithm classifies the problem as a large residual problem, it automatically switches to the Newton method with numerically computed second derivatives.

PROGRAMMING LANGUAGE: Standard Fortran (Fortran 66)

COMPUTER SYSTEMS WHERE THE CODE HAS BEEN IMPLEMENTED:
Control Data Cyber 170/730

SPECIAL PROGRAM FEATURES: The user can either give the 1:st derivatives analytically or let the package compute them. The default values of some of the parameters in these easy-to-use versions are easily changed by the user.

PRACTICAL APPLICATION PROBLEMS SOLVED BY THE CODE: Roentgen Stereophotogrammetry

CONDITIONS OF AVAILABILITY: Report unsatisfactory behaviour

CHARGE: 200 $ (or Skr: 1000 or DM 350) for ENLSIP and ELSUNC together

ADDRESS FROM WHERE THE CODE CAN BE ORDERED: University of Umea, Institute of Information Processing, S-901 87 Umea, Sweden

REFERENCES:

Lindström, P. Algorithms for nonlinear least squares - particularly problems with constraints. (Dissertation) Report UMINF-106.83 (1983).

Lindström, P. Preliminary versions of two user guides. One (ENLSIP), for constrained - one (ELSUNC) for unconstrained non-linear least squares problems.

NAME OF THE CODE: MULTEX/ISAAC

AUTHOR(S): H.G. Bock, J. Schlöder

MATHEMATICAL PROBLEM: determine $p, x_1, \ldots, x_m$, $p \in \mathbb{R}^{n_p}$, $x_j \in \mathbb{R}^{n_j}$

$$\Sigma_{j=1}^{m} \| F_j(p, x_j) \|_2^2 = \min$$

s.t.:

$$G_j(p, x_j) = 0 \text{ or } \geq 0$$

DOMAIN OF APPLICATIONS: Nonlinear constrained least squares problems with global (p) and local (x_j) variables. Problem class includes high-dimensional identification problems from MULTiple EXperiment data, implicitly defined models, special separable least squares problems, nonlinear constraints, ill-conditioned or singular problems.

MATHEMATICAL METHOD: Solution by generalized Gauss-Newton method (convergence: global and locally linear). Linear subproblem solver exploits sparse structure of Jacobian. Statistical a posteriori sensitivity analysis. For very hard problems: continuation techniques.

PROGRAMMING LANGUAGE: FORTRAN IV

COMPUTER SYSTEMS WHERE THE CODE HAS BEEN IMPLEMENTED:
IBM 370/168, 4331, 3081

SPECIAL PROGRAM FEATURES: ISAAC: for general nonlinear constrained least squares. MULTEX: for multiple experiment structure including equality constraints (inequalities in test). Suited for parallel processing. Special version for local variables appearing linearly.

PRACTICAL APPLICATION PROBLEMS SOLVED BY THE CODE: E.g. rate constant identification for oscillating Belousov-Zhabotinskij-reaction from measured bistability points (894 Variables), identification of titration constants.

CONDITIONS OF AVAILABILITY: On request

CHARGE:

ADDRESS FROM WHERE THE CODE CAN BE ORDERED: Institut für Angewandte Mathematik, Universität Bonn, SFB 72, D-5300 Bonn 1, FRG

REFERENCES:

H.G. Bock (1980): Numerical treatment of inverse problems in chemical reaction systems, Springer Series Chemical Physics 18, 102-125

J. Schlöder, H.G. Bock (1983): Identification of rate constants in bistable chemical reactions, Progress in Scientific Computing 2, Birkhäuser, Boston, 27-47.

NAME OF THE CODE: PQ1.

AUTHOR(S): Robert Mifflin

MATHEMATICAL PROBLEM: Minimize f(x) subject to c(x) ≦ 0 where x is a single variable and c is a single constraint function. A subgradient of f(x) is required at each feasible (infeasible) x. For example, f(x) may be the maximum of several differentiable functions $f_i(c_i)$ and a subgradient at x is the derivative of any one of the f_i's (c_i's) that attain the maximum at x. Note that several constraints $c_i(x) \leqq 0$ are equivalent to one constraint $c(x) = \max_i c_i(x) \leqq 0$.

DOMAIN OF APPLICATIONS: Single variable constrained minimization and line search. A line search function value decrease requirement may be imposed via an appropriately defined constraint function c as in [3].
Constrained root finding, i.e. finding an $x \in [a,b]$ such that r(x)=0, can be approached by minimizing $f(x) = |r(x)| = \max[r(x),-r(x)]$ subject to $c(x) = \max[a-x,x-b] \leqq 0$.

MATHEMATICAL METHOD: The method is described and analyzed in [3]. It combines polyhedral and quadratic approximation, an automatic scale-free penalty technique, and a safeguard. The safeguard insures convergence and keeps apart points used in divided difference formulas for approximating second derivatives, but it does not detract from rapid convergence of the better than linear type [4].

PROGRAMMING LANGUAGE: FORTRAN (an older BASIC program is also available [2]).

COMPUTER SYSTEMS WHERE THE CODE HAS BEEN IMPLEMENTED: VAX 11/750 (IBM personal computer).

SPECIAL PROGRAM FEATURES: Minimization is done one step at a time by calling PQ1 sequentially . The calling program should include the user's stopping test, the call to the user's function evaluation subroutine and input/output. PQ1 is written so that it can be used in "nested" optimization applications such as the one described next.

PRACTICAL APPLICATION PROBLEMS SOLVED BY THE CODE: A practical single resource allocation problem [1] with five bounded decision variables has been solved by a dual technique that uses PQ1 in a nested manner. This technique involves solving a single variable "outer" dual problem where each function evaluation requires solving a five variable "inner" Lagrangian problem that separates into five single variable problems.

CONDITIONS OF AVAILABILITY: A forthcoming paper [5] contains a brief description of the method, some information for users, a description of the nested application and a listing of PQ1 (less than 100 lines of code excluding comments).

CHARGE: None for the paper. A tape version might be produced and supplied with a charge to cover costs if there is sufficient demand.

ADDRESS FROM WHERE THE CODE CAN BE ORDERED: Mathematics 2930, Washington State University, Pullman, WA 99164, U.S.A.

REFERENCES:

1. K.W. Heiner, M. Kupferschmidt and J.G. Ecker, Maximizing Restitution for Erroneous Medical Payments when Auditing Samples from More than One Provider, INTERFACES 13 (1983), 12-17
2. I. Mifflin and R. Mifflin, A Basic Program for Solving Univariate Constrained Minimization Problems, Dept. of Pure and Applied Mathematics, WSU, Pullman, 1983.
3. R. Mifflin, Stationarity and Superlinear Convergence of an Algorithm for Univariate Locally Lipschitz Constrained Minimization, Mathematical Programming 28 (1984), 50-71.
4. R. Mifflin, Better than Linear Convergence and Safeguarding in Nonsmooth Minimization, in P. Thoft-Christensen, ed., System Modelling and Optimization, Springer-Verlag, Berlin, 1984.
5. R. Mifflin, A Computational Algorithm for Univariate Minimization and a Nested Application, forthcoming.

NAME OF THE CODE: M1FC1

AUTHOR(S): C. Lemarechal

MATHEMATICAL PROBLEM: Unconstrained minimization of an objective function f on the $\mathbb{R}^n$

DOMAIN OF APPLICATIONS: When the problem is difficult because of some of the following characteristics:
- many variables
- bad conditioning (the code accepts non-continuous derivatives)
- f expensive (in CPU)

MATHEMATICAL METHOD: Bundle method

PROGRAMMING LANGUAGE: FORTRAN IV

COMPUTER SYSTEMS WHERE THE CODE HAS BEEN IMPLEMENTED:
HB68, IBM 3033, CRAY 1, AMDAHL, CDC 6600, Burrough

SPECIAL PROGRAM FEATURES: Documentation in French
Bad final rate of convergence - otherwise, very reliable

PRACTICAL APPLICATION PROBLEMS SOLVED BY THE CODE: Air to air combat, optimum design, crystallography (not solved because not solvable), inresion of seismic data, identification of constitution laws with large deformation.

CONDITIONS OF AVAILABILITY: Free for academic use, otherwise for sale

CHARGE: 6000 ff

ADDRESS FROM WHERE THE CODE CAN BE ORDERED:
INRIA, BP 105, 78153 Le Chesnay, France

REFERENCES: Lemarechal, Strodiot, Bihain
in: Nonlinear Programming 4, eds. Mangasarian, Meyer, Robinson
Academic Press, 1981.

NAME OF THE CODE: UNCMIN

AUTHOR(S): Robert B. Schnabel, John E. Koontz, Barry E. Weiss

MATHEMATICAL PROBLEM:

Unconstrained minimization: minimize $f: \mathbb{R} \to \mathbb{R}$,

$$x \in \mathbb{R}^n$$

where $f \in C^2$.

DOMAIN OF APPLICATIONS: Any unconstrained minimization problems with small or moderate dimension. (An n×n matrix is stored, an an n×n linear system is solved at each iteration).

MATHEMATICAL METHOD: Various choices are available:
1. Line search or trust region/dogleg or trust region/"optimal step"
2. Analytic or finite-difference gradient
3. Analytic, finite-difference gradient

PROGRAMMING LANGUAGE: FORTRAN 66 (Standard)

COMPUTER SYSTEMS WHERE THE CODE HAS BEEN IMPLEMENTED: Many including CDC, VAX, IBM (?)

SPECIAL PROGRAM FEATURES:
1. Many methods available
2. User may provide scaling information
3. Simple and complex calling sequences available
4. Automatic checking of user-supplied derivatives

PRACTICAL APPLICATION PROBLEMS SOLVED BY THE CODE:
Many including: maximum likelihood problems, time series related problems, various problems from engineering

CONDITIONS OF AVAILABILITY: No restrictions

CHARGE: $ 100 (U.S.)

ADDRESS FROM WHERE THE CODE CAN BE ORDERED:
Robert B. Schnabel, Department of Computer Science, University of Colorado, Boulder, Colorado 80309, USA

REFERENCES: "A modular system of algorithms for unconstrained minimization", by Schnabel, Koontz and Weiss, Technical Report CU-CS-240-82, Department of Computer Science, University of Colorado, Boulder

NAME OF THE CODE: QNBD

AUTHOR(S): J.F. Bonnans

MATHEMATICAL PROBLEM: Minimization of a smooth function, the variables being subject to bound constraints

DOMAIN OF APPLICATIONS: As a quasi-Newton method is used, the number of variables should not be too large

MATHEMATICAL METHOD: The algorithm consists in a projection of a direction given by a variable metric method, as in (1). However, the line search is an extension of one due to P. Wolfe (3), and the parameters of the algorithm are modified to insure a fast convergence even if the strict complementary conditions do not hold. The method is explained in details in (2).

PROGRAMMING LANGUAGE: Standard Fortran

COMPUTER SYSTEMS WHERE THE CODE HAS BEEN IMPLEMENTED:
DPS 68 Multics of INRIA

SPECIAL PROGRAM FEATURES: Uses the subroutine MC11A of the Harwell library and a modified version of the subroutine MLIS1 of C. Lemaréchal (INRIA).

PRACTICAL APPLICATION PROBLEMS SOLVED BY THE CODE:
The control of a thermical system. See (2).

CONDITIONS OF AVAILABILITY: Free for academic use.

CHARGE: to be discussed for commercial use.

ADDRESS FROM WHERE THE CODE CAN BE ORDERED:
J.F. Bonnans, INRIA, Domaine de Voluceau, B.P. 105, Rocquencourt 78153 Le Chesnay Cedex, France

REFERENCES:
1. D.P. Bertsekas, Projected Newton methods for optimization problems with simple constraints, SIAM J. Cont. Opt. 20, pp. 221-246, 1982
2. J.F. Bonnans, A variant of a projected variable metric method for bound constrained optimization problems. INRIA Report n°242, 1983
3. P. Wolfe, Convergence conditions for ascent methods, SIAM Review 11, pp. 226-235, 1969.

NAME OF THE CODE: GQOPT3
AUTHOR(S): Richard E. Quandt and others
MATHEMATICAL PROBLEM: Numerical Optimization
DOMAIN OF APPLICATIONS: Statistics, Econometrics
MATHEMATICAL METHOD: various
PROGRAMMING LANGUAGE: FORTRAN
COMPUTER SYSTEMS WHERE THE CODE HAS BEEN IMPLEMENTED: IBM, VAX
SPECIAL PROGRAM FEATURES: Many special features
PRACTICAL APPLICATION PROBLEMS SOLVED BY THE CODE: Likelihood function maximization, Sum of squares minimization
CONDITIONS OF AVAILABILITY: Agreement not to resell or distribute
CHARGE: $ 150
ADDRESS FROM WHERE THE CODE CAN BE ORDERED: Prof. R.E. Quandt, Department of Economics, Princeton University, Princeton, NJ 08544
REFERENCES:

NAME OF THE CODE: GPKTC
AUTHOR(S): M. Rijckaert, E. Walraven
MATHEMATICAL PROBLEM: Generalized geometric programs
DOMAIN OF APPLICATIONS: Constrained nonlinear programming
MATHEMATICAL METHOD: "Condensation Technique"
PROGRAMMING LANGUAGE: FORTRAN (77)
COMPUTER SYSTEMS WHERE THE CODE HAS BEEN IMPLEMENTED: IBM 3033
SPECIAL PROGRAM FEATURES:
PRACTICAL APPLICATION PROBLEMS SOLVED BY THE CODE: Mainly engineering problems
CONDITIONS OF AVAILABILITY: Special conditions for academic use
CHARGE:
ADDRESS FROM WHERE THE CODE CAN BE ORDERED: M. Rijckaert, K.U. Leuven, B-3030 Heverlee, Belgium
REFERENCES:

NAME OF THE CODE: QPSOL Version 3.1

AUTHOR(S): Philip E. Gill, Walter Murray, Michael A. Saunders, Margaret H. Wright

MATHEMATICAL PROBLEM:

minimize $\frac{1}{2} x^T H x + c^T x$

subject to $\ell \leqslant \{ {x \atop Ax} \} \leqslant u.$

DOMAIN OF APPLICATIONS: The code is intended for in-core dense problems. The code has been designed to be used efficiently within an SOP method for nonlinear programming

MATHEMATICAL METHOD: Version 3.1 is a two-phase, null-space active-set method. An orthogonal factorization is updated as constraints enter or leave the predicted active set.

PROGRAMMING LANGUAGE: ANSI Fortran 66

COMPUTER SYSTEMS WHERE THE CODE HAS BEEN IMPLEMENTED:
All major US/European mainframe/mini/super-mini computers.

SPECIAL PROGRAM FEATURES:
1. The code takes advantage of upper/lower bounds.
2. The user may specify the accuracy to which the constraints are satisfied.
3. The matrix H may be indefinite or semi-definite.

PRACTICAL APPLICATION PROBLEMS SOLVED BY THE CODE:
The code has been distributed to 50 sites world wide. The object code version will be available at Mark 12 of the NAG library.

CONDITIONS OF AVAILABILITY: Copyright is held by Stanford University. A simple license agreement (limiting redistribution) must be signed.

CHARGE: At cost to universities/nonprofit organizations. $1000 for commercial licenses.

ADDRESS FROM WHERE THE CODE CAN BE ORDERED: Office of Technology Licensing, Room 105 Encina Hall, Stanford University, Stanford, California 94305, USA

REFERENCES: Gill, Murray, Saunders and Wright (1983), "User's Guide for SOL/QPSOL: A Fortran Package for Quadratic Programming," Report SOL 83-7, Department of Operations Research, Stanford University, California

NAME OF THE CODE: ZQPCVX
AUTHOR(S): M.J.D. Powell, University of Cambridge
MATHEMATICAL PROBLEM: Convex Quadratic Programming
DOMAIN OF APPLICATIONS: Convex Quadratic Programming Problems
MATHEMATICAL METHOD: An extension of the primal-dual method proposed by Goldfarb and Idnani (1983). It is described in detail in Powell (1983a).
PROGRAMMING LANGUAGE: FORTRAN
COMPUTER SYSTEMS WHERE THE CODE HAS BENN IMPLEMENTED: IBM 3081
SPECIAL PROGRAM FEATURES: Most suitable for small dense problems.
PRACTICAL APPLICATION PROBLEMS SOLVED BY THE CODE: None yet known to author. Some testing is reported in Powell (1983b) <u>NB</u> This code has been offered to the Harwell and IMSL subroutine libraries.
CONDITIONS OF AVAILABILITY: None
CHARGE: None
ADDRESS FROM WEHRE THE CODE CAN BE ORDERED: Listing (636 lines) from Prof. M.J.D. Powell, DAMTP, Silver Street, Cambridge CB3 9EW, England
REFERENCES: D. Goldfarb and A. Idnani (1983) Math.Prog., Vol. 27, pp. 1-33. M.J.D. Powell (1983a) Report DAMTP 1983/NA 17. M.J.D. Powell (1983b) Report DAMTP 1983/NA 19.

NAME OF THE CODE: DUAL QP

AUTHOR(S): Ashok Idnani

MATHEMATICAL PROBLEM: min $a^T x + \frac{1}{2} x^T G x$

subject to $C_e^T x = b_e$

$C^T x \geq b$

where G is positive definite.

DOMAIN OF APPLICATIONS: Non-linear Programs, Portfolio selection, etc.

MATHEMATICAL METHOD: Dual Projection method described in Mathematical Programming dated Aug. 83.

PROGRAMMING LANGUAGE: FORTRAN IV and ASSEMBLER, FORTRAN 77.

COMPUTER SYSTEMS WHERE THE CODE HAS BEEN IMPLEMENTED:
IBM 370, 4300 series, 3000 series, IBM PC with and without 8087 chip

SPECIAL PROGRAM FEATURES: Has its own interface for Powell's NLP algorithms and also can be used as an independent QP solver.

PRACTICAL APPLICATION PROBLEMS SOLVED BY THE CODE:
Non-linear Programs and Portfolio selections.

CONDITIONS OF AVAILABILITY:

CHARGE:

ADDRESS FROM WHERE THE CODE CAN BE ORDERED:
OCI
P.O. Box 144
Park Ridge, New Jersey 07656 USA

REFERENCES:

NAME OF THE CODE: CONQUA, START
AUTHOR(S): W.J. Kribbe, Drs.
MATHEMATICAL PROBLEM: Min $c^T x + 5\ x^T H\ x$ s.t. $Ax=b$
DOMAIN OF APPLICATIONS:
MATHEMATICAL METHOD: Reduced Conjugate Gradient
PROGRAMMING LANGUAGE: FORTRAN
COMPUTER SYSTEMS WHERE THE CODE HAS BEEN IMPLEMENTED: DEC 2060
SPECIAL PROGRAM FEATURES:
PRACTICAL APPLICATION PROBLEMS SOLVED BY THE CODE: Used in banking-environment
CONDITIONS OF AVAILABILITY: See reference
CHARGE: 10000 Dutch guilders
ADDRESS FROM WHERE THE CODE CAN BE ORDERED: Nassaukade 57, 2281 XH Rijswijk, The Netherlands
REFERENCES: (1) Documentation of the FORTRAN-Subroutines for quadratic programming CONQUA and START. (Report 8231/I Econometric Institute EUR)

NAME OF THE CODE: LINONL
AUTHOR(S): J. Hald
MATHEMATICAL PROBLEM: Minimization of a sum $F(x) = \sum_{i=1}^{m} \|f_i(x)\|$ subject to linear constraints $a_i + g_i^T x \geq 0$. The functions f_i are supposed to be smooth.
DOMAIN OF APPLICATIONS: Estimation of parameters of a mathematical model. Optimal design, centering, tolerancing and tuning.
MATHEMATICAL METHOD: The method is a combination of an LP method and a Quasi-Newton method. The Quasi-Newton algorithm is used only if an irregular solution is delected in which case second order derivative information is estimated from the user supplied first order information. The algorithm can converge only to stationary points and normally the final convergence is at least superlinear.
PROGRAMMING LANGUAGE: FORTRAN
COMPUTER SYSTEMS WHERE THE CODE HAS BEEN IMPLEMENTED: IBM 3033, UNIVAC 1100, CDC
SPECIAL PROGRAM FEATURES:
PRACTICAL APPLICATION PROBLEMS SOLVED BY THE CODE:
CONDITIONS OF AVAILABILITY: We can send a tape
CHARGE: $ 100
ADDRESS FROM WHERE THE CODE CAN BE ORDERED: Kaj Madsen, Building 303, Technical University, DK-2800 Lyngby, Denmark
REFERENCES: J. Hald and K. Madsen: "Combined LP and Quasi-Newton methods for nonlinear ℓ_1 Optimization". Submitted to SIAM J. Num. Anal. 1983. J. Hald: "A2-stage algorithm for nonlinear ℓ_1 Optimization". Report NI-81-03, Institute for Numerical Analysis, Technical University, DK-2800 Lyngby, Denmark.

NAME OF THE CODE: MMLA1Q

AUTHOR(S): J. Hald

MATHEMATICAL PROBLEM:

Minimization of the maximum of a finite set of smooth functions, $F(x) = \max f_j(x)$
subject to linear constraints, $a_i + g_i^T x \geq 0 \quad i=1,\ldots,k$

DOMAIN OF APPLICATIONS:
Estimation of parameters of a mathematical model.
Optimal design, centering, tolerancing and tuning.

MATHEMATICAL METHOD: The method is a combination of an LP method and a Quasi-Newton method. The Quasi-Newton algorithm is used only if an irregular solution is delected in which case secon order derivative information is estimated from the user supplied first order information. The algorithm can converge only to stationary points and normally the final convergence is at least superlinear.

PROGRAMMING LANGUAGE: FORTRAN

COMPUTER SYSTEMS WHERE THE CODE HAS BEEN IMPLEMENTED:
IBM 3033, UNIVAC 1100, CDC.

SPECIAL PPOGRAM FEATURES:

PRACTICAL APPLICATION PROBLEMS SOLVED BY THE CODE:
Optimization of antenna arrays. Active microwave circuit design problems. Refinement of molecular vibrational force fields of molecules from infrared spectrascopic data. And many others.

CHARGE: $ 100

ADDRESS FROM WHERE THE CODE CAN BE ORDERED:
Kaj Madsen, Building 303, Technical University, DK-2800 Lyngby, Denmark

REFERENCES:
J. Hald and K. Madsen: "Combined LP and Quasi-Newton methods for minimax Optimization". Math.Prog. 20 (1981), pp. 49-62.
J. Hald: "MMLA1Q, a FORTRAN subroutine for linearly constrained minimax Optimization". Report NI-81-01, Institute for Numerical Analysis, Technical University, DK-2800 Lyngby, Denmark

NAME OF THE CODE: TOMP	
AUTHOR(S): Dieter Kraft	
MATHEMATICAL PROBLEM: Trajectory optimization by mathematical programming (*) $\min\{f(x(t_f)):\dot{x}(t) = g(x(t),u(t));\ r(x(t_o),x(t_f)) = 0\}$ $u(t),\ t \in [t_o,t_f]$	
DOMAIN OF APPLICATIONS: Optimal control of dynamical systems governed by ordinary differential equations	
MATHEMATICAL METHOD: Direct shooting: approximation of control vector by cubic spline functions, solution of initial value problems, application of sequential quadratic programming	
PROGRAMMING LANGUAGE: FORTRAN	
COMPUTER SYSTEMS WHERE THE CODE HAS BEEN IMPLEMENTED: IBM 3081, PDP 11/65	
SPECIAL PROGRAM FEATURES: Solution (*) without adjoint differential equations	
PRACTICAL APPLICATION PROBLEMS SOLVED BY THE CODE: Optimal trajectories of aircraft & spacecraft Optimal control of a cryogenic wind-tunnel	
CONDITIONS OF AVAILABILITY:	
CHARGE:	
ADDRESS FROM WHERE THE CODE CAN BE ORDERED: Dieter Kraft c/o DFVLR, D-8031 Oberpfaffenhofen	
REFERENCES: FORTRAN-Programme zu numerischen Lösung optimaler Steuerungsprobleme, DFVLR-Mitt. 80-03, 1980.	

NAME OF THE CODE:	PARFIT (PARMEX, COLFIT)

AUTHOR(S): H.G. Bock

MATHEMATICAL PROBLEM: Determine $x \in PC[t_o,t_f]^{n_x}$, $p \in \mathbb{R}^{n_p}$

$$\| r_1(x(t_1),\ldots,x(t_k),p) \|_2 = \min$$

s.t.: $r_2(x(t_1),\ldots,x(t_k),p) \quad = 0 \quad \text{or} \geqslant 0,\ t_j$ fixed or variable

$$\dot{x} = f(t,x,p;\ \text{sign}\ Q(t,x,p)),\ x(t_j+) = c_j(t_j,x(t_j-),p)$$

DOMAIN OF APPLICATIONS: Parameter identification in systems of nonlinear differential equations. PARFIT allows: nonlinear equality and inequality constraints, o.d.e. systems with jump and switching conditions, systems with difficult stability properties (stiff, unstable, singularly perturbed), statistical aposteriori sensitivity analysis, singular problems.

MATHEMATICAL METHOD: Stable transformation to large constrained nonlinear least squares problem by multiple shooting. Solution by generalized Gauss-Newton method (convergence: global and locally linear). Derivative-free linearization by internal numerical differentiation.

PROGRAMMING LANGUAGE: FORTRAN IV

COMPUTER SYSTEMS WHERE THE CODE HAS BEEN IMPLEMENTED:
IBM 370/168, 4331, 3081, VAX 11/780, UNIVAC

SPECIAL PROGRAM FEATURES: Additional routines for automatic generation of right hand sides from chemical reaction systems (CREATE) and treatment of multiple experiment data (PARMEX). Interface for interactive modelling in preparation.

PRACTICAL APPLICATION PROBLEMS SOLVED BY THE CODE: Various parameter identification problems in chemical and biochemical reaction systems, in pharmaco-kinetics and biology; geophysical problems: tracing of seismic rays for layer identification (inversion of travel times).

CONDITIONS OF AVAILABILITY: On request

CHARGE:

ADDRESS FROM WHERE THE CODE CAN BE ORDERED:
Institut für Angewandte Mathematik, SFB 72, Universität Bonn, D-5300 Bonn 1, FRG

REFERENCES:
H.G. Bock (1980): Numerical treatment of inverse problems in chemical reaction systems, Modelling of chemical reactions (eds. Ebert, Deuflhard, Jäger), Springer, Berlin, 102-125
H.G. Bock (1983): Recent advances in parameter identification techniques for o.d.e., Numerical treatment of inverse problems (eds. Deuflhard, Hairer) Birkhäuser, Boston, 95-121.

NAME OF THE CODE: MUSCOD

AUTHOR(S): H.G. Bock, K.J. Plitt

MATHEMATICAL PROBLEM:

Determine $x \in AC[t_o,t_f]^{n_x}$, $u \in L_\infty[t_o,t_f]^{n_u}$, $p \in \mathbb{R}^{n_p}$

$$\int_{t_o}^{t_f} l(t,x,p,u)dt = \min$$

s.t.: $\dot{x} = f(t,x,p,u;\ \text{sign}\ Q(t,x,p))$

$r(x(t_o),x(t_f),t_f,p) = 0$ or ≥ 0, $u(t) \in U$, $p \in P$.

DOMAIN OF APPLICATIONS: Direct solution of *constrained optimal control and parameter optimization problems for* dynamic processes.

Problem features allowed: Control and parameter constraints, discontinuous dynamics, boundary conditions, unstable and/or stiff systems, singular and bang-bang control. State inequality constraints and other cost functions by transformations. Extensions to multipoint boundary conditions.

MATHEMATICAL METHOD: *Stable* transformation of control problem to large constrained nonlinear programming problem by multiple shooting parameterization of state and control variables. Solution by successive quadratic programming method. Convergence: global and locally superlinear. High rank variable metric updates.

PROGRAMMING LANGUAGE: FORTRAN IV

COMPUTER SYSTEMS WHERE THE CODE HAS BEEN IMPLEMENTED:
IBM 370/168, 4331, 3081, Andahl 470/V6

SPECIAL PROGRAM FEATURES: Completely derivative-free (derivative generation by internal numerical differentiation), no adjoint differential equations necessary. Performance insensitive to poor initial guess of control. No deterioration of convergence for refined parameterization due to high rank updates. *Remark*: SQP/QPSOLV modules for nonlinear programming in preparation.

PRACTICAL APPLICATION PROBLEMS SOLVED: E.g. aerospace problems (trajectory optimization for aircrafts, interplanetary low thrust satellites, re-entry manoevres), optimal robot manipulator control, optimal transition from exhaustible to reproducible resources.

CONDITIONS OF AVAILABILITY: On request

CHARGE:

ADDRESS FROM WHERE THE CODE CAN BE ORDERED: Institut für Angewandte Mathematik, SFB 72, Universität Bonn, D-5300 Bonn 1, FRG

REFERENCES:

H.G. Bock, K.J. Plitt (1983): A multiple shooting algorithm for direct solution of constrained optimal control problems, Proc. 9th IFAC World Congress Automatic Control, Pergamon Press.

NAME OF THE CODE: OPCON

AUTHOR(S): H.G. Bock

MATHEMATICAL PROBLEM:

Determine $x \in PC[t_o,t_f]^{n_x}$, $u \in L_\infty[t_o,t_f]^{n_u}$, $p \in \mathbb{R}^{n_p}$

$$\Phi(t_f,x(t_f),p) + \int_{t_o}^{t_f} L(t,x,p,u)dt + \max_{t_o \leq t \leq t_f} M(t,x,p,u) = \min$$

$$\dot{x} = f(t,x,p,u, \text{sign } Q(t,x,p)),$$

s.t.: $r(x(t_1),\ldots,x(t_k),p) = 0$ or ≥ 0,

$$x(t_j+) = c_j(t_j,x(t_j-),p) \quad \text{if} \quad Q_j(t_j,x(t_j-),p) = 0,$$

$$g(t,x(t),p,u(t)) \geq 0, \; u(t) \in U, \; p \in P.$$

DOMAIN OF APPLICATIONS: Indirect solution of *constrained optimal control and parameter optimization problems* for dynamic processes. Problem class includes discontinuous state variables and right hand sides, state and control inequality constraints, disconnected control regions, minimax or Chebyshev control problems, singular and bang-bang control, multipoint boundary conditions (eg. rendezvous and flybys).

MATHEMATICAL METHOD: Transformation of optimal control problem by means of the maximum-principle to associated *multipoint boundary value problem with jump and switching conditions*. Stable solution by multiple shooting. Globally convergent generalized Gauss-Newton method for nonlinear problem solution.

PROGRAMMING LANGUAGE: FORTRAN IV

COMPUTER SYSTEMS WHERE THE CODE HAS BEEN IMPLEMENTED:
IBM 370/168, 4331, 3081

SPECIAL PROGRAM FEATURES: Derivative-free (gradient generation by internal numerical differentiation). Adaptive integration accuracy scheme. Generation of initial data by inverse problem algorithms or continuation techniques. Refined switching procedures ("competing Hamiltonians" COMPH). Feedback approximations by → FAUST.

PRACTICAL APPLICATION PROBLEMS SOLVED BY THE CODE: E.g. trajectory optimization for aircrafts, low thrust satellites, robot manipulators, minimization of maximum temperature for space shuttle reentry, minimum energy strategies for New York subway system.

CONDITIONS OF AVAILABILITY: on request

CHARGE:

ADDRESS FROM WHERE THE CODE CAN BE ORDERED: Institut für Angewandte Mathematik, SFB 72, Universität Bonn, D-53 Bonn 1, FRG

REFERENCES: H.G. Bock (1978), Numerical solution of nonlinear multipoint boundary value problems with applications to optimal control, ZAMM 58.
H.G. Bock (1981): *Numerical treatment of state-constrained and Chebyshev control problems* (in German), Carl-Cranz-Gesellschaft, Heidelberg.

H.G. Bock, R. Longman (1980): Optimal control of velocity profiles for minimization of energy consumption in the New York subway system, Proc. 2nd IFAC Workshop Control Applics. of Nonlinear Progr., (ed. Well).

NAME OF THE CODE: FAUST

AUTHOR(S): H.G. Bock, P. Krämer-Eis

MATHEMATICAL PROBLEM:

$\forall (t_o, x^o, q^o) \in G$ determine $u(t_o, x^o, q^o) \in L_\infty[t_o, t_f]^{n_u}$

$$\Phi(t_f, x(t_f), q) = \min_{x,u}$$

s.t.: $r(t_f, x(t_f), q) = 0$, $g(t, x(t), q, u(t)) \geq 0$, $u(t) \in U$

$$\dot{x} = f(t, x, q, u, \text{sign } Q(t, x, q)), x(t_o) = x^o, q = q^o.$$

$$x(t_j+) = C_j(t_j, x(t_j-), q) \quad \text{if} \quad Q_j(t_j, x(t_j-), q) = 0.$$

DOMAIN OF APPLICATIONS: Approximation of *optimal feedback control strategies* which adapt optimal open loop solutions to perturbations of state variables or process parameters.

MATHEMATICAL METHOD: (to be used in combination with program system OPCON). Computation of maximum principle solution by multiple shooting. Expansion of discretized (nondifferentiable) multipoint boundary value problem with jump and switching conditions. Recursive determination of coefficient matrices for feedback control law.

PROGRAMMING LANGUAGE: FORTRAN IV

COMPUTER SYSTEMS WHERE THE CODE HAS BEEN IMPLEMENTED:
IBM 370/168, 4331, 3081

SPECIAL PROGRAM FEATURES: Algorithm is derivative-free, stable due to multiple shooting, avoids laborious on-line computations. Suited for discontinuous dynamical systems, control constraints, unstable and/or stiff systems. Direct treatment of perturbed system parameters in dynamical system or side conditions.

PRACTICAL APPLICATION PROBLEMS SOLVED: E.g. optimal orbit-orbit low thrust transfers, reentry manoeuvre of spacecrafts, energy optimal control of vehicles (New York subway).

CONDITIONS OF AVAILABILITY: on request

CHARGE:

ADDRESS FROM WHERE THE CODE CAN BE ORDERED: Institut für Angewandte Mathematik, SFB 72, Universität Bonn, D-53 Bonn 1, FRG

REFERENCES:

H.G. Bock, P. Krämer-Eis (1981): An efficient algorithm for approximate computation of feedback control laws in nonlinear processes. ZAMM 61.

H.G. Bock, P. Krämer-Eis (1983): A multiple shooting algorithm for numerical computation of open and closed loop controls in nonlinear systems, Proc. 9th IFAC World Congress Automatic Control, Pergamon Press.

NATO ASI Series F

Vol. 1: Issues in Acoustic Signal – Image Processing and Recognition. Edited by C. H. Chen. VIII, 333 pages. 1983.

Vol. 2: Image Sequence Processing and Dynamic Scene Analyisi. Edited by T. S. Huang. IX, 749 pages. 1983.

Vol. 3: Electronic Systems Effectiveness and Life Cycle Costing. Edited by J. K. Skwirzynski. XVII, 732 pages. 1983.

Vol. 4: Pictorial Data Analysis. Edited by R. M. Haralick. VIII, 468 pages. 1983.

Vol. 5: International Calibration Study of Traffic Conflict Techniques. Edited by E. Asmussen VII, 229 pages. 1984.

Vol. 6: Information Technology and the Computer Network. Edited by K. G. Beauchamp. VIII, 271 pages. 1984.

Vol. 7: High-Speed Computation. Edited by J. S. Kowalik. IX, 441 pages. 1984.

Vol. 8: Program Transformation and Programming Environments. Report on an Workshop directed by F. L. Bauer and H. Remus. Edited by P. Pepper. XIV, 378 pages. 1984.

Vol. 9: Computer Aided Analysis and Optimization of Mechanical System Dynamics. Edited by E. J. Haug. XXII, 700 pages. 1984.

Vol. 10: Simulation and Model-Based Methodologies: An Integrative View. Edited by T. I. Ören, B. P. Zeigler, M. S. Elzas. XIII, 651 pages. 1984.

Vol. 11: Robotics and Artificial Intelligence. Edited by M. Brady, L. A. Gerhardt, H. F. Davidson. XVII, 693 pages. 1984.

Vol. 12: Combinatorial Algorithms on Words. Edited by A. Apostolico, Z. Galil. VIII, 361 pages. 1985.

Vol. 13: Logics and Models of Concurrent Systems. Edited by K. R. Apt. VIII, 498 pages. 1985.

Vol. 14: Combinatorial Algorithms on Words. Edited by A. Apostolico and Z. Galil. VIII, 361 pages. 1985.

Vol. 15: Computational Mathematical Programming. Edited by K. Schittkowski. VIII, 451 pages. 1985.